TAKE
NOTE!

TO ACCOMPANY

FUNDAMENTALS
OF BIOCHEMISTRY

Second Edition

Donald Voet
University of Pennsylvania

Judith G. Voet
Swarthmore College

Charlotte W. Pratt
Seattle, Washington

WILEY

JOHN WILEY & SONS, INC.

To order books or for customer service call 1-800-CALL-WILEY (225-5945).

ISBN-13 978- 0-471-73930-2
ISBN-10 0-471-73930-8

Printed in the United States of America

10 9 8 7 6 5 4 3 2 1

Printed and bound by Courier Kendallville, Inc.

CONTENTS

CHAPTER 1

Introduction to the Chemistry of Life

Table 1-2 Key to Structure. Common Functional Groups and Linkages in Biochemistry

Compound Name	Structure[a]	Functional Group or Linkage
Amine[b]	RNH_2 or $R\overset{+}{N}H_3$ R_2NH or $R_2\overset{+}{N}H_2$ R_3N or $R_3\overset{+}{N}H$	$-N{<}$ or $-\overset{+}{\underset{\vert}{N}}-$ (amino group)
Alcohol	ROH	$-OH$ (hydroxyl group)
Thiol	RSH	$-SH$ (sulfhydryl group)
Ether	ROR	$-O-$ (ether linkage)
Aldehyde	$R-\overset{O}{\overset{\|}{C}}-H$	$-\overset{O}{\overset{\|}{C}}-$ (carbonyl group)
Ketone	$R-\overset{O}{\overset{\|}{C}}-R$	$-\overset{O}{\overset{\|}{C}}-$ (carbonyl group)
Carboxylic acid[b]	$R-\overset{O}{\overset{\|}{C}}-OH$ or $R-\overset{O}{\overset{\|}{C}}-O^-$	$-\overset{O}{\overset{\|}{C}}-OH$ (carboxyl group) or $-\overset{O}{\overset{\|}{C}}-O^-$ (carboxylate group)
Ester	$R-\overset{O}{\overset{\|}{C}}-OR$	$-\overset{O}{\overset{\|}{C}}-O-$ (ester linkage) $R-\overset{O}{\overset{\|}{C}}-$ (acyl group)[c]
Thioester	$R-\overset{O}{\overset{\|}{C}}-SR$	$-\overset{O}{\overset{\|}{C}}-S-$ (thioester linkage) $R-\overset{O}{\overset{\|}{C}}-$ (acyl group)[c]
Amide	$R-\overset{O}{\overset{\|}{C}}-NH_2$ $R-\overset{O}{\overset{\|}{C}}-NHR$ $R-\overset{O}{\overset{\|}{C}}-NR_2$	$-\overset{O}{\overset{\|}{C}}-N{<}$ (amido group) $R-\overset{O}{\overset{\|}{C}}-$ (acyl group)[c]
Imine (Schiff base)[b]	$R{=}NH$ or $R{=}\overset{+}{N}H_2$ $R{=}NR$ or $R{=}\overset{+}{N}HR$	${>}C{=}N-$ or ${>}C{=}\overset{+}{N}{<}$ (imino group)
Disulfide	$R-S-S-R$	$-S-S-$ (disulfide linkage)
Phosphate ester[b]	$R-O-\overset{O}{\overset{\|}{\underset{\underset{OH}{\vert}}{P}}}-O^-$	$-\overset{O}{\overset{\|}{\underset{\underset{OH}{\vert}}{P}}}-O^-$ (phosphoryl group)
Diphosphate ester[b]	$R-O-\overset{O}{\overset{\|}{\underset{\underset{O^-}{\vert}}{P}}}-O-\overset{O}{\overset{\|}{\underset{\underset{OH}{\vert}}{P}}}-O^-$	$-\overset{O}{\overset{\|}{\underset{\underset{O^-}{\vert}}{P}}}-O-\overset{O}{\overset{\|}{\underset{\underset{OH}{\vert}}{P}}}-O^-$ (phosphoanhydride group)
Phosphate diester[b]	$R-O-\overset{O}{\overset{\|}{\underset{\underset{O^-}{\vert}}{P}}}-O-R$	$-O-\overset{O}{\overset{\|}{\underset{\underset{O^-}{\vert}}{P}}}-O-$ (phosphodiester linkage)

[a]R represents any carbon-containing group. In a molecule with more than one R group, the groups may be the same or different.

[b]Under physiological conditions, these groups are ionized and hence bear a positive or negative charge.

[c]If attached to an atom other than carbon.

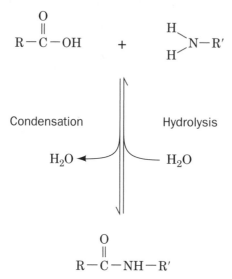

Condensation   Hydrolysis

H_2O ←   ← H_2O

Figure 1-3 Reaction of a carboxylic acid with an amine.

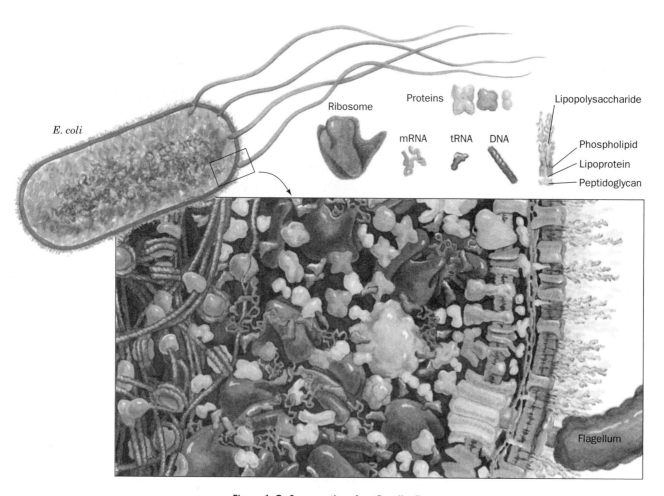

E. coli

Ribosome   Proteins   Lipopolysaccharide

mRNA   tRNA   DNA   Phospholipid

          Lipoprotein

          Peptidoglycan

Flagellum

Figure 1-6 Cross section of an *E. coli* cell.

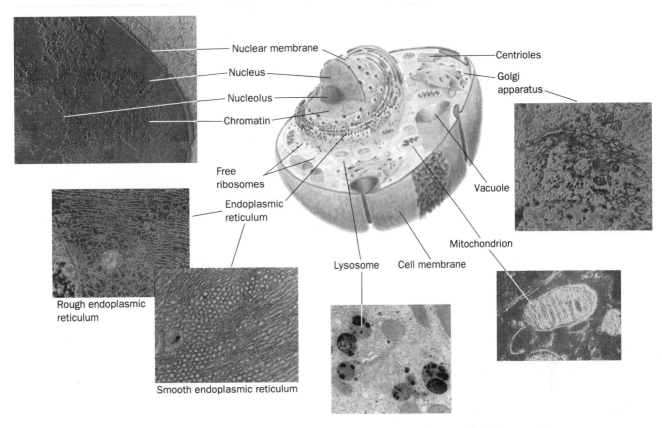

Nuclear membrane

Nucleus

Nucleolus

Chromatin

Free ribosomes

Endoplasmic reticulum

Centrioles

Golgi apparatus

Vacuole

Mitochondrion

Lysosome

Cell membrane

Rough endoplasmic reticulum

Smooth endoplasmic reticulum

Figure 1-8 Diagram of a typical animal cell accompanied by electron micrographs of its organelles.

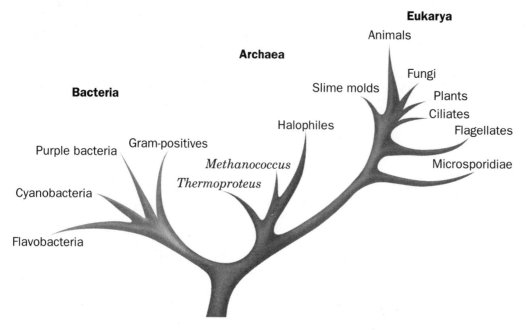

Figure 1-9 Phylogenetic tree showing three domains of organisms.

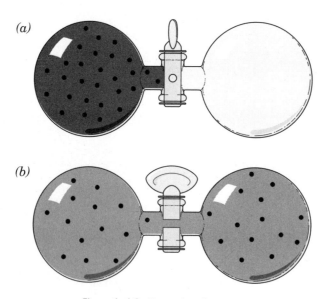

Figure 1-10 Illustration of entropy.

$$S = k_B \ln W \qquad [1\text{-}5]$$

$$\Delta S_{system} + \Delta S_{surroundings} = \Delta S_{universe} > 0 \qquad [1\text{-}6]$$

$$\Delta S \geq \frac{q}{T} \qquad [1\text{-}7]$$

$$\Delta S \geq \frac{q_P}{T} = \frac{\Delta H}{T} \qquad [1\text{-}8]$$

$$\Delta H - T\Delta S \leq 0 \qquad [1\text{-}9]$$

$$G = H - TS \qquad [1\text{-}10]$$

$$\boxed{\Delta G = \Delta H - T\Delta S < 0} \qquad [1\text{-}11]$$

Table 1-3 Variation of Reaction Spontaneity (Sign of ΔG) with the Signs of ΔH and ΔS

ΔH	ΔS	$\Delta G = \Delta H - T\Delta S$
−	+	The reaction is both enthalpically favored (exothermic) and entropically favored. It is spontaneous (exergonic) at all temperatures.
−	−	The reaction is enthalpically favored but entropically opposed. It is spontaneous only at temperatures *below* $T = \Delta H/\Delta S$.
+	+	The reaction is enthalpically opposed (endothermic) but entropically favored. It is spontaneous only at temperatures *above* $T = \Delta H/\Delta S$.
+	−	The reaction is both enthalpically and entropically opposed. It is unspontaneous (endergonic) at all temperatures.

$$\overline{G}_A = \overline{G}_A^\circ = RT \ln[A] \qquad [1\text{-}12]$$

$$a\mathrm{A} + b\mathrm{B} \rightleftharpoons c\mathrm{C} + d\mathrm{D}$$

$$\Delta G = c\overline{G}_C + d\overline{G}_D - a\overline{G}_A - b\overline{G}_B \qquad [1\text{-}13]$$

$$\Delta G^\circ = c\overline{G}_C^\circ + d\overline{G}_D^\circ - a\overline{G}_A^\circ - b\overline{G}_B^\circ \qquad [1\text{-}14]$$

$$\Delta G = \Delta G^\circ + RT \ln\left(\frac{[C]^c[D]^d}{[A]^a[B]^b}\right) \qquad [1\text{-}15]$$

$$\boxed{\Delta G^\circ = -RT \ln K_{eq}} \qquad [1\text{-}16]$$

$$K_{eq} = \frac{[C]_{eq}^c[D]_{eq}^d}{[A]_{eq}^a[B]_{eq}^b} = e^{-\Delta G^\circ/RT} \qquad [1\text{-}17]$$

$$\ln K_{eq} = \frac{-\Delta H^\circ}{R}\left(\frac{1}{T}\right) + \frac{\Delta S^\circ}{R} \qquad [1\text{-}18]$$

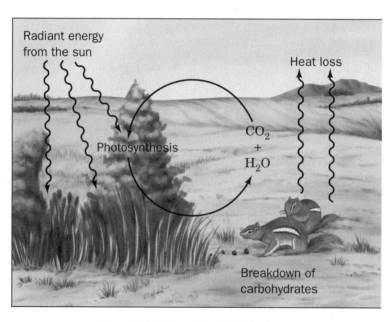

Figure 1-11 Energy flow in the biosphere.

CHAPTER 2

Water

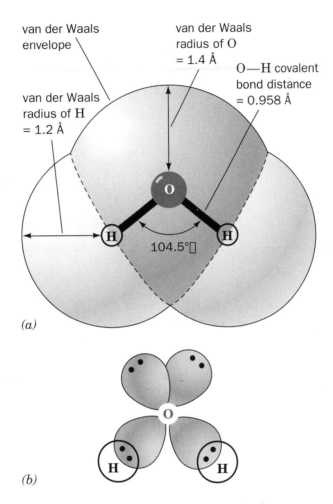

van der Waals envelope

van der Waals radius of O = 1.4 Å

O—H covalent bond distance = 0.958 Å

van der Waals radius of H = 1.2 Å

104.5°

(a)

(b)

Figure 2-1 Structure of the water molecule.

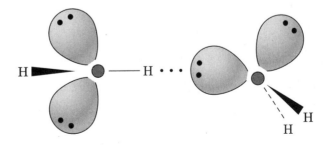

H ⸺ H · · ·

H

H

Figure 2-2 A hydrogen bond in water.

Table 2-1 Bond Energies in Biomolecules

Type of Bond	Example	Bond Strength $(kJ \cdot mol^{-1})$
Covalent	O—H	460
	C—H	414
	C—C	348
Noncovalent		
Ionic interaction	$-COO^-\cdots{}^+H_3N-$	86
van der Waals forces		
Hydrogen bond	$-O-H\cdots O\diagdown$	20
Dipole–dipole interaction	$\diagdown\!C=O\cdots\diagdown\!C=O$	9.3
London dispersion forces	(see structure)	0.3

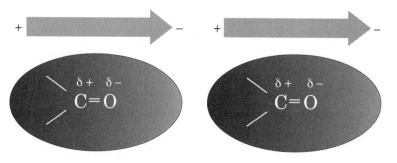

(a) Interactions between permanent dipoles

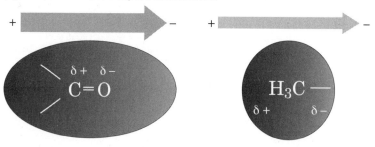

(b) Dipole–induced dipole interactions

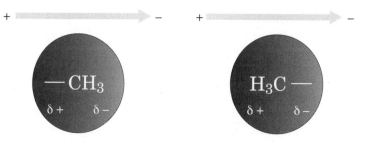

(c) London dispersion forces

Figure 2-5 Dipole–dipole interactions.

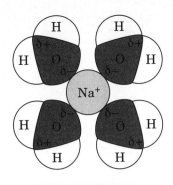

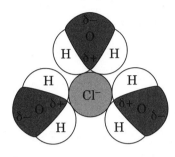

Figure 2-6 Solvation of ions.

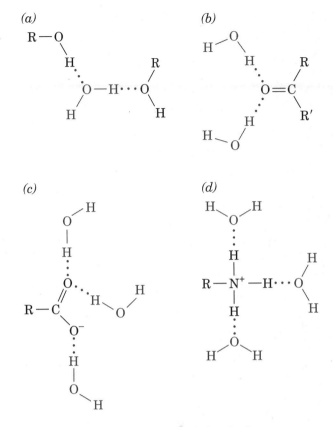

Figure 2-7 Hydrogen bonding by functional groups.

Table 2-2 Thermodynamic Changes for Transferring Hydrocarbons from Water to Nonpolar Solvents at 25°C

Process	ΔH $(kJ \cdot mol^{-1})$	$-T\Delta S$ $(kJ \cdot mol^{-1})$	ΔG $(kJ \cdot mol^{-1})$
CH_4 in $H_2O \rightleftharpoons CH_4$ in C_6H_6	11.7	−22.6	−10.9
CH_4 in $H_2O \rightleftharpoons CH_4$ in CCl_4	10.5	−22.6	−12.1
C_2H_6 in $H_2O \rightleftharpoons C_2H_6$ in benzene	9.2	−25.1	−15.9
C_2H_4 in $H_2O \rightleftharpoons C_2H_4$ in benzene	6.7	−18.8	−12.1
C_2H_2 in $H_2O \rightleftharpoons C_2H_2$ in benzene	0.8	−8.8	−8.0
Benzene in $H_2O \rightleftharpoons$ liquid benzene[a]	0.0	−17.2	−17.2
Toluene in $H_2O \rightleftharpoons$ liquid toluene[a]	0.0	−20.0	−20.0

[a]Data measured at 18°C.

Source: Kauzmann, W., *Adv. Protein Chem.* **14,** 39 (1959).

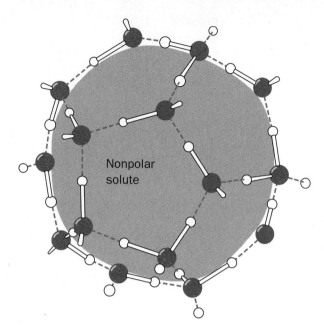

Figure 2-8 Orientation of water molecules around a nonpolar solute.

Nonpolar solute

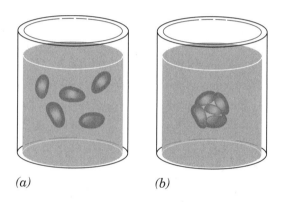

(a)   (b)

Figure 2-9 Aggregation of nonpolar molecules in water.

$$CH_3CH_2CH_2CH_2CH_2CH_2CH_2CH_2CH_2CH_2CH_2CH_2CH_2CH_2-\overset{\overset{\displaystyle O}{\|}}{C}-O^-$$

Palmitate ($C_{15}H_{31}COO^-$)

$$CH_3CH_2CH_2CH_2CH_2CH_2CH_2CH_2-\overset{\overset{\displaystyle H}{|}}{C}=\overset{\overset{\displaystyle H}{|}}{C}-CH_2CH_2CH_2CH_2CH_2CH_2CH_2-\overset{\overset{\displaystyle O}{\|}}{C}-O^-$$

Oleate ($C_{17}H_{33}COO^-$)

Figure 2-10 Fatty acid anions (soaps).

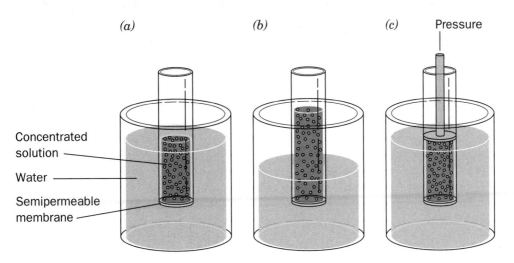

Figure 2-13 Osmotic pressure.

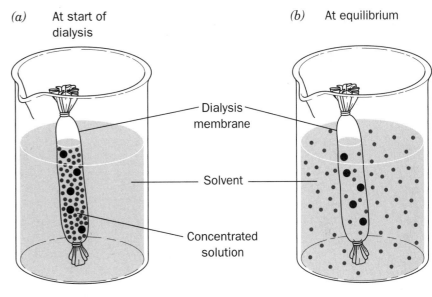

Figure 2-14 Dialysis.

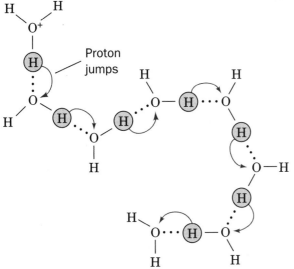

Figure 2-15 Proton jumping.

$$K_w = [\text{H}^+][\text{OH}^-] \qquad\qquad [2\text{-}2]$$

$$\text{pH} = -\log[\text{H}^+] = \log \frac{1}{[\text{H}^+]} \qquad\qquad [2\text{-}3]$$

Table 2-3 pH Values of Some Common Substances

Substance	pH
1 M NaOH	14
Household ammonia	12
Seawater	8
Blood	7.4
Milk	7
Saliva	6.6
Tomato juice	4.4
Vinegar	3
Gastric juice	1.5
1 M HCl	0

$$K = \frac{[\text{H}_3\text{O}^+][\text{A}^-]}{[\text{HA}][\text{H}_2\text{O}]} \qquad\qquad [2\text{-}4]$$

$$K_a = K[\text{H}_2\text{O}] = \frac{[\text{H}^+][\text{A}^-]}{[\text{HA}]} \qquad\qquad [2\text{-}5]$$

$$\text{p}K = -\log K \qquad\qquad [2\text{-}6]$$

$$[\text{H}^+] = K\frac{[\text{HA}]}{[\text{A}^-]} \qquad\qquad [2\text{-}7]$$

$$\text{pH} = -\log K + \log \frac{[\text{A}^-]}{[\text{HA}]} \qquad\qquad [2\text{-}8]$$

$$\boxed{\text{pH} = \text{p}K + \log \frac{[\text{A}^-]}{[\text{HA}]}} \qquad\qquad [2\text{-}9]$$

Table 2-4 Dissociation Constants and pK Values at 25°C of Some Acids

Acid	K	pK
Oxalic acid	5.37×10^{-2}	1.27 (pK_1)
H_3PO_4	7.08×10^{-3}	2.15 (pK_1)
Formic acid	1.78×10^{-4}	3.75
Succinic acid	6.17×10^{-5}	4.21 (pK_1)
Oxalate$^-$	5.37×10^{-5}	4.27 (pK_2)
Acetic acid	1.74×10^{-5}	4.76
Succinate$^-$	2.29×10^{-6}	5.64 (pK_2)
2-(N-Morpholino)ethanesulfonic acid (MES)	8.13×10^{-7}	6.09
H_2CO_3	4.47×10^{-7}	6.35 (pK_1)[a]
Piperazine-N,N'-bis(2-ethanesulfonic acid) (PIPES)	1.74×10^{-7}	6.76
$H_2PO_4^-$	1.51×10^{-7}	6.82 (pK_2)
3-(N-Morpholino)propanesulfonic acid (MOPS)	7.08×10^{-8}	7.15
N-2-Hydroxyethylpiperazine-N'-2-ethanesulfonic acid (HEPES)	3.39×10^{-8}	7.47
Tris(hydroxymethyl)aminomethane (Tris)	8.32×10^{-9}	8.08
NH_4^+	5.62×10^{-10}	9.25
Glycine (amino group)	1.66×10^{-10}	9.78
HCO_3^-	4.68×10^{-11}	10.33 (pK_2)
Piperidine	7.58×10^{-12}	11.12
HPO_4^{2-}	4.17×10^{-13}	12.38 (pK_3)

Source: Dawson, R.M.C., Elliott, D.C., Elliott, W.H., and Jones, K.M., *Data for Biochemical Research* (3rd ed.), pp. 424–425, Oxford Science Publications (1986); *and* Good, N.E., Winget, G.D., Winter, W., Connolly, T.N., Izawa, S., and Singh, R.M.M., *Biochemistry* **5**, 467 (1966).

[a]The pK for the overall reaction $CO_2 + H_2O \rightleftharpoons H_2CO_3 \rightleftharpoons H^+ + HCO_3^-$; see Box 2-2.

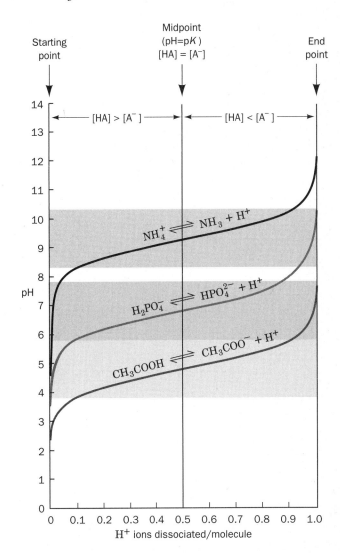

Figure 2-17 Titration curves for acetic acid, phosphate, and ammonia.

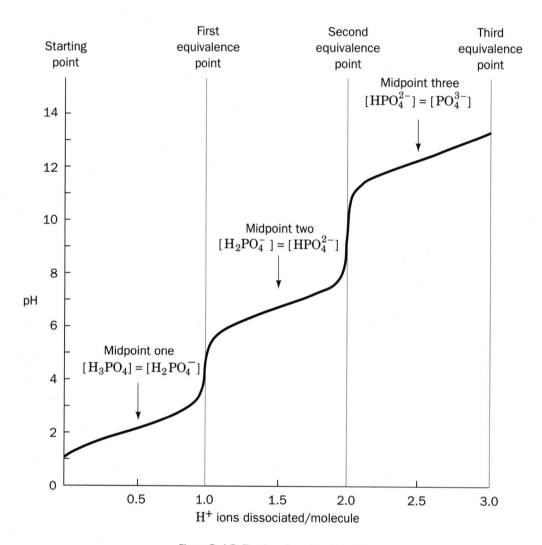

Figure 2-18 Titration of a polyprotic acid.

CHAPTER 3

Nucleotides, Nucleic Acids, and Genetic Information

Purine   **Pyrimidine**

Table 3-1 Names and Abbreviations of Nucleic Acid Bases, Nucleosides, and Nucleotides

Base Formula	Base (X = H)	Nucleoside (X = ribose[a])	Nucleotide[b] (X = ribose phosphate[a])
	Adenine Ade A	Adenosine Ado A	Adenylic acid Adenosine monophosphate AMP
	Guanine Gua G	Guanosine Guo G	Guanylic acid Guanosine monophosphate GMP
	Cytosine Cyt C	Cytidine Cyd C	Cytidylic acid Cytidine monophosphate CMP
	Uracil Ura U	Uridine Urd U	Uridylic acid Uridine monophosphate UMP
	Thymine Thy T	Deoxythymidine dThd dT	Deoxythymidylic acid Deoxythymidine monophosphate dTMP

[a]The presence of a 2′-deoxyribose unit in place of ribose, as occurs in DNA, is implied by the prefixes "deoxy" or "d." For example, the deoxynucleoside of adenine is deoxyadenosine or dA. However, for thymine-containing residues, which rarely occur in RNA, the prefix is redundant and may be dropped. The presence of a ribose unit may be explicitly implied by the prefix "ribo."

[b]The position of the phosphate group in a nucleotide may be explicitly specified as in, for example, 3′-AMP and 5′-GMP.

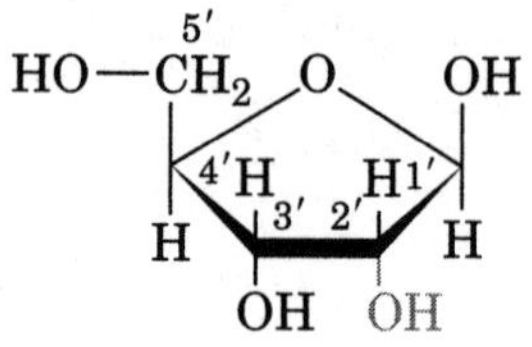

Ribose 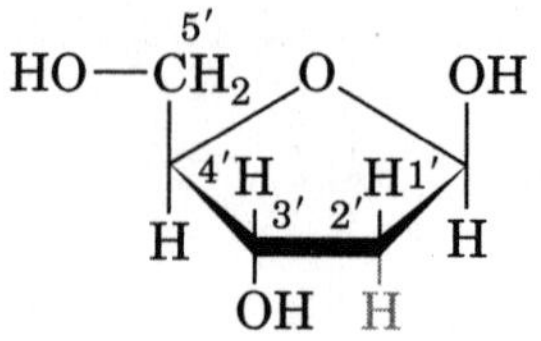   **Deoxyribose**

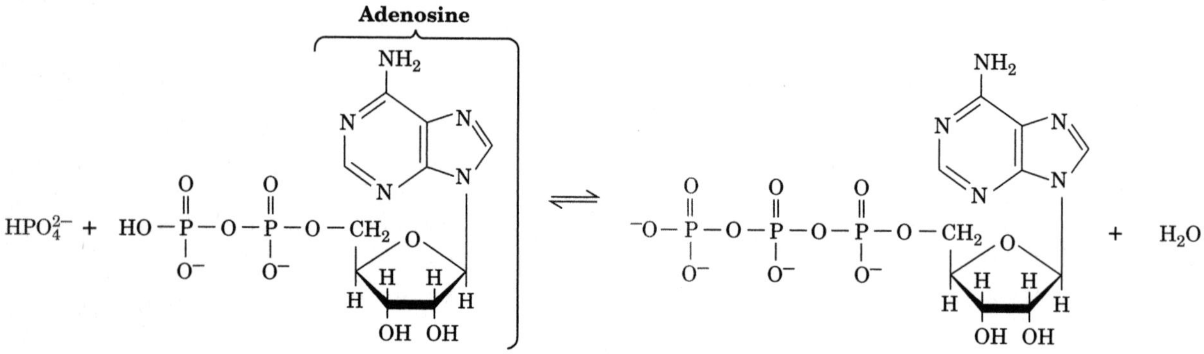

Adenosine diphosphate (ADP)   **Adenosine triphosphate (ATP)**

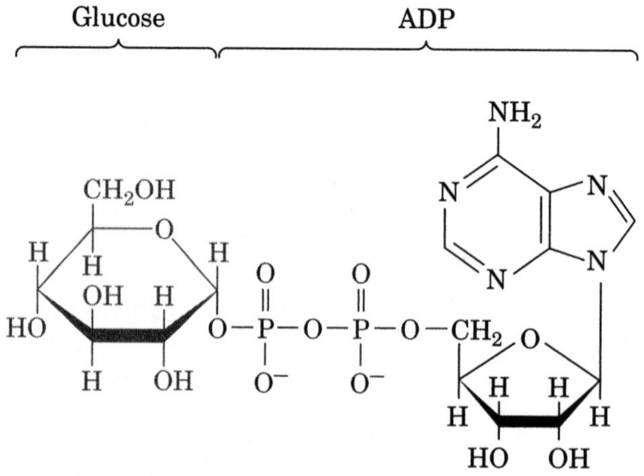

Figure 3-2 ADP–glucose.

(a)

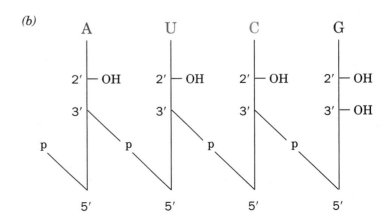

A

U (T)

C

G

Figure 3-3 *Key to Structure.* Chemical structure of a nucleic acid.

(b)

	A		U		C		G
2'	— OH	2'	— OH	2'	— OH	2'	— OH
3'		3'		3'		3'	— OH
p		p		p		p	
	5'		5'		5'		5'

17

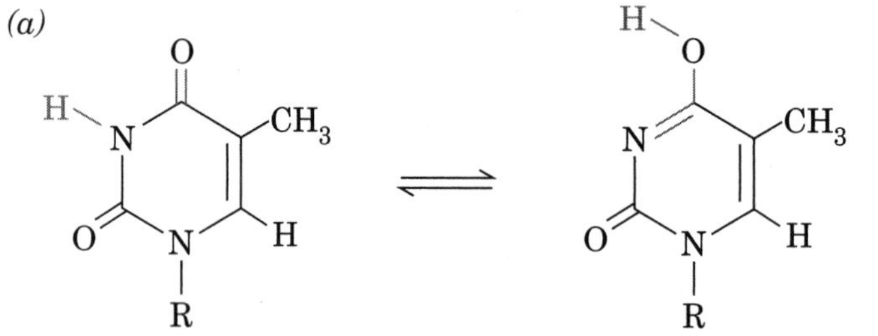

Thymine
(keto *or* lactam form)

Thymine
(enol *or* lactim form)

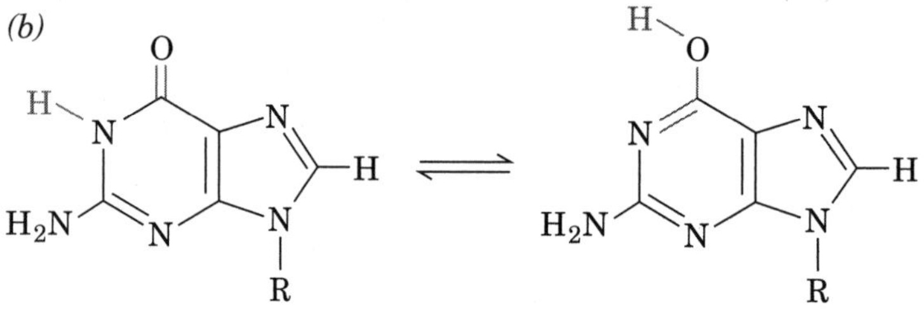

Guanine
(keto *or* lactam form)

Guanine
(enol *or* lactim form)

Figure 3-4 Tautomeric forms of bases.

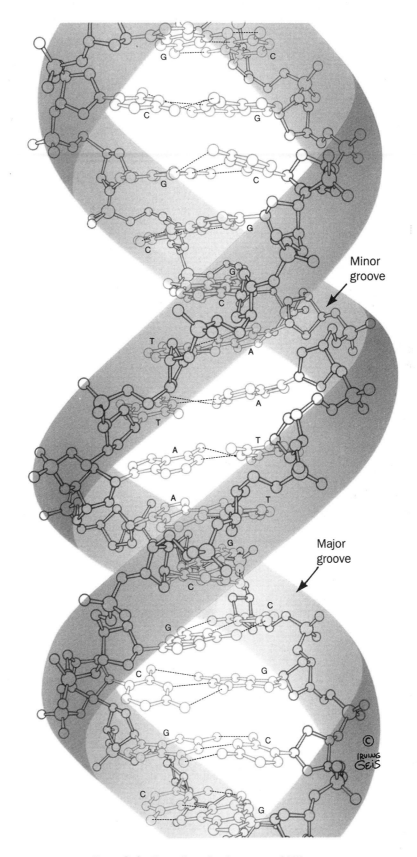

Figure 3-6 Three-dimensional structure of DNA.

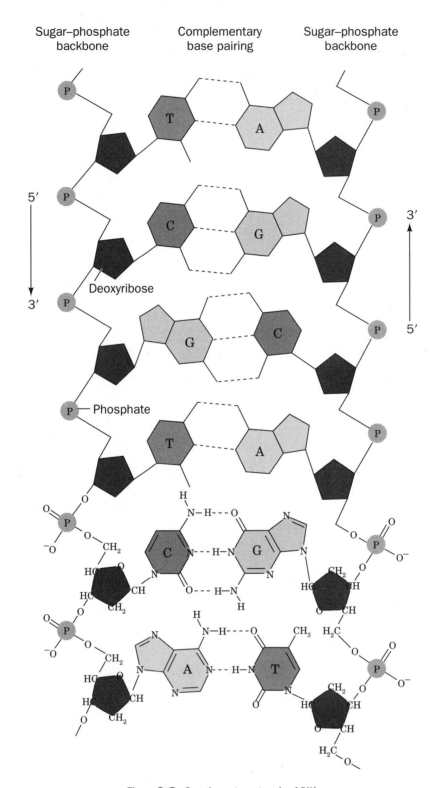

Figure 3-8 Complementary strands of DNA.

Figure 3-9 Formation of a stem–loop structure.

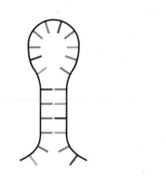

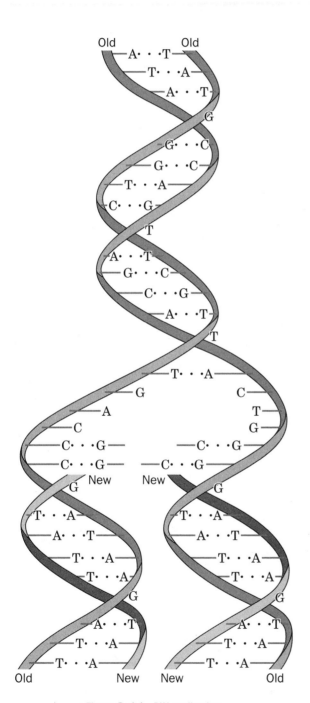

Figure 3-11 DNA replication.

DNA   5′ — A–G–A–G–G–T–G–C–T — 3′
       3′ — T–C–T–C–C–A–C–G–A — 5′

mRNA   5′ — A–G–A–G–G–U–G–C–U — 3′
tRNAs 3′   U–C–U   C–C–A   C–G–A   5′

        Arginine Glycine Alanine

Protein   —Arginine–Glycine–Alanine—

Figure 3-12 Transcription and translation.

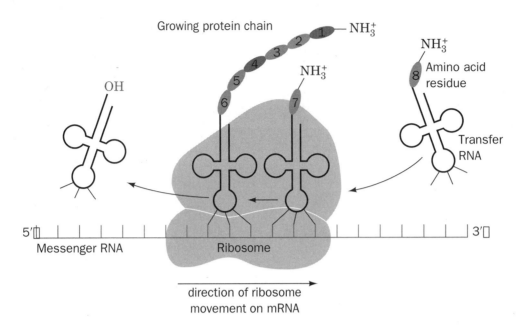

Growing protein chain

direction of ribosome
movement on mRNA

Figure 3-14 Translation.

Table 3-2 Recognition and Cleavage Sites of Some Restriction Enzymes

Enzyme	Recognition Sequence[a]	Microorganism
*Alu*I	AG$\downarrow$CT	*Arthrobacter luteus*
*Bam*HI	G$\downarrow$GATCC	*Bacillus amyloliquefaciens* H
*Bgl*I	GCCNNNNN$\downarrow$NGGC	*Bacillus globigii*
*Bgl*II	A$\downarrow$GATCT	*Bacillus globigii*
*Eco*RI	G$\downarrow$AATTC	*Escherichia coli* RY13
*Eco*RII	$\downarrow$CC(^A_T)GG	*Escherichia coli* R245
*Eco*RV	GAT$\downarrow$ATC	*Escherichia coli* J62 pLG74
*Hae*II	RGCGC$\downarrow$Y	*Haemophilus aegyptius*
*Hae*III	GG$\downarrow$CC	*Haemophilus aegyptius*
*Hind*III	A$\downarrow$AGCTT	*Haemophilus influenzae* R_d
*Hpa*II	C$\downarrow$CGG	*Haemophilus parainfluenzae*
*Msp*I	C$\downarrow$CGG	*Moraxella* species
*Pst*I	CTGCA$\downarrow$G	*Providencia stuartii* 164
*Pvu*II	CAG$\downarrow$CTG	*Proteus vulgaris*
*Sal*I	G$\downarrow$TCGAC	*Streptomyces albus* G
*Taq*I	T$\downarrow$CGA	*Thermus aquaticus*
*Xho*I	C$\downarrow$TCGAG	*Xanthomonas holcicola*

[a]The recognition sequence is abbreviated so that only one strand, reading 5′ to 3′, is given. The cleavage site is represented by an arrow ($\downarrow$). R, Y, and N represent a purine nucleotide, a pyrimidine nucleotide, and any nucleotide, respectively.

Source: Roberts, R.J. and Macelis, D., REBASE—the restriction enzyme database, http://rebase.neb.com.

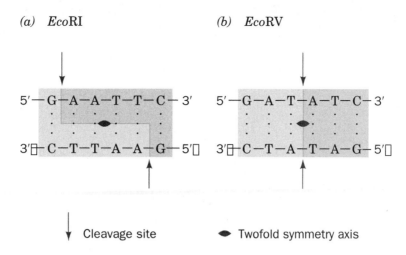

(a) *Eco*RI

5′ —G—A—A—T—T—C— 3′

3′ —C—T—T—A—A—G— 5′

(b) *Eco*RV

5′ —G—A—T—A—T—C— 3′

3′ —C—T—A—T—A—G— 5′

↓ Cleavage site   ● Twofold symmetry axis

Figure 3-16 Restriction sites.

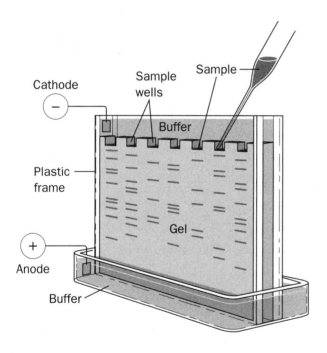

Cathode

Sample wells

Sample

Buffer

Plastic frame

Gel

Anode

Buffer

Figure 3-17 Apparatus for gel electrophoresis.

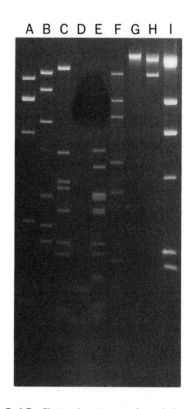

A B C D E F G H I

Figure 3-18 Electrophoretogram of restriction digests.

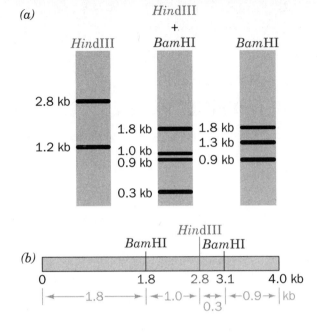

Figure 3-19 Construction of a restriction map.

RESTRICTION FRAGMENT LENGTH POLYMORPHISMS (RFLPs)

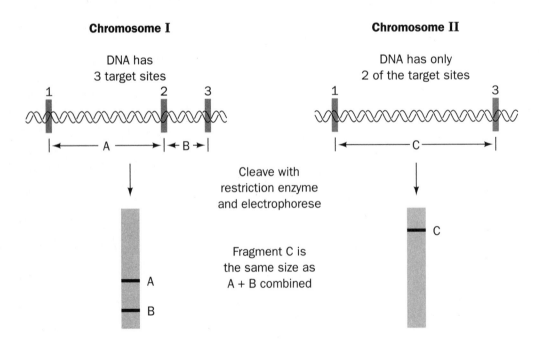

THE CHAIN-TERMINATOR METHOD OF SEQUENCING DNA

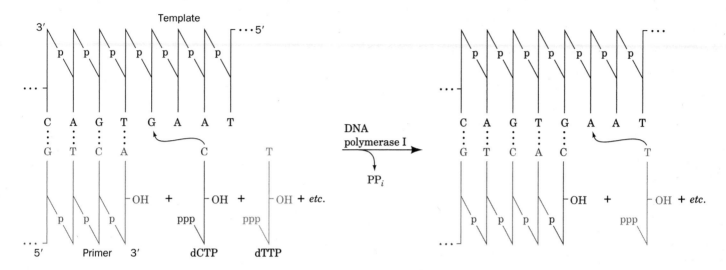

Figure 3-20   Action of DNA polymerase I.

2′,3′-Dideoxynucleoside
triphosphate

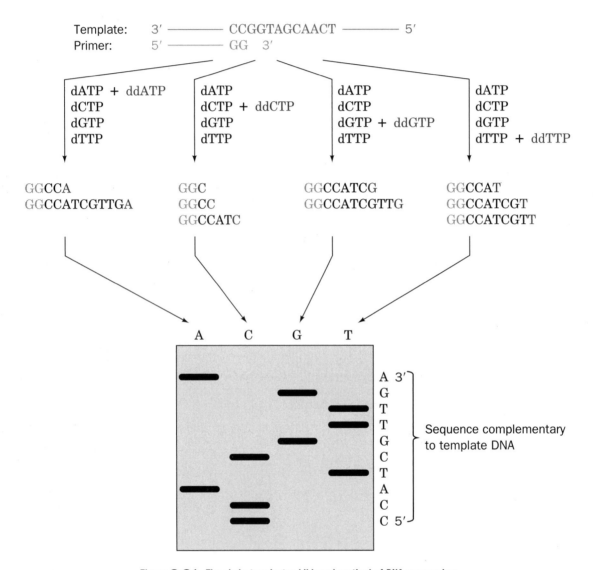

Figure 3-21 The chain-terminator (dideoxy) method of DNA sequencing.

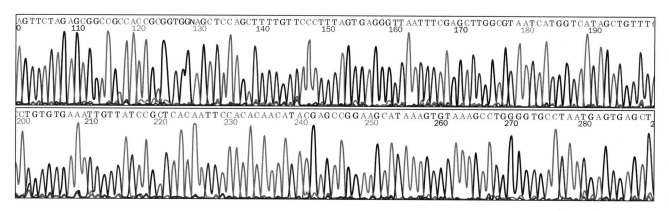

Figure 3-23 Automated DNA sequencing.

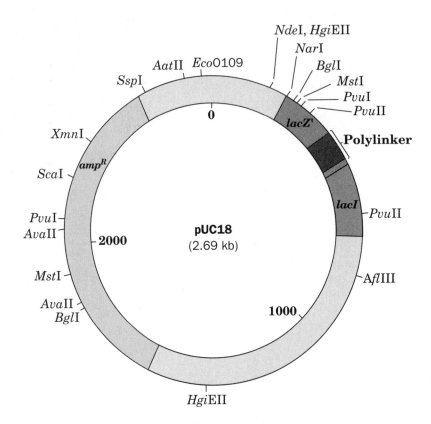

Figure 3-25 The plasmid pUC18.

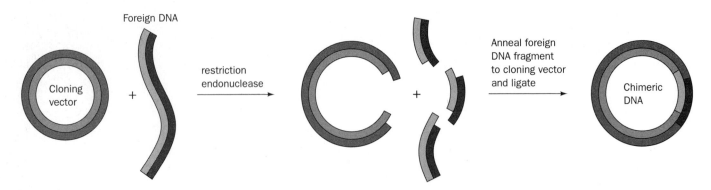

Figure 3-27 Construction of a recombinant DNA molecule.

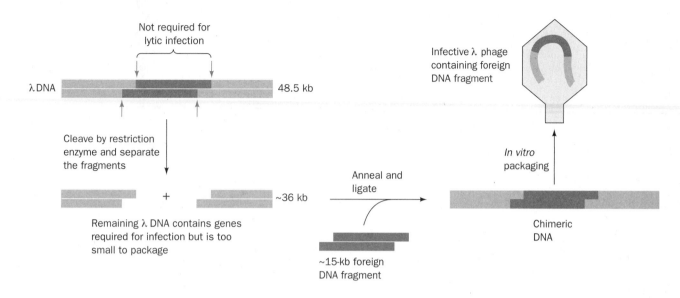

Figure 3-28 Cloning with bacteriophage λ.

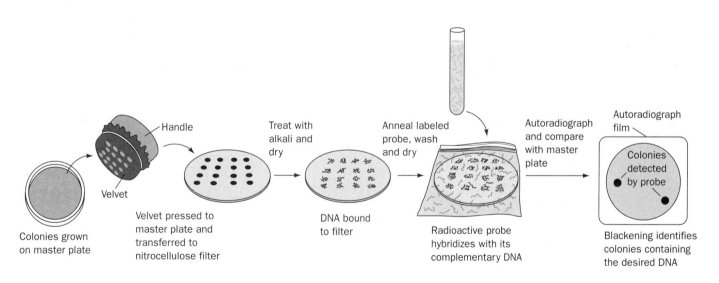

Figure 3-29 Colony (*in situ*) hybridization.

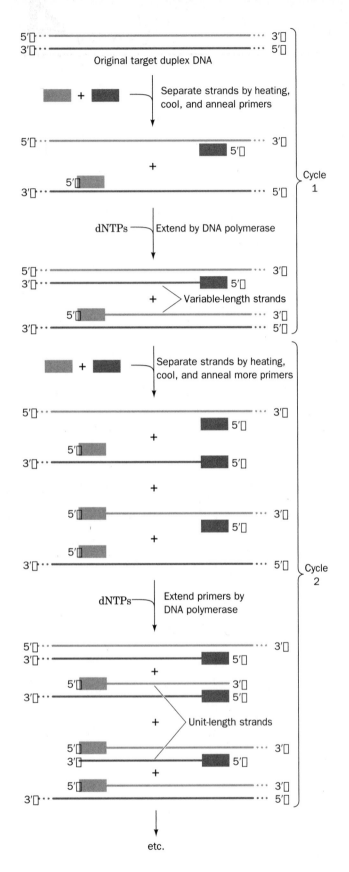

Figure 3-30 The polymerase chain reaction (PCR).

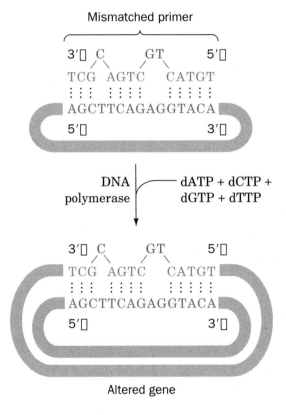

Figure 3-31 Site-directed mutagenesis.

CHAPTER 4

Amino Acids

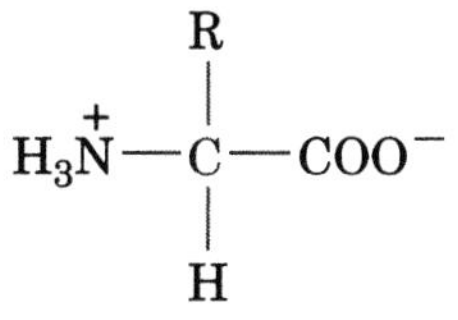

Figure 4-2 A zwitterionic amino acid.

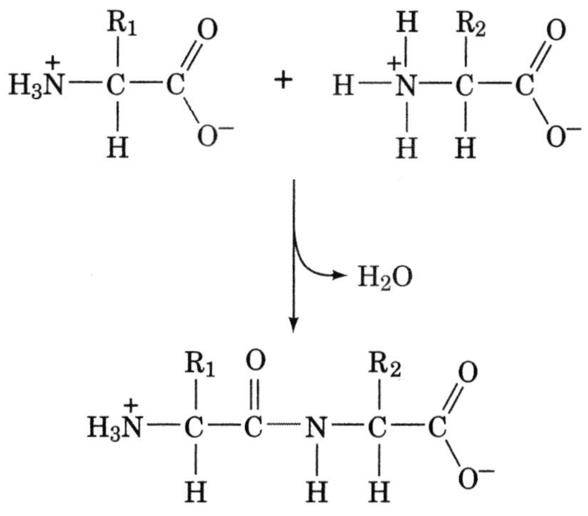

Figure 4-3 Condensation of two amino acids.

Table 4-1 *Key to Structure.* Covalent Structures and Abbreviations of the "Standard" Amino Acids of Proteins, Their Occurrence, and the pK Values of Their Ionizable Groups

Name, Three-letter Symbol, and One-letter Symbol	Structural Formula[a]	Residue Mass (D)[b]	Average Occurrence in Proteins (%)[c]	pK_1 α-COOH[d]	pK_2 α-NH$_3^+$[d]	pK_R Side Chain[d]
Amino acids with nonpolar side chains						
Glycine Gly G		57.0	7.2	2.35	9.78	
Alanine Ala A		71.1	7.8	2.35	9.87	
Valine Val V		99.1	6.6	2.29	9.74	
Leucine Leu L		113.2	9.1	2.33	9.74	
Isoleucine Ile I		113.2	5.3	2.32	9.76	
Methionine Met M		131.2	2.2	2.13	9.28	
Proline Pro P		97.1	5.2	1.95	10.64	
Phenylalanine Phe F		147.2	3.9	2.20	9.31	
Tryptophan Trp W		186.2	1.4	2.46	9.41	

[a]The ionic forms shown are those predominating at pH 7.0 (except for that of histidine[f]) although residue mass is given for the neutral compound. The C_α atoms, as well as those atoms marked with an asterisk, are chiral centers with configurations as indicated according to Fischer projection formulas (Section 4-2). The standard organic numbering system is provided for heterocycles.

[b]The residue masses are given for the neutral residues. For the molecular masses of the parent amino acids, add 18.0 D, the molecular mass of H_2O, to the residue masses. For side chain masses, subtract 56.0 D, the formula mass of a peptide group, from the residue masses.

[c]Calculated from a database of nonredundant proteins containing 300,688 residues as compiled by Doolittle, R.F. *in* Fasman, G.D. (Ed.), *Predictions of Protein Structure and the Principles of Protein Conformation,* Plenum Press (1989).

[d]Data from Dawson, R.M.C., Elliott, D.C., Elliott, W.H., and Jones, K.M., *Data for Biochemical Research* (3rd ed.), pp. 1–31, Oxford Science Publications (1986).

Table 4-1 (continued)

Name, Three-letter Symbol, and One-letter Symbol	Structural Formula[a]	Residue Mass (D)[b]	Average Occurrence in Proteins (%)[c]	pK_1 α-COOH[d]	pK_2 α-NH_3^+[d]	pK_R Side Chain[d]
Amino acids with uncharged polar side chains						
Serine Ser S		87.1	6.8	2.19	9.21	
Threonine Thr T		101.1	5.9	2.09	9.10	
Asparagine[e] Asn N		114.1	4.3	2.14	8.72	
Glutamine[e] Gln Q		128.1	4.3	2.17	9.13	
Tyrosine Tyr Y		163.2	3.2	2.20	9.21	10.46 (phenol)
Cysteine Cys C		103.1	1.9	1.92	10.70	8.37 (sulfhydryl)
Amino acids with charged polar side chains						
Lysine Lys K		128.2	5.9	2.16	9.06	10.54 (ε-NH_3^+)
Arginine Arg R		156.2	5.1	1.82	8.99	12.48 (guanidino)
Histidine[f] His H		137.1	2.3	1.80	9.33	6.04 (imidazole)
Aspartic acid[e] Asp D		115.1	5.3	1.99	9.90	3.90 (β-COOH)
Glutamic acid[e] Glu E		129.1	6.3	2.10	9.47	4.07 (γ-COOH)

[e]The three- and one-letter symbols for asparagine *or* aspartic acid are Asx and B, whereas for glutamine *or* glutamic acid they are Glx and Z. The one-letter symbol for an undetermined or "nonstandard" amino acid is X.

[f]Both neutral and protonated forms of histidine are present at pH 7.0, since its pK_R is close to 7.0.

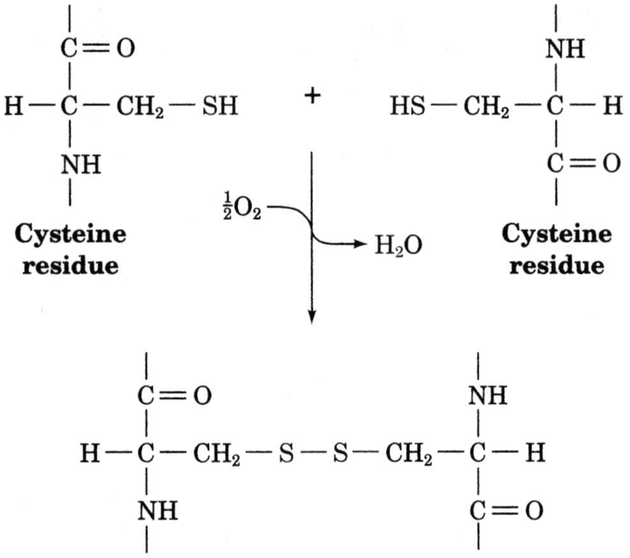

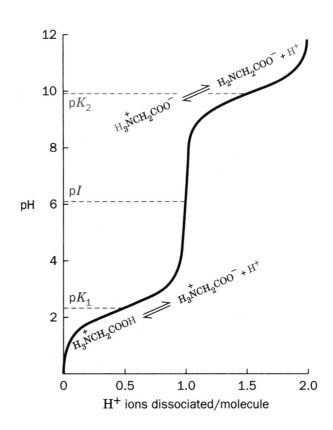

Figure 4-6 Disulfide-bonded cysteine residues.

Figure 4-8 Titration of glycine.

$$pH = pK + \log \frac{[A^-]}{[HA]} \qquad [4\text{-}1]$$

$$pI = \tfrac{1}{2}(pK_i + pK_j) \qquad [4\text{-}2]$$

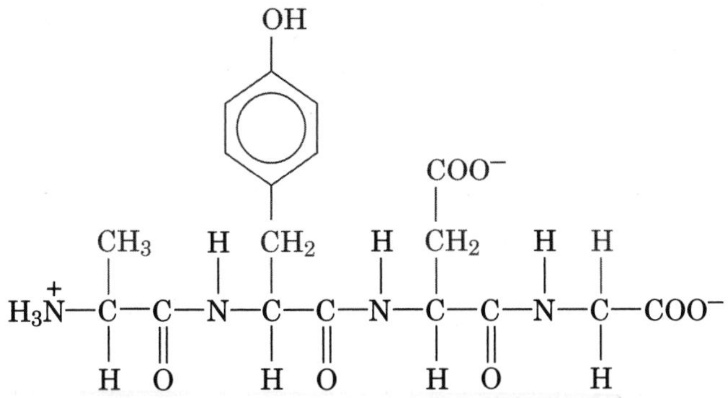

OH

COO⁻

CH_3 H CH_2 H CH_2 H   H

$H_3\overset{+}{N}$—C—C—N—C—C—N—C—C—N—C—COO⁻

H   O   H   O   H   O   H

Ala — Tyr — Asp — Gly

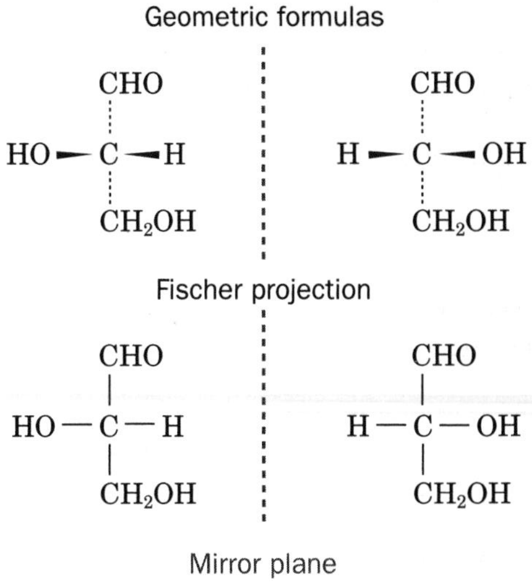

Geometric formulas

CHO   CHO

HO—C—H   H—C—OH

CH_2OH CH_2OH

Fischer projection

CHO   CHO

HO—C—H   H—C—OH

CH_2OH CH_2OH

Mirror plane

L-Glyceraldehyde   D-Glyceraldehyde

Figure 4-12 The Fischer convention.

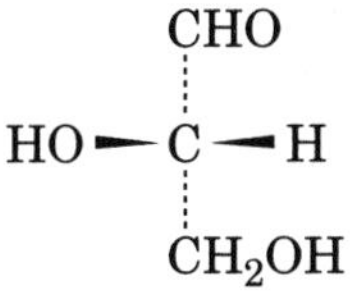

CHO   COO⁻

HO—C—H $H_3\overset{+}{N}$—C—H

CH_2OH R

L-Glyceraldehyde 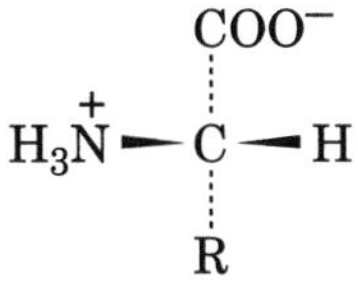   L-α-Amino acid

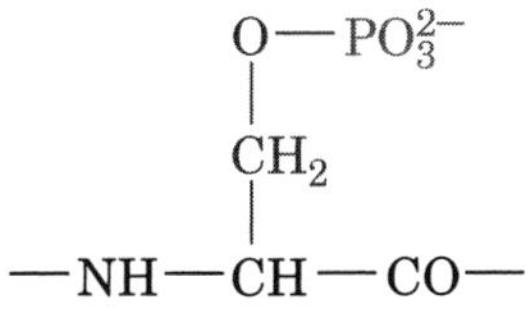

O—PO_3^{2-}

CH_2

—NH—CH—CO—

***O*-Phosphoserine**

⁻OOC   COO⁻

CH γ

βCH₂

—NH—αCH—CO—

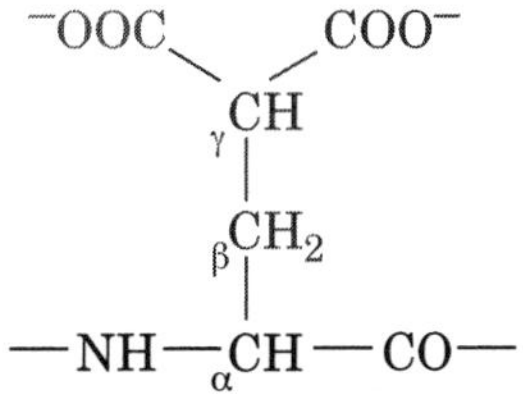

γ-Carboxyglutamate

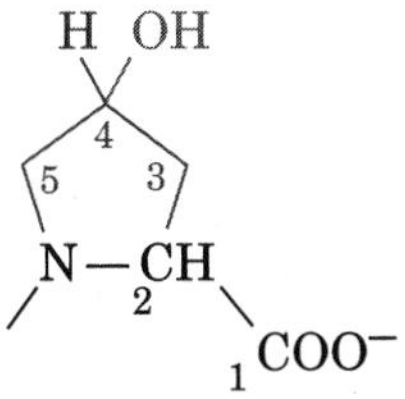

H OH

4

5   3

N—CH

2   COO⁻ 1

4-Hydroxyproline

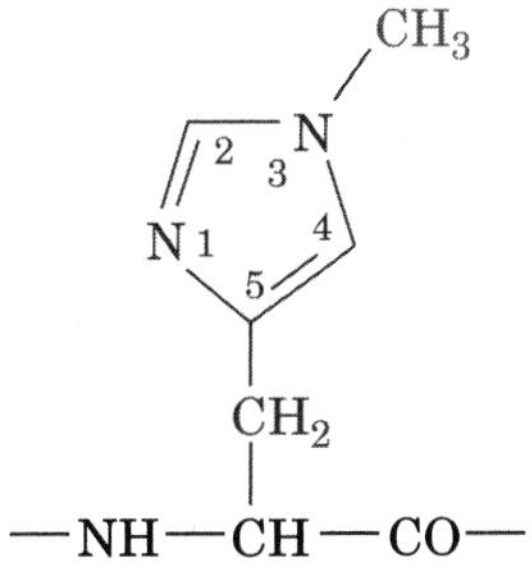

CH₃

N

2   3

N 1   4

5

CH₂

—NH—CH—CO—

3-Methylhistidine

O
‖
HN—C—CH₃

εCH₂

δCH₂

γCH₂

βCH₂

—NH—αCH—CO—

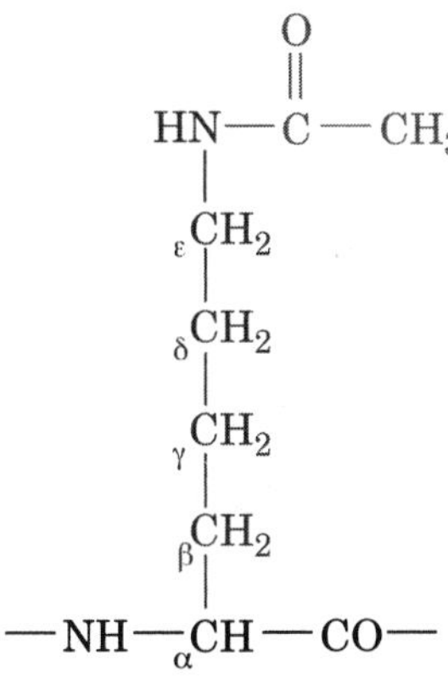

ε-*N*-Acetyllysine

Figure 4-15 Some modified amino acid residues in proteins.

CHAPTER 5

Proteins: Primary Structure

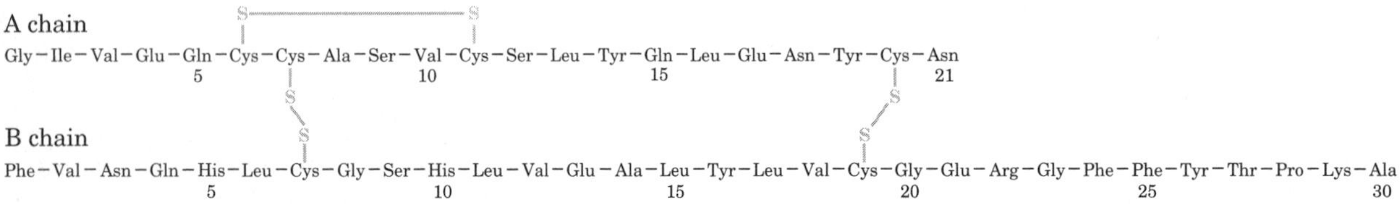

A chain

Gly−Ile−Val−Glu−Gln−Cys−Cys−Ala−Ser−Val−Cys−Ser−Leu−Tyr−Gln−Leu−Glu−Asn−Tyr−Cys−Asn

          5   10   15   21

B chain

Phe−Val−Asn−Gln−His−Leu−Cys−Gly−Ser−His−Leu−Val−Glu−Ala−Leu−Tyr−Leu−Val−Cys−Gly−Glu−Arg−Gly−Phe−Phe−Tyr−Thr−Pro−Lys−Ala

     5   10   15   20   25   30

Figure 5-1 The primary structure of bovine insulin.

Table 5-1 Compositions of Some Proteins

Protein	Amino Acid Residues	Subunits	Polypeptide Molecular Mass (D)
Proteinase inhibitor III (bitter gourd)	30	1	3,427
Cytochrome c (human)	104	1	11,617
Myoglobin (horse)	153	1	16,951
Interferon-γ (rabbit)	288	2	33,842
Chorismate mutase (*Bacillus subtilis*)	381	3	43,551
Triose phosphate isomerase (*E. coli*)	510	2	53,944
Hemoglobin (human)	574	4	61,986
RNA polymerase (bacteriophage T7)	883	1	98,885
Nucleoside diphosphate kinase (*Dictyostelium discoideum*)	930	6	100,764
Pyruvate decarboxylase (yeast)	2,252	4	245,456
Glutamine synthetase (*E. coli*)	5,616	12	621,264
Titin (human)	26,926	1	2,993,428

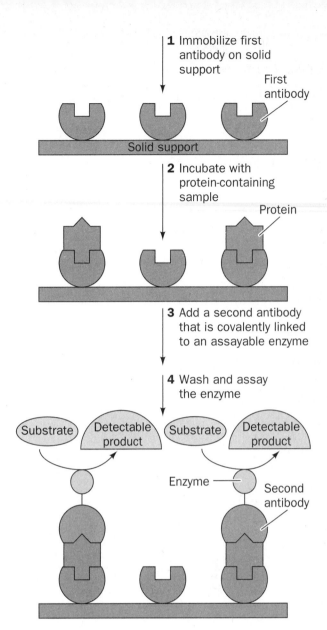

Figure 5-3 Enzyme-linked immunosorbent assay.

SEPARATION TECHNIQUES

Protein Characteristic	*Purification Procedure*
Solubility	Salting out
Ionic Charge	Ion exchange chromatography
	Electrophoresis
	Isoelectric focusing
Polarity	Hydrophobic interaction chromatography
Size	Gel filtration chromatography
	SDS-PAGE
	Ultracentrifugation
Binding specificity	Affinity chromatography

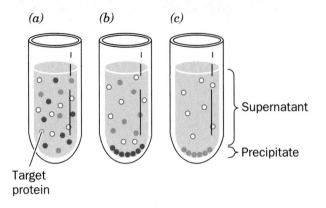

Target
protein

Supernatant

Precipitate

Figure 5-5 Fractionation by salting out.

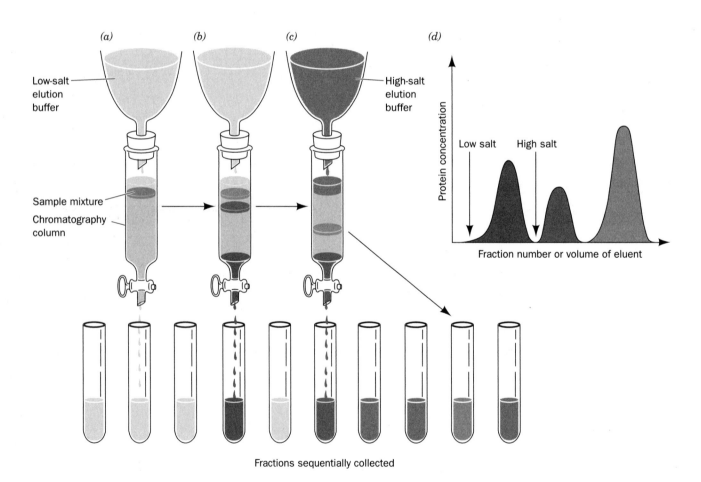

Low-salt
elution
buffer

High-salt
elution
buffer

Sample mixture

Chromatography
column

Protein concentration

Low salt   High salt

Fraction number or volume of eluent

Fractions sequentially collected

Figure 5-6 Ion exchange chromatography.

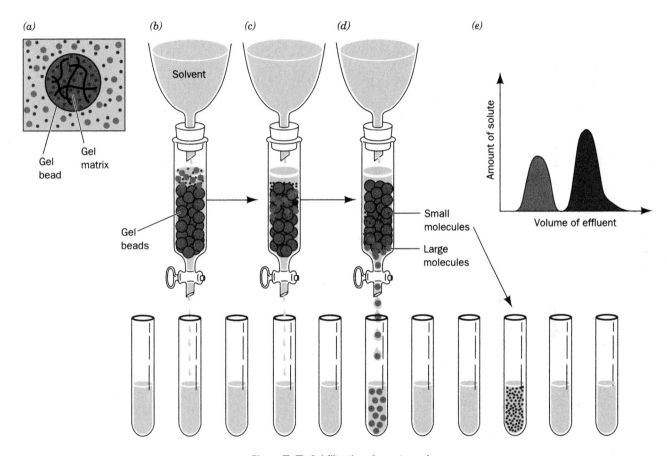

(a)

Gel
bead

Gel
matrix

(b)

Solvent

Gel
beads

(c)

(d)

Small
molecules

Large
molecules

(e)

Amount of solute

Volume of effluent

Figure 5-7 Gel filtration chromatography.

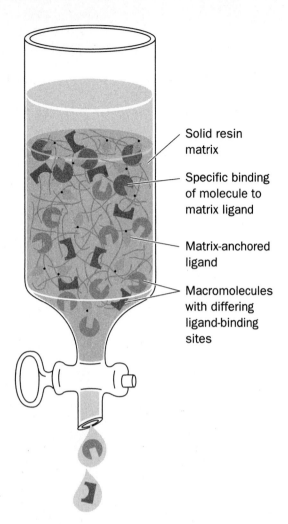

Figure 5-8 Affinity chromatography.

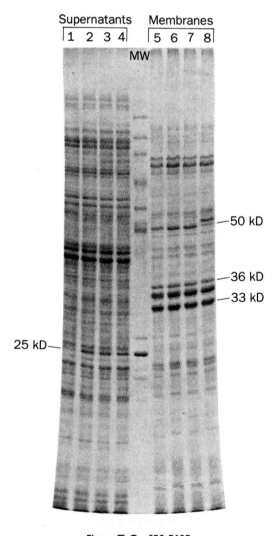

Figure 5-9 SDS-PAGE.

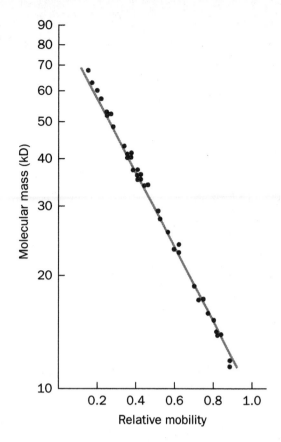

Figure 5-10 Logarithmic relationship between the molecular mass of a protein and its electrophoretic mobility in SDS-PAGE.

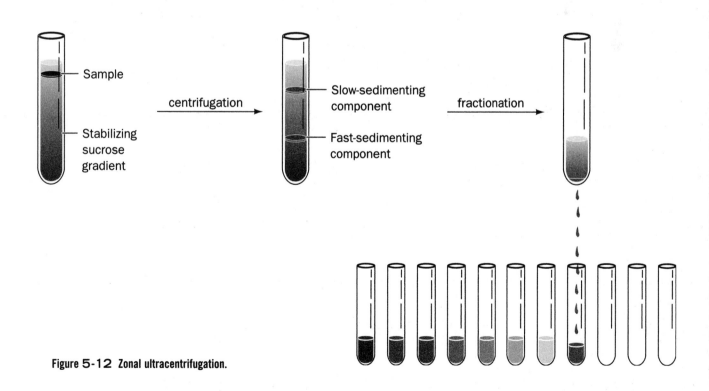

Figure 5-12 Zonal ultracentrifugation.

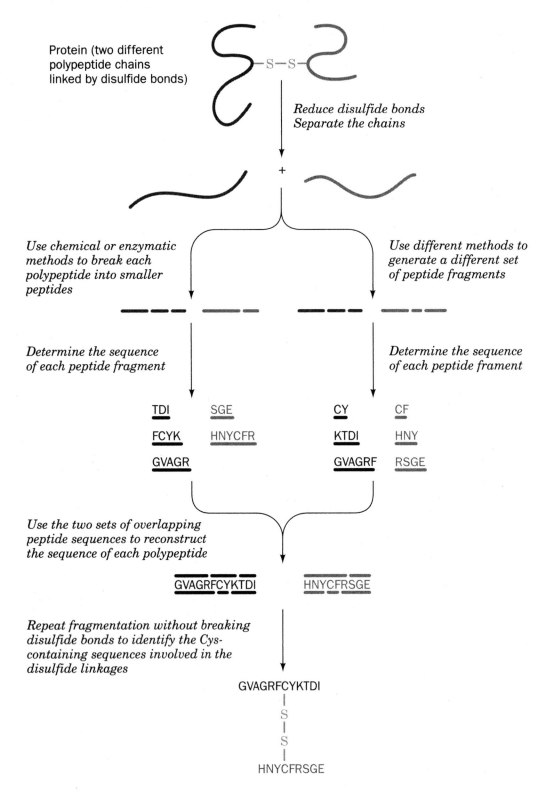

Figure 5-13 Overview of protein sequencing.

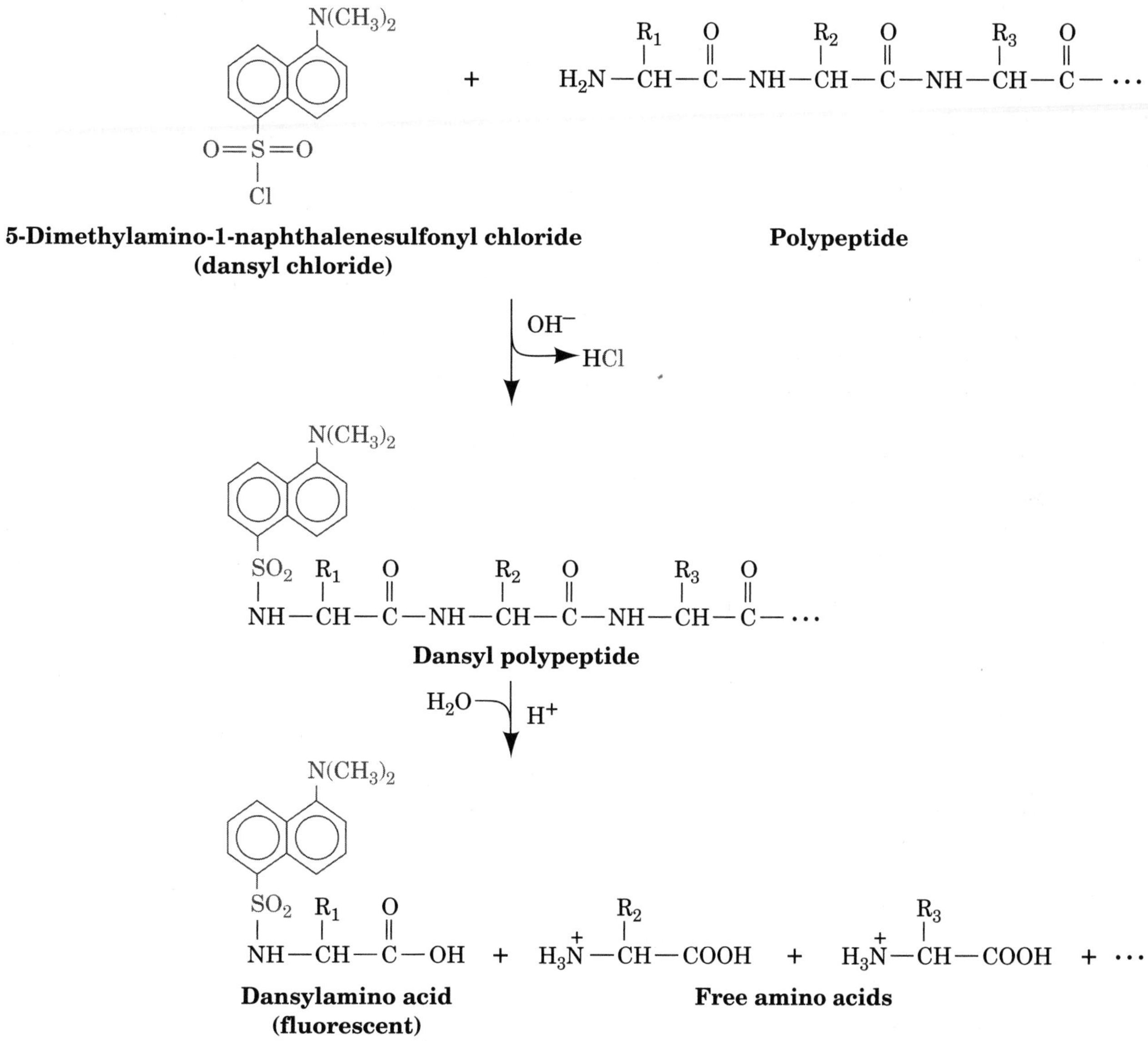

5-Dimethylamino-1-naphthalenesulfonyl chloride (dansyl chloride)

Polypeptide

Dansyl polypeptide

Dansylamino acid (fluorescent)

Free amino acids

Figure 5-14 The dansyl chloride reaction.

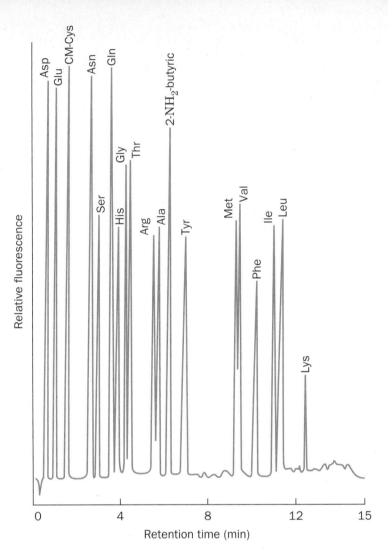

Figure 5-15 Amino acid analysis.

Table 5-3 Specificities of Various Endopeptidases

$$-NH-CH-\overset{\overset{R_{n-1}}{|}}{C}-\overset{\overset{O}{\parallel}}{\underset{\uparrow}{C}}-NH-CH-\overset{\overset{R_n}{|}}{C}-\overset{\overset{O}{\parallel}}{C}-$$

Scissile
peptide bond

Enzyme	Source	Specificity	Comments
Trypsin	Bovine pancreas	R_{n-1} = positively charged residues: Arg, Lys; $R_n \neq$ Pro	Highly specific
Chymotrypsin	Bovine pancreas	R_{n-1} = bulky hydrophobic residues: Phe, Trp, Tyr; $R_n \neq$ Pro	Cleaves more slowly for R_{n-1} = Asn, His, Met, Leu
Elastase	Bovine pancreas	R_{n-1} = small neutral residues: Ala, Gly, Ser, Val; $R_n \neq$ Pro	
Thermolysin	*Bacillus thermoproteolyticus*	R_n = Ile, Met, Phe, Trp, Tyr, Val; $R_{n-1} \neq$ Pro	Occasionally cleaves at R_n = Ala, Asp, His, Thr; heat stable
Pepsin	Bovine gastric mucosa	R_n = Leu, Phe, Trp, Tyr; $R_{n-1} \neq$ Pro	Also others; quite nonspecific; pH optimum = 2
Endopeptidase V8	*Staphylococcus aureus*	R_{n-1} = Glu	

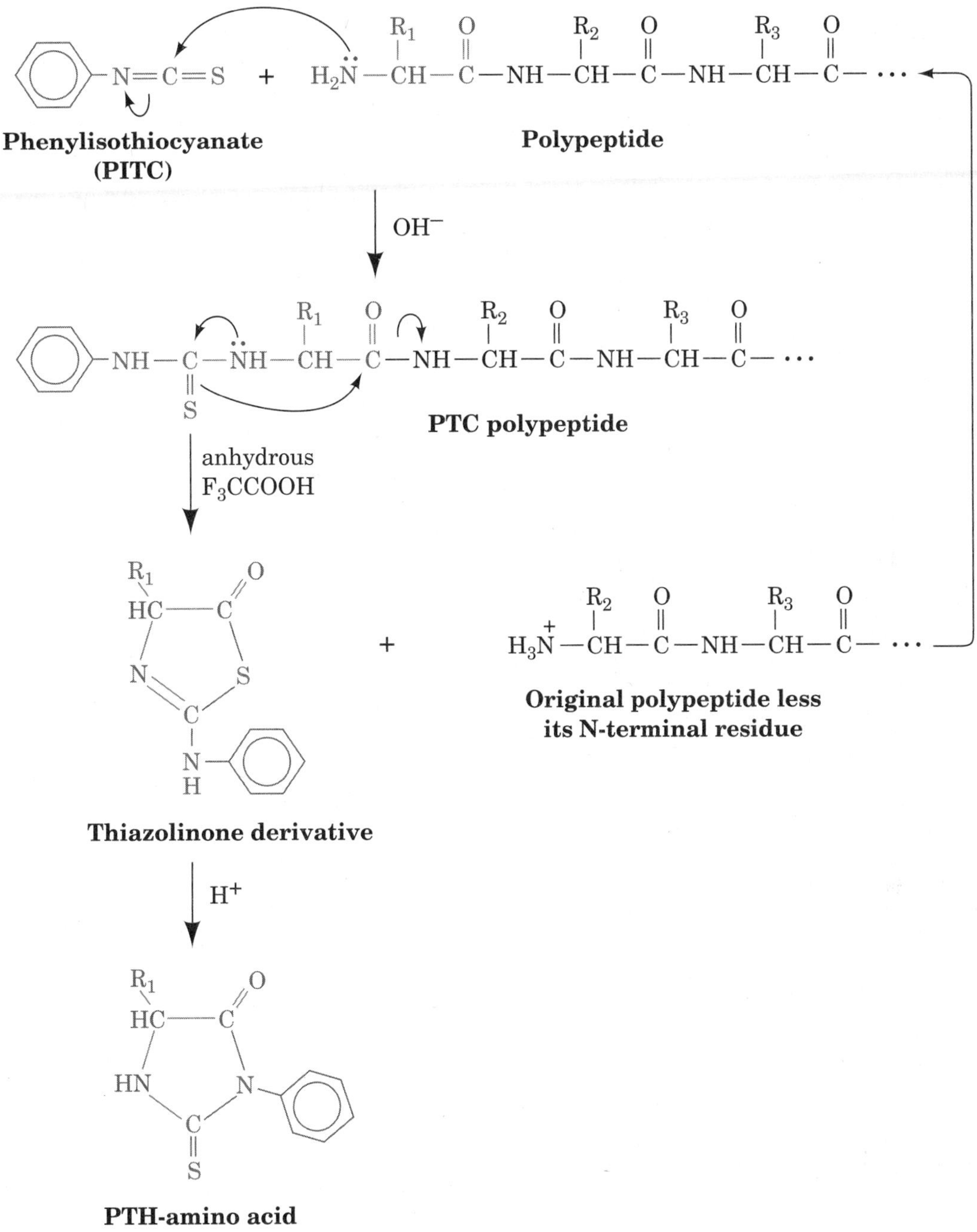

Figure 5-17 Edman degradation.

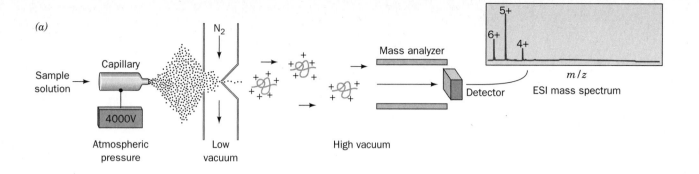

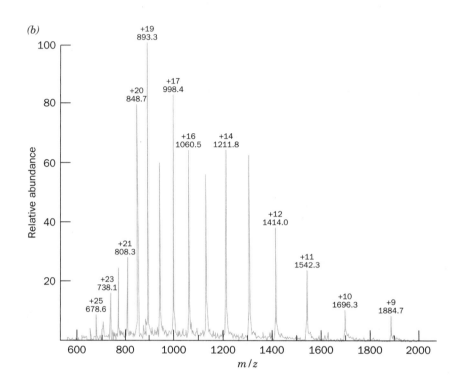

Figure 5-18 Electrospray ionization mass spectrometry (ESI).

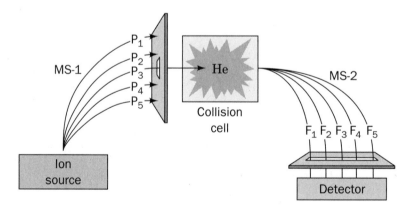

Figure 5-19 Tandem mass spectrometry in amino acid sequencing.

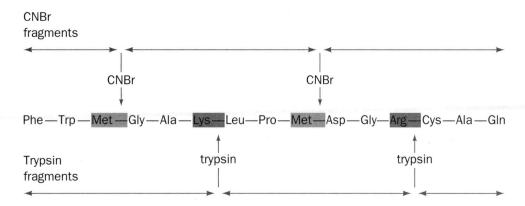

Figure 5-20 Generating overlapping fragments to determine the amino acid sequence of a polypeptide.

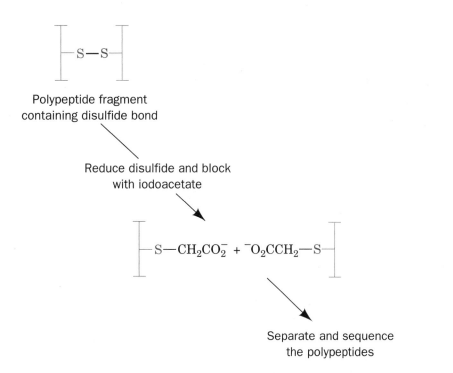

Figure 5-21 Determining the positions of disulfide bonds.

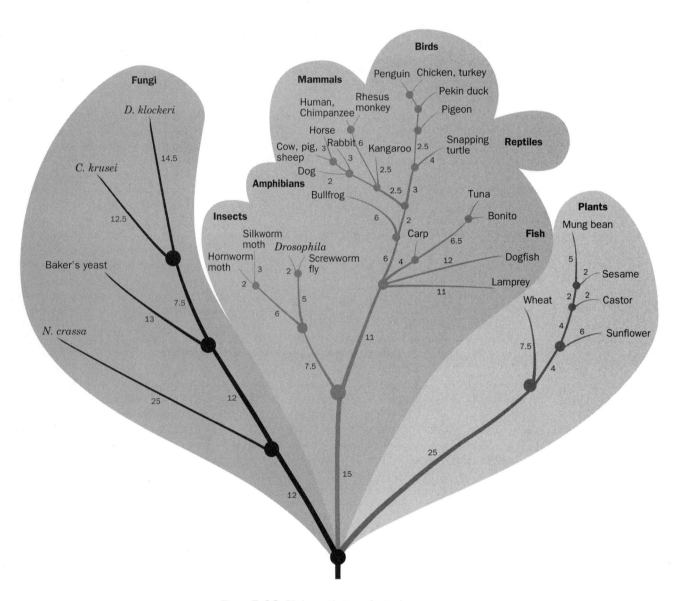

Figure 5-23 Phylogenetic tree of cytochrome *c*.

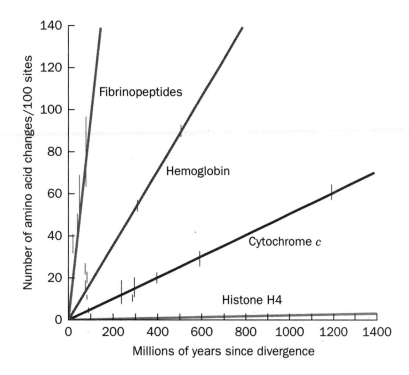

Figure 5-24 Rates of evolution of four proteins.

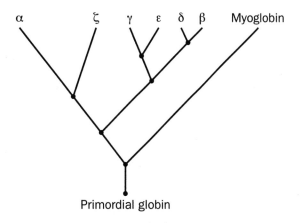

Figure 5-25 Genealogy of the globin family.

CHAPTER 6

Proteins: Three-Dimensional Structure

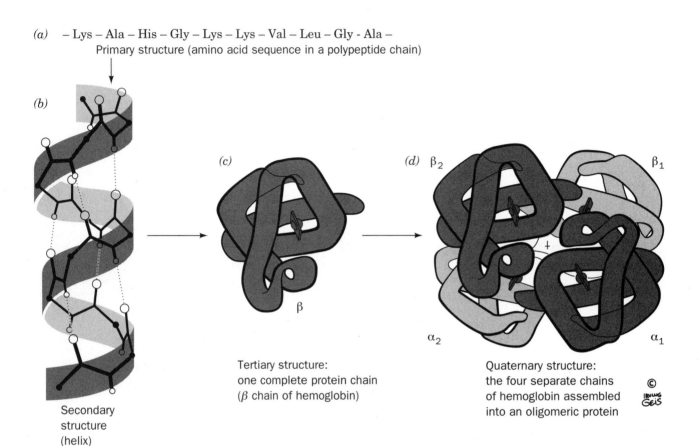

(a)   – Lys – Ala – His – Gly – Lys – Lys – Val – Leu – Gly - Ala –

Primary structure (amino acid sequence in a polypeptide chain)

(b)

(c)

(d) β₂   β₁

α₂   α₁

Secondary
structure
(helix)

Tertiary structure:
one complete protein chain
(β chain of hemoglobin)

Quaternary structure:
the four separate chains
of hemoglobin assembled
into an oligomeric protein

© IRVING GEIS

Figure 6-1 Levels of protein structure.

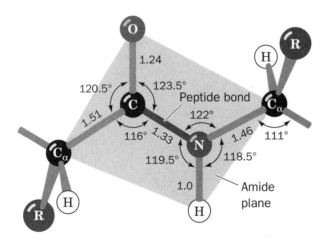

Figure 6-2 The trans peptide group.

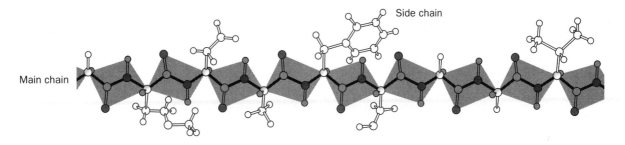

Figure 6-3 Extended conformation of a polypeptide.

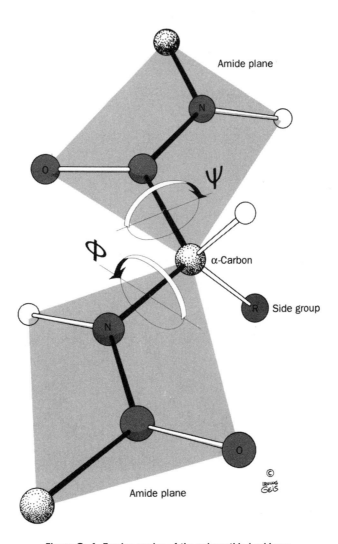

Figure 6-4 Torsion angles of the polypeptide backbone.

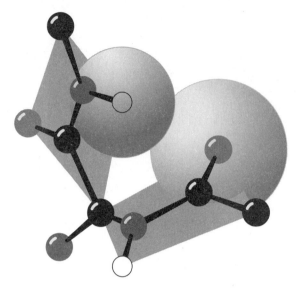

Figure 6-5 Steric interference between adjacent peptide groups.

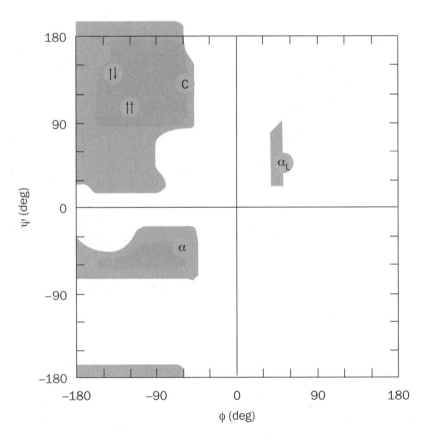

Figure 6-6 The Ramachandran diagram.

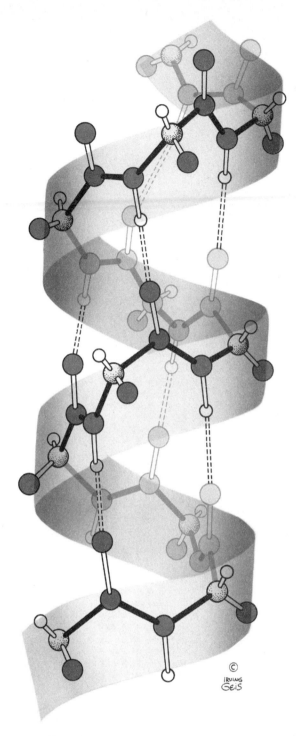

Figure 6-7 *Key to Structure.* The α helix.

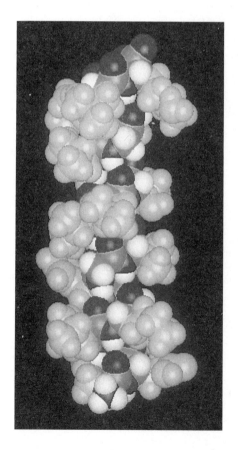

Figure 6-8 Space-filling model of an α helix.

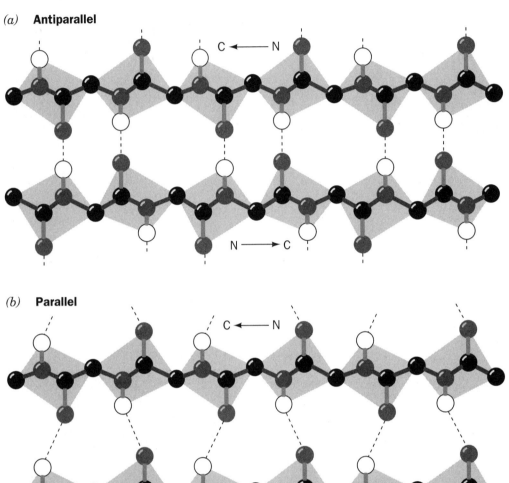

(a) **Antiparallel**

C ← N

N → C

(b) **Parallel**

C ← N

C ← N

Figure 6-9 *Key to Structure.* β Sheets.

Figure 6-12 Diagram of a β sheet in bovine carboxypeptidase A.

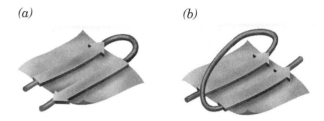

Figure 6-13 Connections between adjacent strands in β sheets.

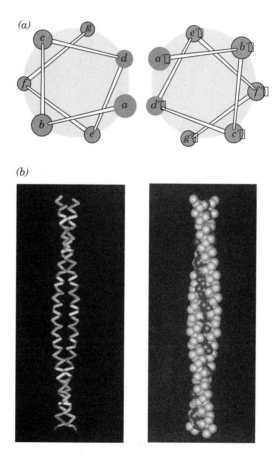

Figure 6-14 A coiled coil.

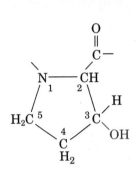

4-Hydroxyprolyl residue (Hyp)

3-Hydroxyprolyl residue

5-Hydroxylysyl residue (Hyl)

Figure 6-16 The collagen triple helix.

Table 6-1 Propensities of Amino Acid Residues for α Helical and β Sheet Conformations

Residue	P_α	P_β
Ala	1.42	0.83
Arg	0.98	0.93
Asn	0.67	0.89
Asp	1.01	0.54
Cys	0.70	1.19
Gln	1.11	1.10
Glu	1.51	0.37
Gly	0.57	0.75
His	1.00	0.87
Ile	1.08	1.60
Leu	1.21	1.30
Lys	1.16	0.74
Met	1.45	1.05
Phe	1.13	1.38
Pro	0.57	0.55
Ser	0.77	0.75
Thr	0.83	1.19
Trp	1.08	1.37
Tyr	0.69	1.47
Val	1.06	1.70

Source: Chou, P.Y. and Fasman, G.D., *Annu. Rev. Biochem.* **47,** 258 (1978).

56

(a) **Type I**   (b) **Type II**

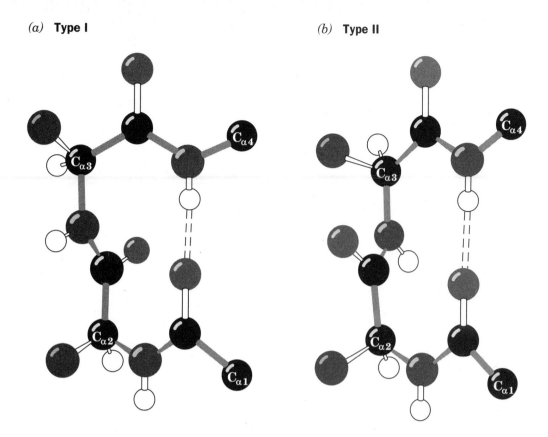

Figure 6-19 Reverse turns in polypeptide chains.

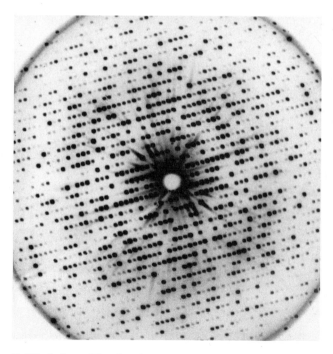

Figure 6-22 An X-ray diffraction photograph of a crystal of sperm whale myoglobin.

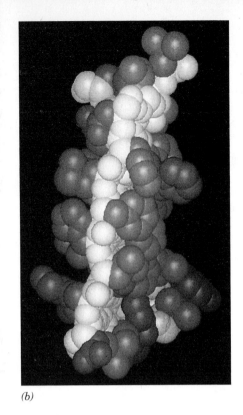

Figure 6-26 Side chain locations in an α helix and a β sheet.

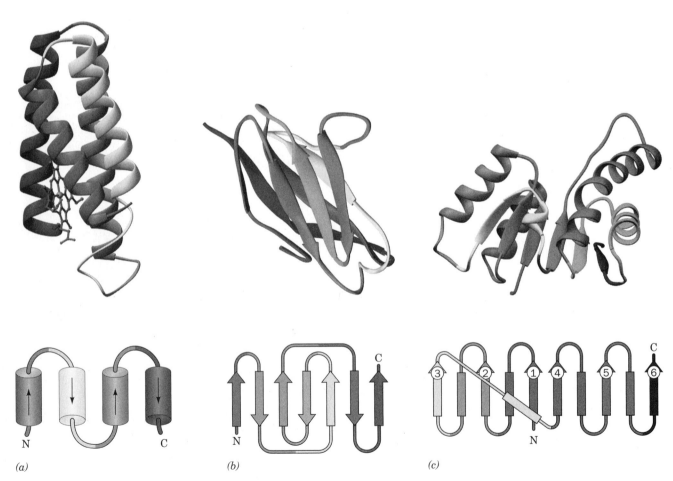

(a)

(b)

(c)

Figure 6-28 A selection of protein structures.

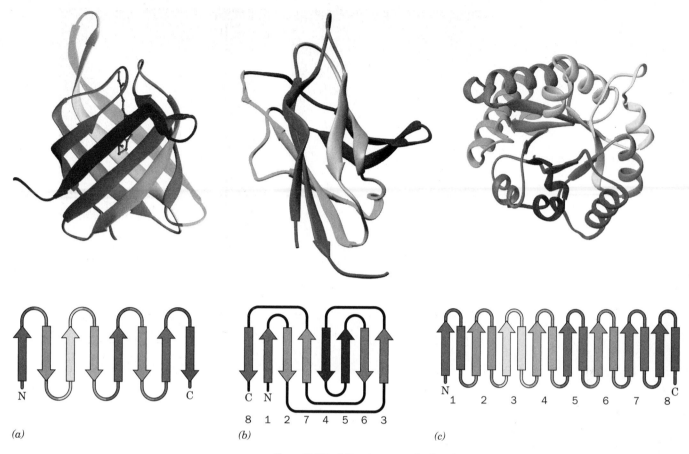

(b)

(c)

Figure 6-30 X-Ray structures of β barrels.

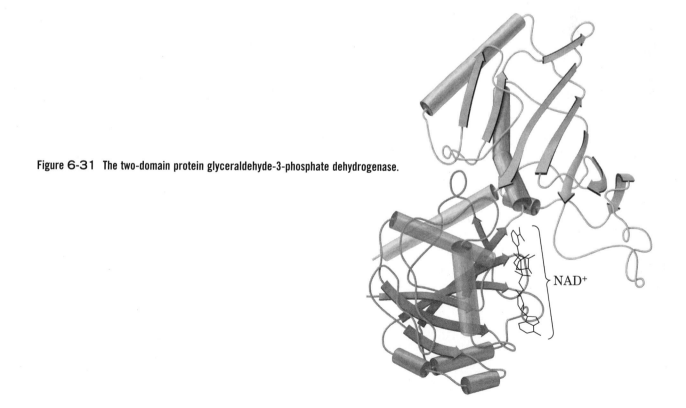

Figure 6-31 The two-domain protein glyceraldehyde-3-phosphate dehydrogenase.

NAD⁺

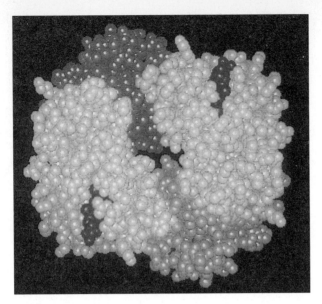

Figure 6-33 Quaternary structure of hemoglobin.

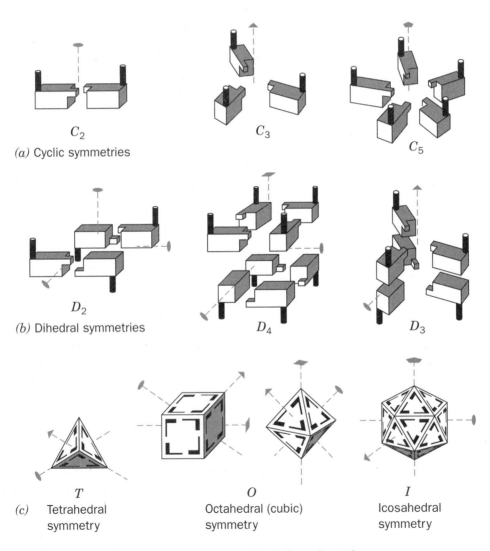

C_2

(a) Cyclic symmetries

C_3

C_5

D_2

(b) Dihedral symmetries

D_4

D_3

T
Tetrahedral
symmetry

O
Octahedral (cubic)
symmetry

I
Icosahedral
symmetry

Figure 6-34 Some symmetries of oligomeric proteins.

Table 6-2 Hydropathy Scale for Amino Acid Side Chains

Side Chain	Hydropathy
Ile	4.5
Val	4.2
Leu	3.8
Phe	2.8
Cys	2.5
Met	1.9
Ala	1.8
Gly	−0.4
Thr	−0.7
Ser	−0.8
Trp	−0.9
Tyr	−1.3
Pro	−1.6
His	−3.2
Glu	−3.5
Gln	−3.5
Asp	−3.5
Asn	−3.5
Lys	−3.9
Arg	−4.5

Source: Kyte, J. and Doolittle, R.F., *J. Mol. Biol.* **157,** 110 (1982).

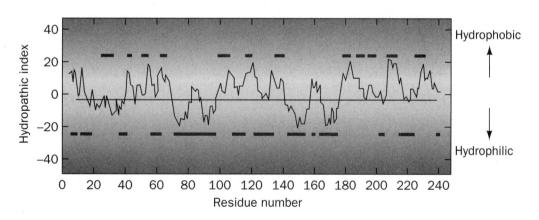

Figure 6-35 A hydropathic index plot for bovine chymotrypsinogen.

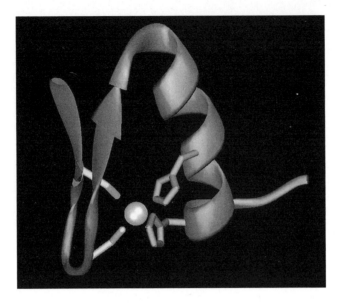

Figure 6-37 A zinc finger motif.

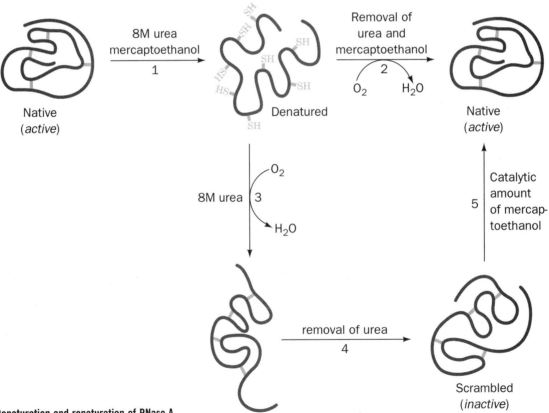

Native
(*active*)

8M urea
mercaptoethanol
1

SH SH SH
HS SH
HS SH
SH
SH
Denatured

Removal of
urea and
mercaptoethanol
2
O_2 H_2O

Native
(*active*)

8M urea

O_2
3

H_2O

removal of urea
4

Scrambled
(*inactive*)

5

Catalytic
amount
of mercap-
toethanol

Figure 6-39 Denaturation and renaturation of RNase A.

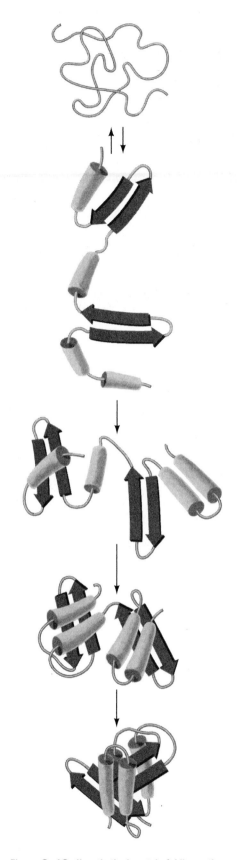

Figure 6-40 Hypothetical protein folding pathway.

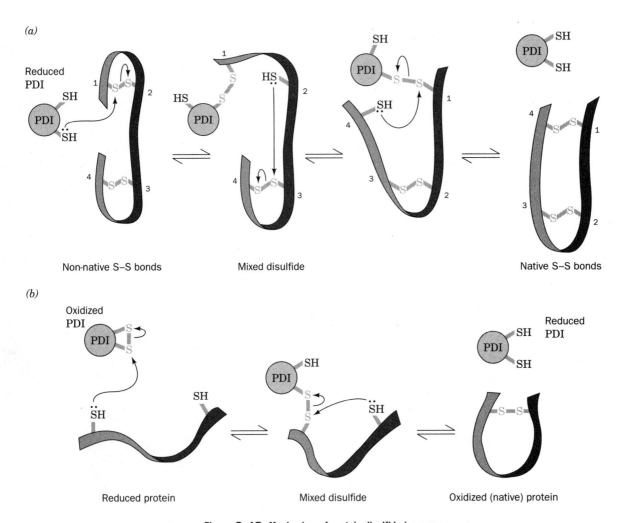

(a)

Reduced PDI

Non-native S–S bonds   Mixed disulfide   Native S–S bonds

(b)

Oxidized PDI   Reduced PDI

Reduced protein   Mixed disulfide   Oxidized (native) protein

Figure 6-42 Mechanism of protein disulfide isomerase.

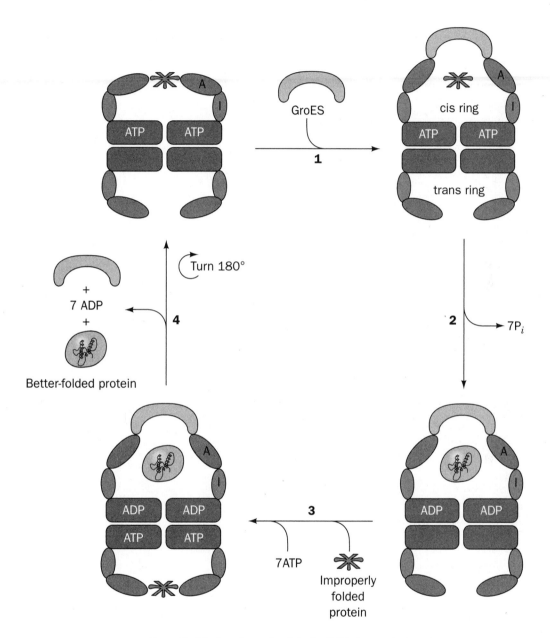

Figure 6-46 Reaction cycle of the GroEL/ES chaperonin.

CHAPTER 7

Protein Function: Myoglobin and Hemoglobin

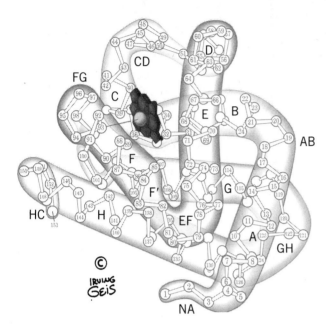

Figure 7-1 Structure of sperm whale myoglobin.

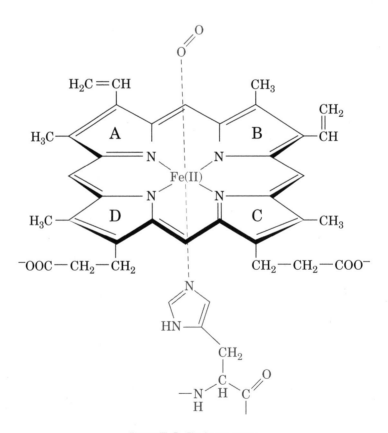

Figure 7-2 The heme group.

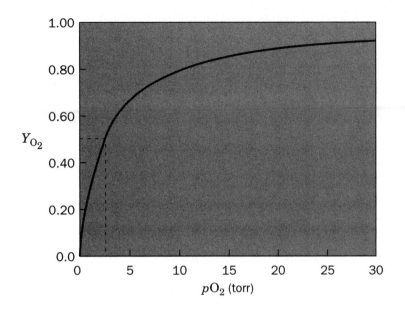

Figure 7-4 Oxygen-binding curve of myoglobin.

$$Mb + O_2 \rightleftharpoons MbO_2$$

$$K = \frac{[Mb][O_2]}{[MbO_2]} \qquad [7\text{-}1]$$

$$Y_{O_2} = \frac{[MbO_2]}{[Mb] + [MbO_2]} \qquad [7\text{-}2]$$

$$Y_{O_2} = \frac{[O_2]}{K + [O_2]} \qquad [7\text{-}5]$$

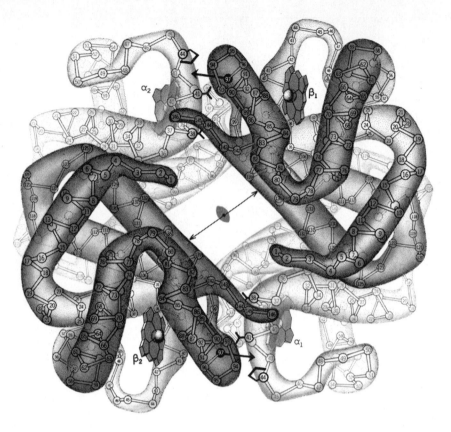

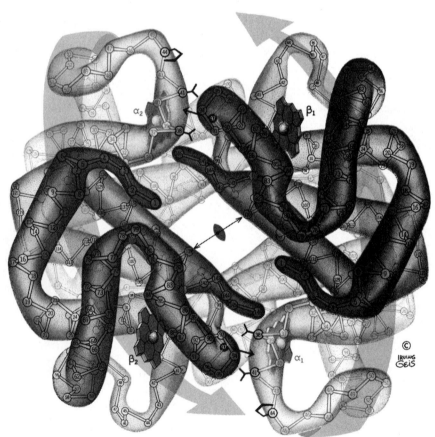

Figure 7-5 Hemoglobin structure.

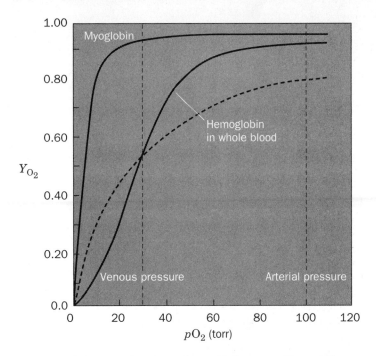

Figure 7-7 *Key to Function.* Oxygen-binding curve of hemoglobin.

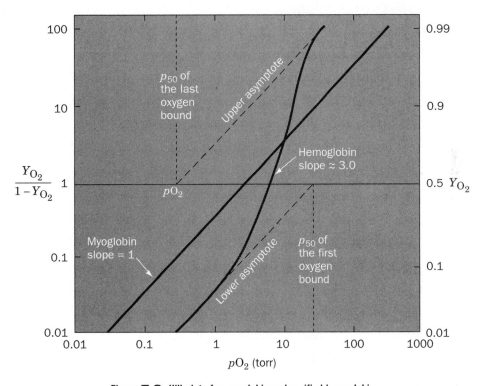

Figure 7-8 Hill plots for myoglobin and purified hemoglobin.

$$\log \left(\frac{Y_{O_2}}{1 - Y_{O_2}} \right) = n \log pO_2 - n \log p_{50}$$

[7-11]

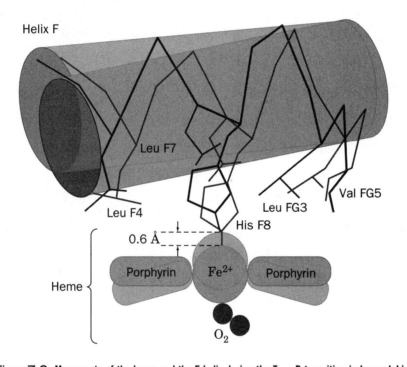

Figure 7-9 Movements of the heme and the F helix during the T → R transition in hemoglobin.

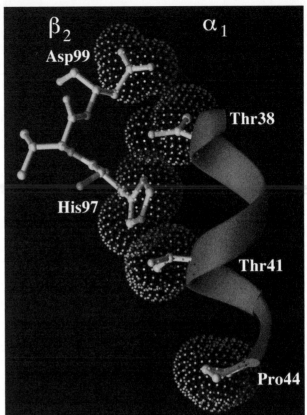

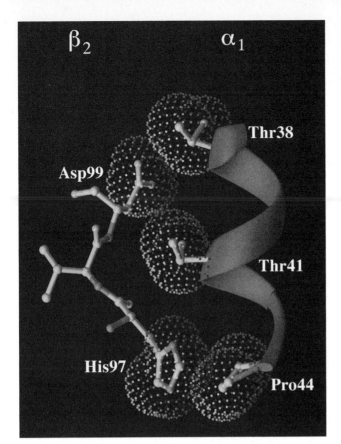

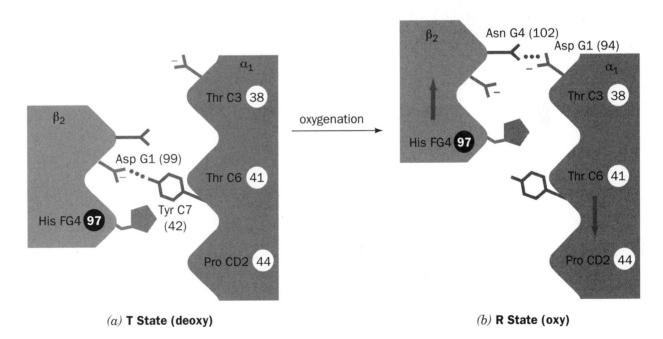

(a) **T State (deoxy)**

oxygenation

(b) **R State (oxy)**

Figure 7-10 Changes at the α_1–β_2 interface during T → R transition in hemoglobin.

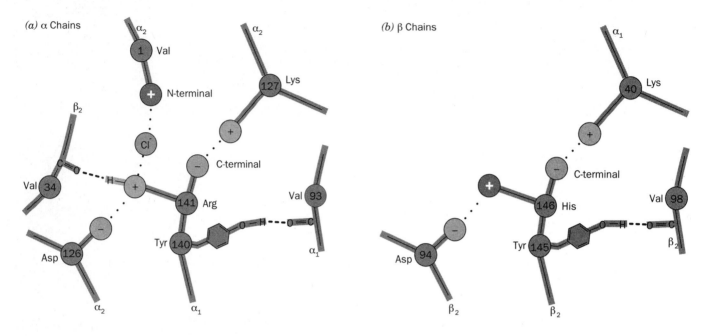

Figure **7-11** Networks of ion pairs and hydrogen bonds in deoxyhemoglobin.

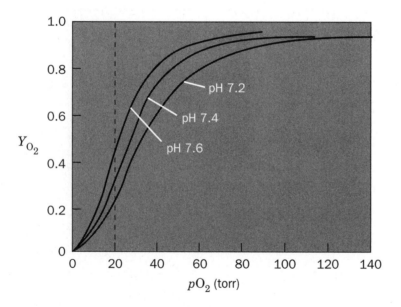

Figure **7-12** The Bohr effect.

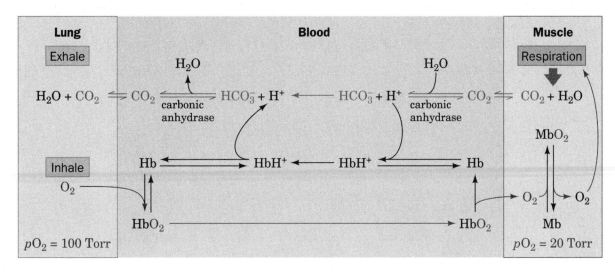

Figure 7-13 *Key to Function.* The roles of hemoglobin and myoglobin in transporting O_2 from the lungs to respiring tissues and CO_2 (as HCO_3^-) from the tissues to the lungs.

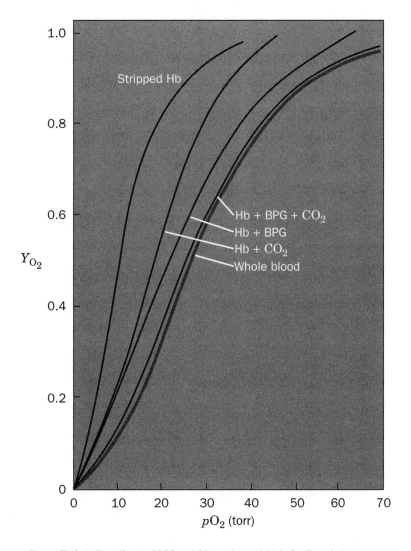

Figure 7-14 The effects of BPG and CO_2 on hemoglobin's O_2 dissociation curve.

$$\text{D-2,3-Bisphosphoglycerate (BPG)}$$

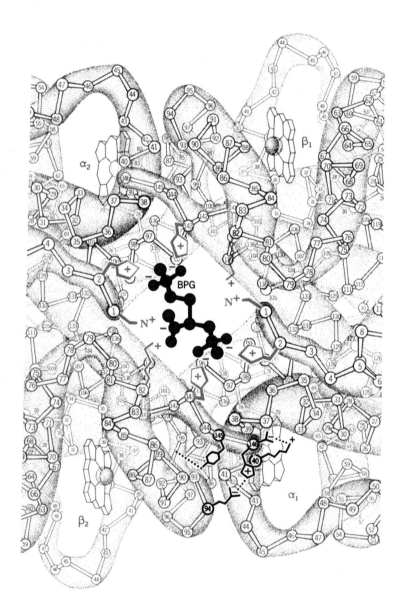

Figure 7-15 Binding of BPG to deoxyhemoglobin.

Table 7-1 Some Hemoglobin Variants

Name[a]	Mutation	Effect
Hammersmith	Phe CD1(42)β → Ser	Weakens heme binding
Bristol	Val E11(67)β → Asp	Weakens heme binding
Bibba	Leu H19(136)α → Pro	Disrupts the H helix
Savannah	Gly B6(24)β → Val	Disrupts the B–E helix interface
Philly	Tyr C1(35)α → Phe	Disrupts hydrogen bonding at the α_1–β_1 interface
Boston	His E7(58)α → Tyr	Promotes methemoglobin formation
Milwaukee	Val E11(67)β → Glu	Promotes methemoglobin formation
Iwate	His F8(87)α → Tyr	Promotes methemoglobin formation
Yakima	Asp G1(99)β → His	Disrupts a hydrogen bond that stabilizes the T conformation
Kansas	Asn G4(102)β → Thr	Distrupts a hydrogen bond that stabilizes the R conformation

[a]Hemoglobin variants are usually named after the place where they were discovered (e.g., hemoglobin Boston).

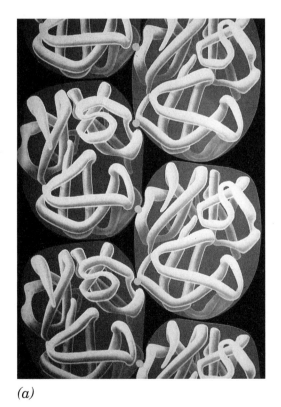

(a)

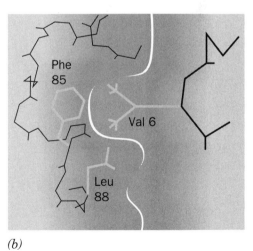

(b)

Figure 7-19 Structure of a deoxyhemoglobin S fiber.

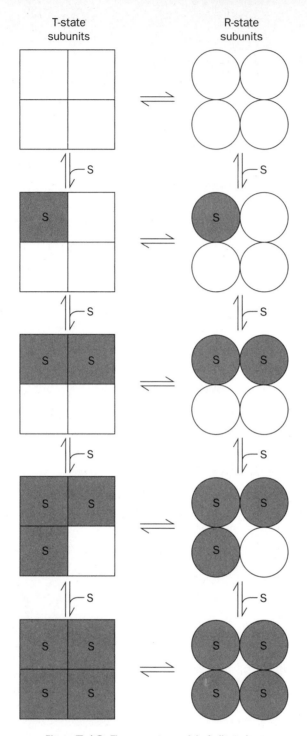

Figure 7-16 The symmetry model of allosterism.

Figure 7-17 The sequential model of allosterism.

CHAPTER 8

Carbohydrates

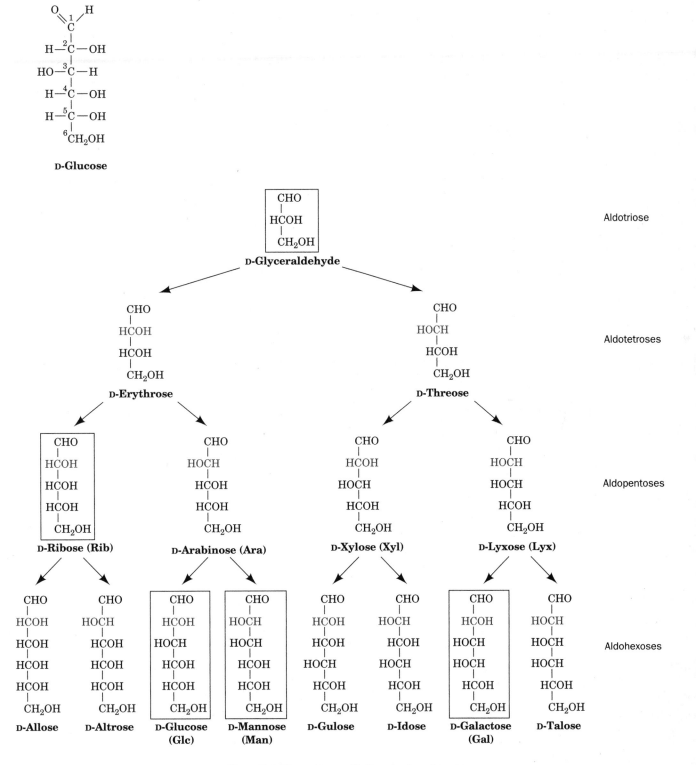

D-Glucose

Figure 8-1 The D-aldoses with three to six carbon atoms.

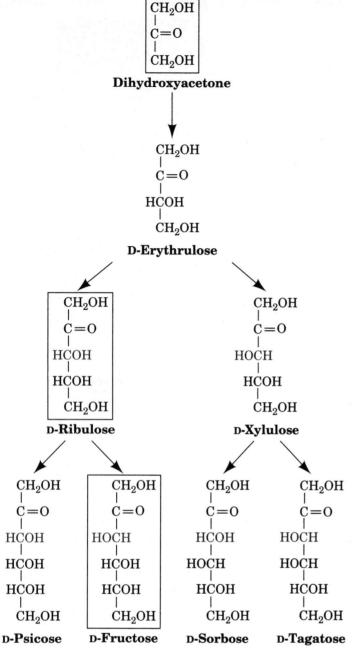

Figure 8-2 The D-ketoses with three to six carbon atoms.

R—OH + R'—C(H)(=O) ⇌ R—O—C(H)(R')(OH)

Alcohol　**Aldehyde**　　**Hemiacetal**

R—OH + R'—C(R'')(=O) ⇌ R—O—C(R'')(R')(OH)

Alcohol　**Ketone**　　**Hemiketal**

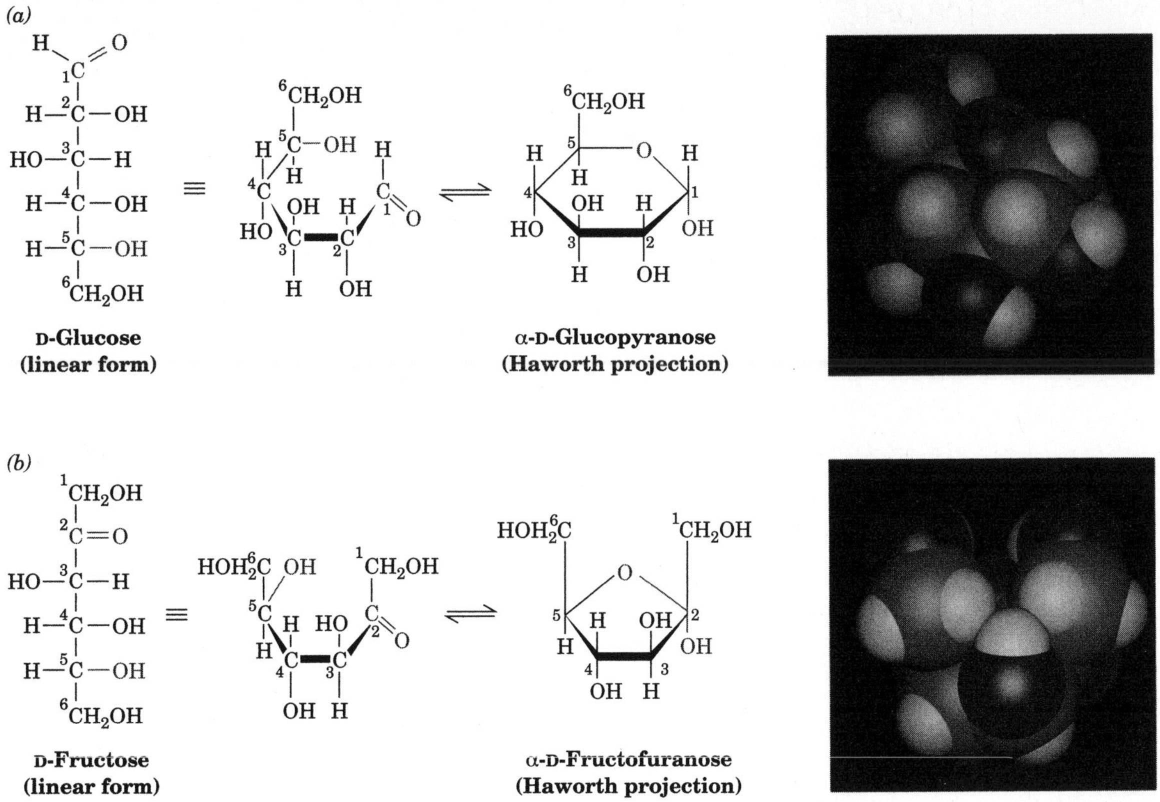

(a)

D-Glucose
(linear form)

α-D-Glucopyranose
(Haworth projection)

(b)

D-Fructose
(linear form)

α-D-Fructofuranose
(Haworth projection)

Figure 8-3 *Key to Structure.* Cyclization of glucose and fructose.

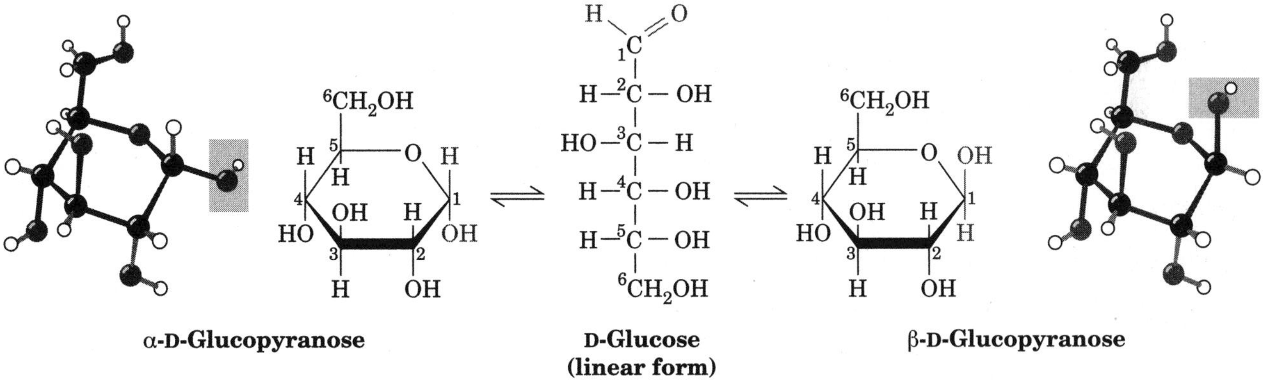

α-D-Glucopyranose

D-Glucose
(linear form)

β-D-Glucopyranose

Figure 8-4 α and β anomers.

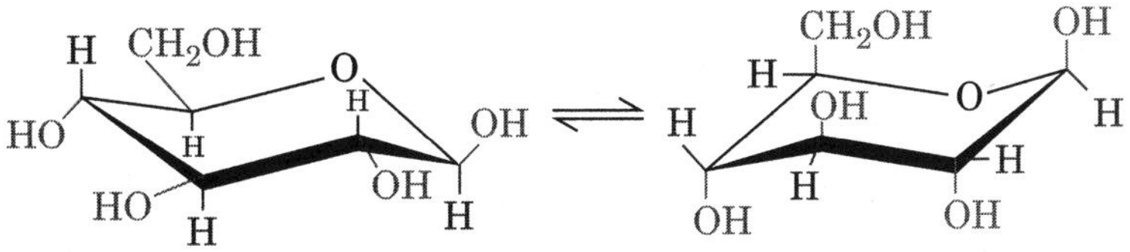

Figure 8-5 The two chair conformations of β-D-glucopyranose.

COOH

H—C—OH

HO—C—H

H—C—OH

H—C—OH

CH₂OH

D-Gluconic acid

O   H

C

H—C—OH

HO—C—H

H—C—OH

H—C—OH

COOH

D-Glucuronic acid

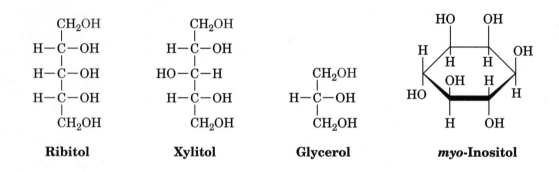

Ribitol

Xylitol

Glycerol

myo-**Inositol**

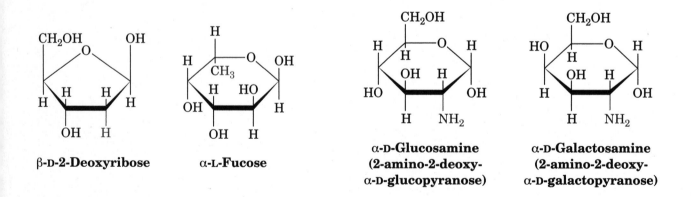

β-D-2-Deoxyribose

α-L-Fucose

**α-D-Glucosamine
(2-amino-2-deoxy-
α-D-glucopyranose)**

**α-D-Galactosamine
(2-amino-2-deoxy-
α-D-galactopyranose)**

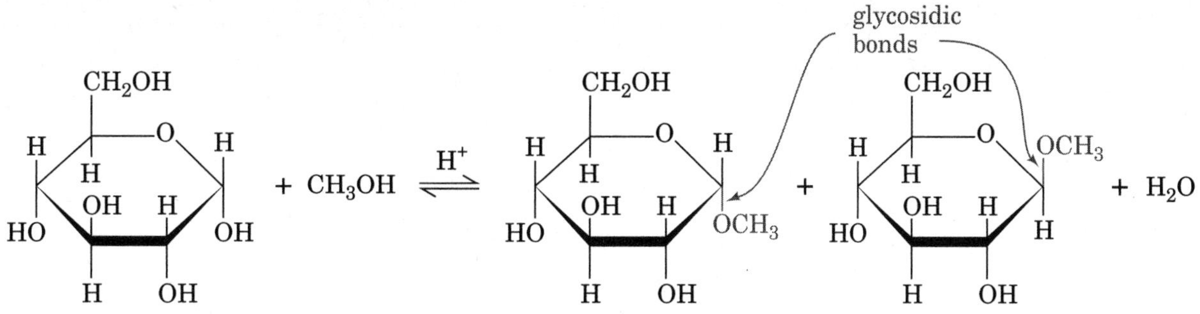

Figure 8-7 Formation of glycosides.

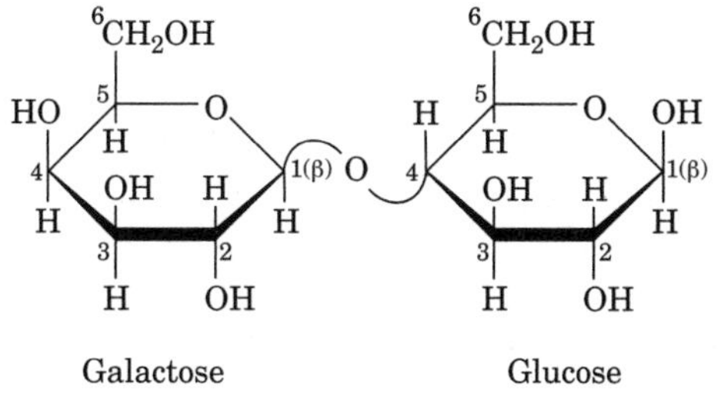

Galactose Glucose

Lactose

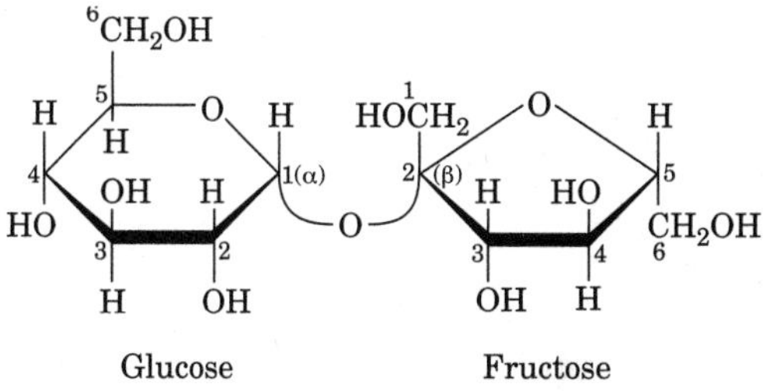

Glucose Fructose

Sucrose

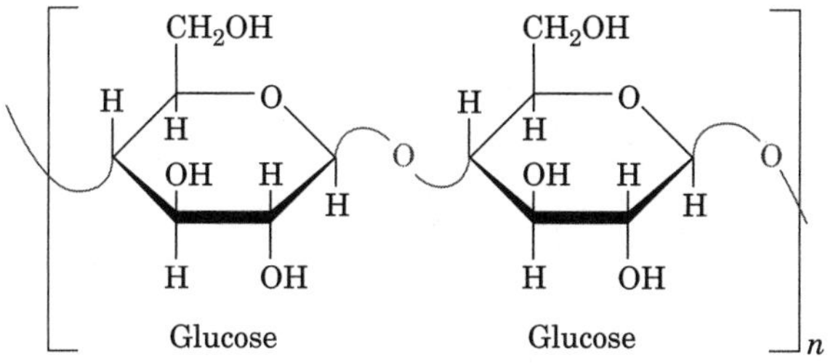

Glucose Glucose

Cellulose

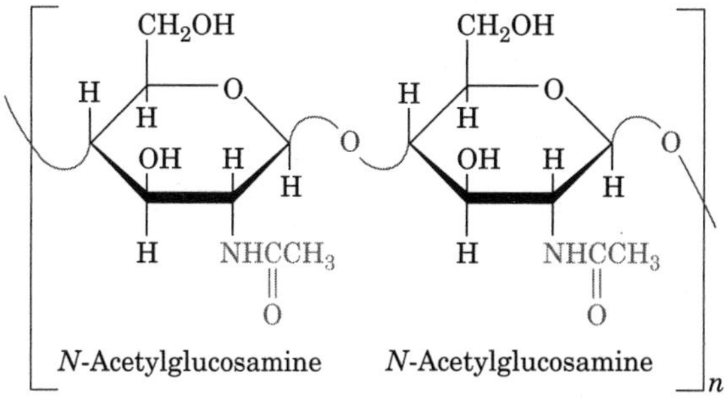

N-Acetylglucosamine N-Acetylglucosamine

Chitin

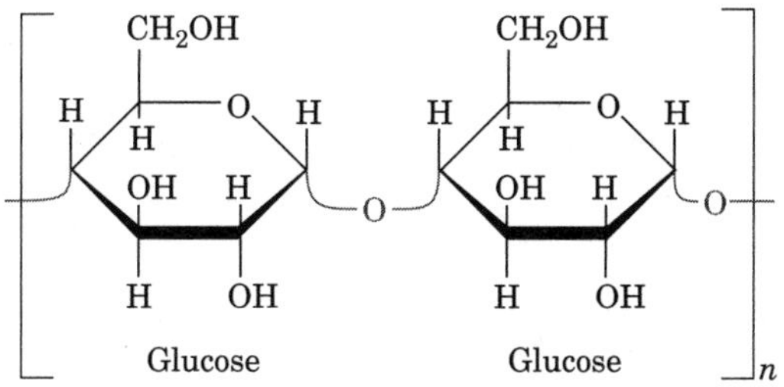

Glucose Glucose

α-Amylose

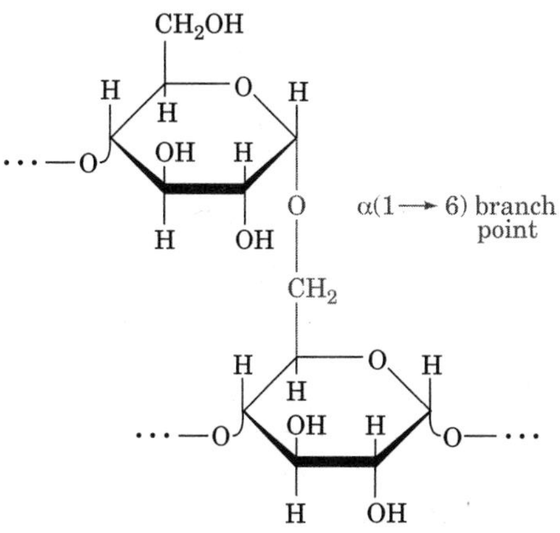

α(1 → 6) branch point

Amylopectin

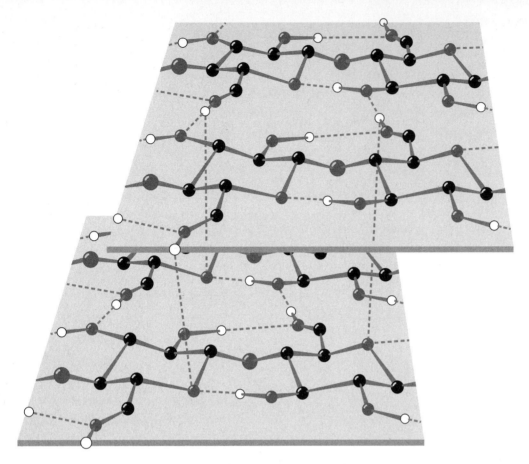

Figure 8-9 Model of cellulose.

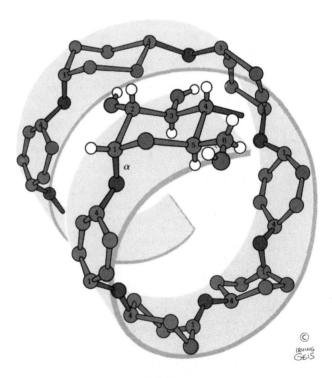

Figure 8-10 a-Amylose.

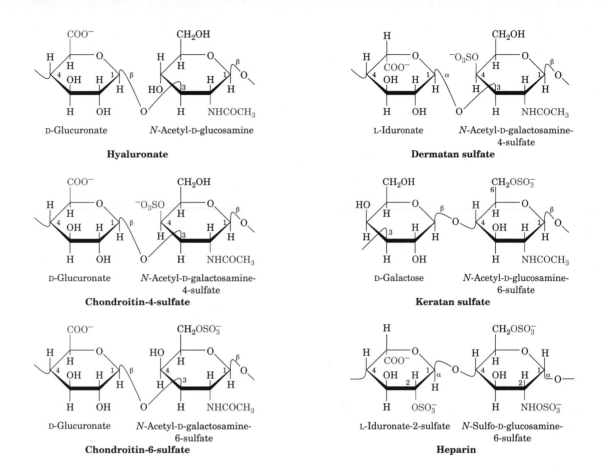

Figure 8-12 Repeating disaccharide units of some glycosaminoglycans.

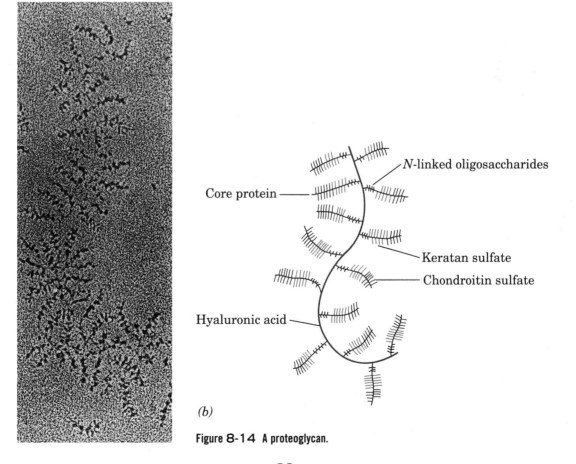

(a)

(b)

Figure 8-14 A proteoglycan.

83

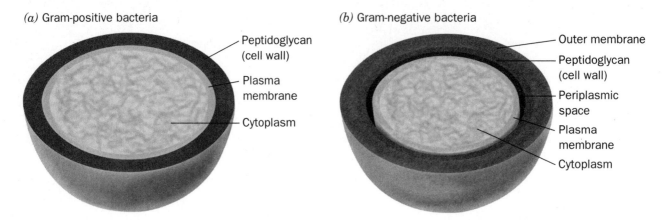

(a) Gram-positive bacteria

Peptidoglycan (cell wall)

Plasma membrane

Cytoplasm

(b) Gram-negative bacteria

Outer membrane

Peptidoglycan (cell wall)

Periplasmic space

Plasma membrane

Cytoplasm

Figure 8-15 Bacterial cell walls.

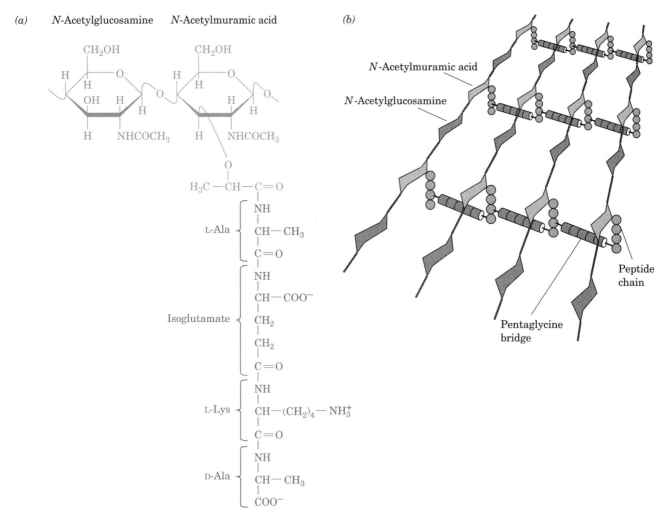

(a) *N*-Acetylglucosamine   *N*-Acetylmuramic acid

L-Ala

Isoglutamate

L-Lys

D-Ala

(b)

N-Acetylmuramic acid

N-Acetylglucosamine

Peptide chain

Pentaglycine bridge

Figure 8-16 Peptidoglycan.

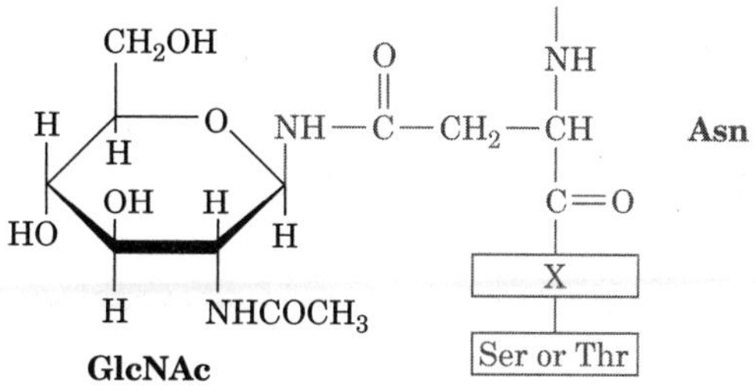
CH$_2$OH
O
NH
H
O
NH—C—CH$_2$—CH
Asn
H
OH
H
C=O
HO
H
X
H
NHCOCH$_3$
Ser or Thr
GlcNAc

Man α(1→6)
Man β(1→4) GlcNAc β(1→4) GlcNAc—
Man α(1→3)

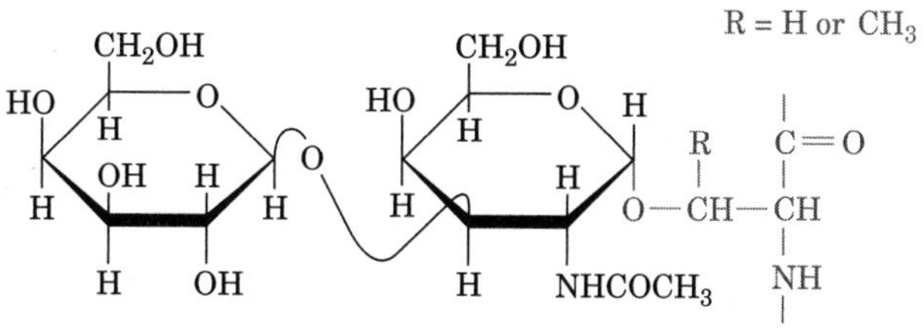
R = H or CH$_3$
CH$_2$OH
CH$_2$OH
HO
O
HO
O
H
H
H
R
C=O
O
OH
H
H
H
H
H
O—CH—CH
NH
H
OH
H
NHCOCH$_3$

β-Galactosyl-(1→3)-α-*N*-acetylgalactosaminyl-Ser/Thr

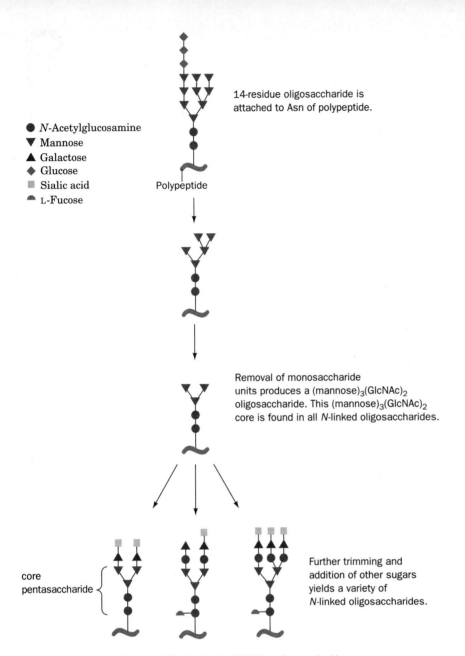

- ● N-Acetylglucosamine
- ▼ Mannose
- ▲ Galactose
- ◆ Glucose
- ▉ Sialic acid
- ◖ L-Fucose

14-residue oligosaccharide is attached to Asn of polypeptide.

Polypeptide

Removal of monosaccharide units produces a (mannose)$_3$(GlcNAc)$_2$ oligosaccharide. This (mannose)$_3$(GlcNAc)$_2$ core is found in all N-linked oligosaccharides.

core pentasaccharide

Further trimming and addition of other sugars yields a variety of N-linked oligosaccharides.

Figure 8-17 Synthesis of N-linked oligosaccharides.

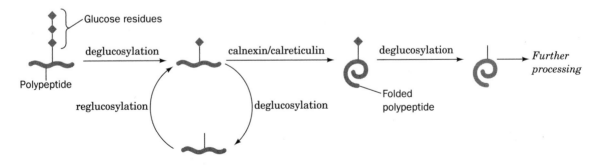

Glucose residues

Polypeptide

deglucosylation

reglucosylation

calnexin/calreticulin

deglucosylation

deglucosylation

Folded polypeptide

Further processing

Figure 8-19 Quality control in oligosaccharide processing.

CHAPTER 9

Lipids and Biological Membranes

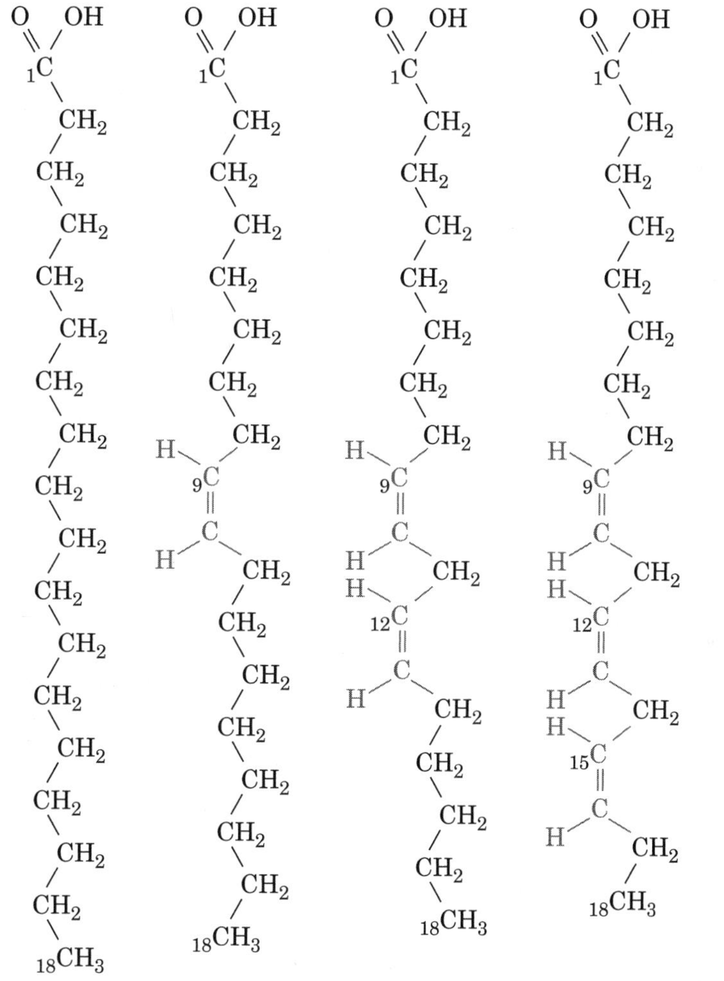

Stearic acid   Oleic acid   Linoleic acid   □-Linolenic acid

Figure 9-1 The structural formulas of some C_{18} fatty acids.

Table 9-1 The Common Biological Fatty Acids

Symbol[a]	Common Name	Systematic Name	Structure	mp (°C)
Saturated fatty acids				
12:0	Lauric acid	Dodecanoic acid	$CH_3(CH_2)_{10}COOH$	44.2
14:0	Myristic acid	Tetradecanoic acid	$CH_3(CH_2)_{12}COOH$	52
16:0	Palmitic acid	Hexadecanoic acid	$CH_3(CH_2)_{14}COOH$	63.1
18:0	Stearic acid	Octadecanoic acid	$CH_3(CH_2)_{16}COOH$	69.1
20:0	Arachidic acid	Eicosanoic acid	$CH_3(CH_2)_{18}COOH$	75.4
22:0	Behenic acid	Docosanoic acid	$CH_3(CH_2)_{20}COOH$	81
24:0	Lignoceric acid	Tetracosanoic acid	$CH_3(CH_2)_{22}COOH$	84.2
Unsaturated fatty acids (all double bonds are cis)				
16:1$n-7$	Palmitoleic acid	9-Hexadecenoic acid	$CH_3(CH_2)_5CH=CH(CH_2)_7COOH$	−0.5
18:1$n-9$	Oleic acid	9-Octadecenoic acid	$CH_3(CH_2)_7CH=CH(CH_2)_7COOH$	13.2
18:2$n-6$	Linoleic acid	9,12-Octadecadienoic acid	$CH_3(CH_2)_4(CH=CHCH_2)_2(CH_2)_6COOH$	−9
18:3$n-3$	α-Linolenic acid	9,12,15-Octadecatrienoic acid	$CH_3CH_2(CH=CHCH_2)_3(CH_2)_6COOH$	−17
18:3$n-6$	γ-Linolenic acid	6,9,12-Octadecatrienoic acid	$CH_3(CH_2)_4(CH=CHCH_2)_3(CH_2)_3COOH$	
20:4$n-6$	Arachidonic acid	5,8,11,14-Eicosatetraenoic acid	$CH_3(CH_2)_4(CH=CHCH_2)_4(CH_2)_2COOH$	−49.5
20:5$n-3$	EPA	5,8,11,14,17-Eicosapentaenoic acid	$CH_3CH_2(CH=CHCH_2)_5(CH_2)_2COOH$	−54
22:6$n-3$	DHA	4,7,10,13,16,19-Docosohexenoic acid	$CH_3CH_2(CH=CHCH_2)_6CH_2COOH$	
24:1$n-9$	Nervonic acid	15-Tetracosenoic acid	$CH_3(CH_2)_7CH=CH(CH_2)_{13}COOH$	39

[a]Number of carbon atoms: Number of double bonds. For unsaturated fatty acids, the quantity "$n-x$" indicates the position of the last double bond in the fatty acid, where n is its number of C atoms, and x is the position of the last double-bonded C atom counting from the methyl terminal (ω) end.

Source: Dawson, R.M.C., Elliott, D.C., Elliott, W.H., and Jones, K.M., *Data for Biochemical Research* (3rd ed.), Chapter 8, Clarendon Press (1986).

$$^1CH_2-OH$$
$$^2CH-OH$$
$$^3CH_2-OH$$

Glycerol

$$^1CH_2-O-\overset{\overset{O}{\|}}{C}-R_1$$
$$^2CH-O-\overset{\overset{O}{\|}}{C}-R_2$$
$$^3CH_2-O-\overset{\overset{O}{\|}}{C}-R_3$$

Triacylglycerol

Table 9-2 The Common Classes of Glycerophospholipids

$$R_2-\underset{\underset{O}{\parallel}}{C}-O-CH \quad \begin{array}{l} CH_2-O-\underset{\underset{O}{\parallel}}{C}-R_1 \\ \\ CH_2-O-\underset{\underset{O^-}{\parallel}}{\overset{\overset{O}{\parallel}}{P}}-O-X \end{array}$$

Name of X—OH	Formula of —X	Name of Phospholipid
Water	—H	Phosphatidic acid
Ethanolamine	$-CH_2CH_2NH_3^+$	Phosphatidylethanolamine
Choline	$-CH_2CH_2N(CH_3)_3^+$	Phosphatidylcholine (lecithin)
Serine	$-CH_2CH(NH_3^+)COO^-$	Phosphatidylserine
myo-Inositol		Phosphatidylinositol
Glycerol	$-CH_2CH(OH)CH_2OH$	Phosphatidylglycerol
Phosphatidylglycerol		Diphosphatidylglycerol (cardiolipin)

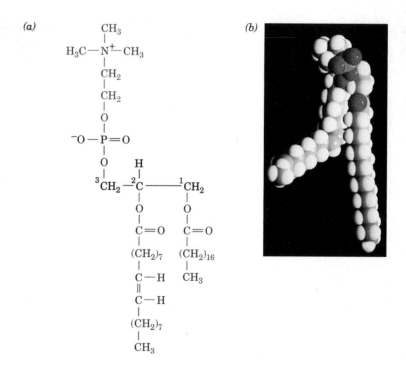

(a)

1-Stearoyl-2-oleoyl-3-phosphatidylcholine

(b)

Figure 9-4 The glycerophospholipid 1-stearoyl-2-oleoyl-3-phosphatidylcholine.

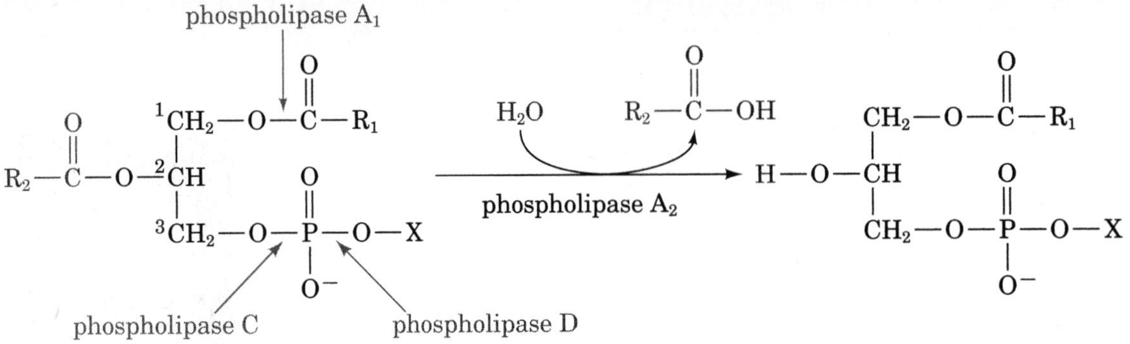

Figure 9-5 Action of phospholipases.

(a)

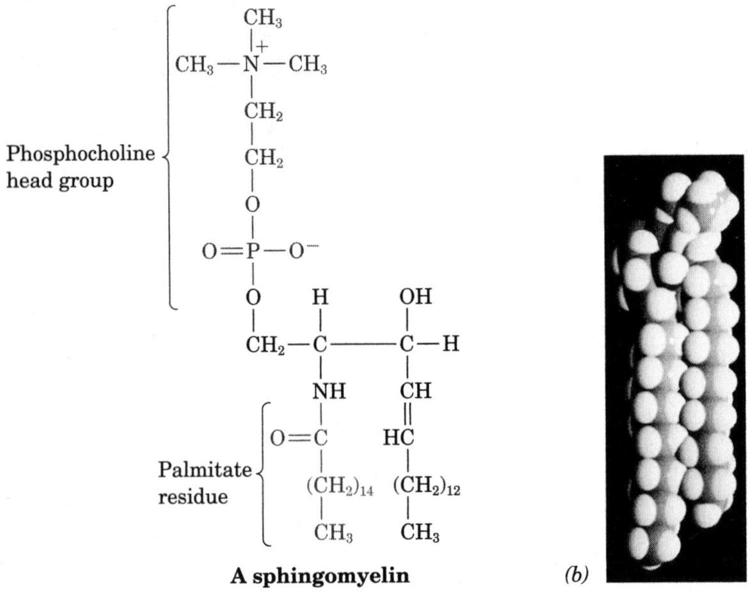

Phosphocholine head group

Palmitate residue

A sphingomyelin

(b)

Figure 9-7 A sphingomyelin.

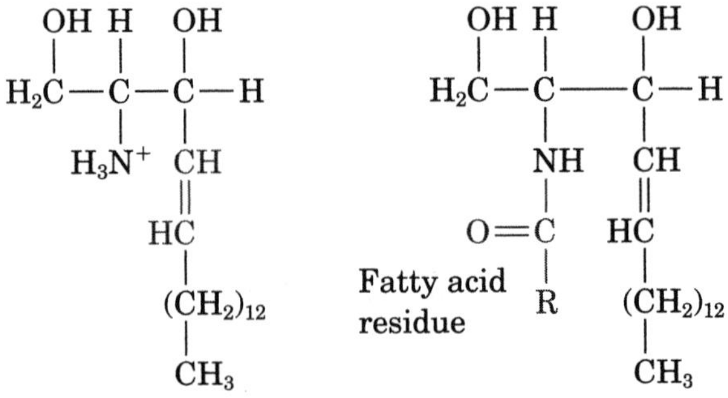

Sphingosine

A ceramide

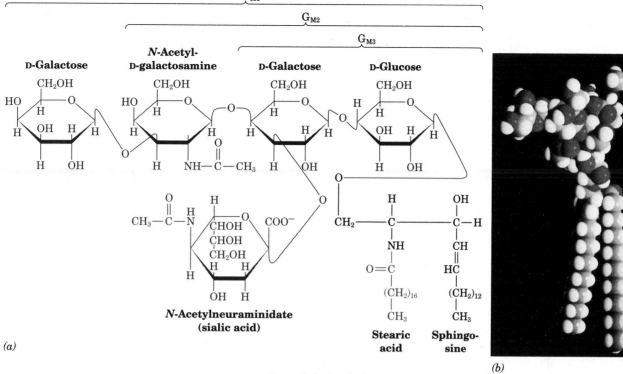

(a)

Figure 9-9 Gangliosides.

(b)

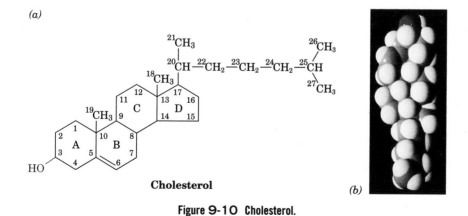

(a)

Cholesterol

(b)

Figure 9-10 Cholesterol.

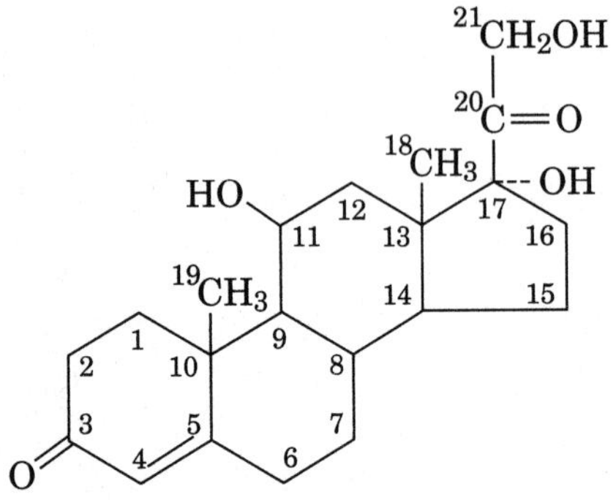

Cortisol (hydrocortisone)
(a glucocorticoid)

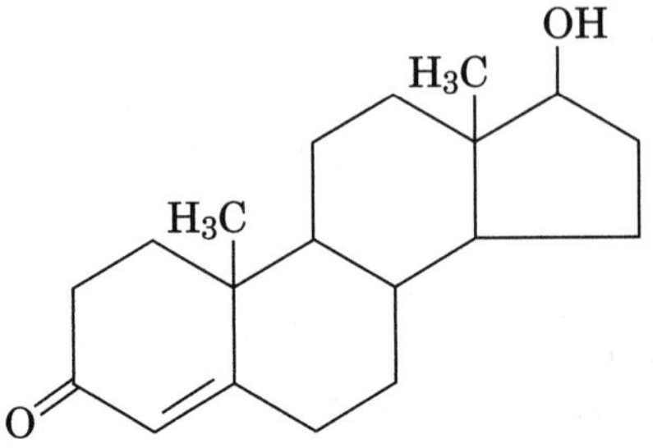

Testosterone
(an androgen)

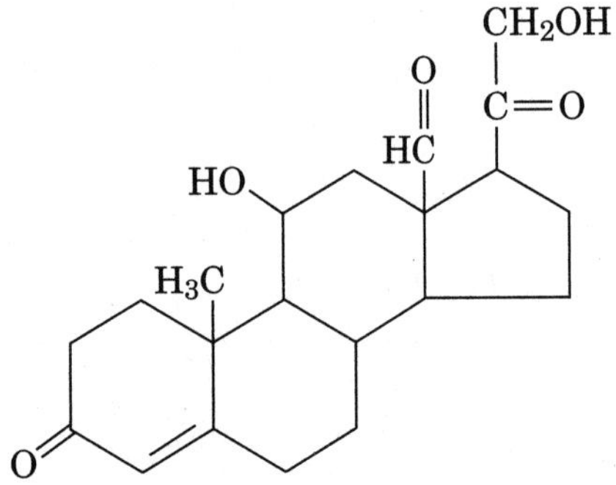

Aldosterone
(a mineralocorticoid)

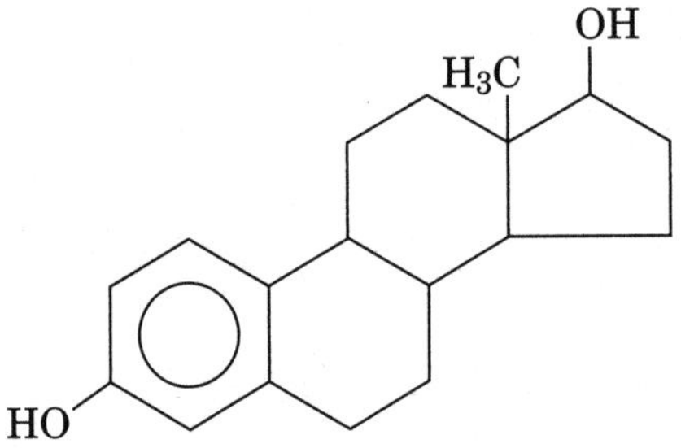

β-Estradiol
(an estrogen)

Figure 9-11 Some representative steroid hormones.

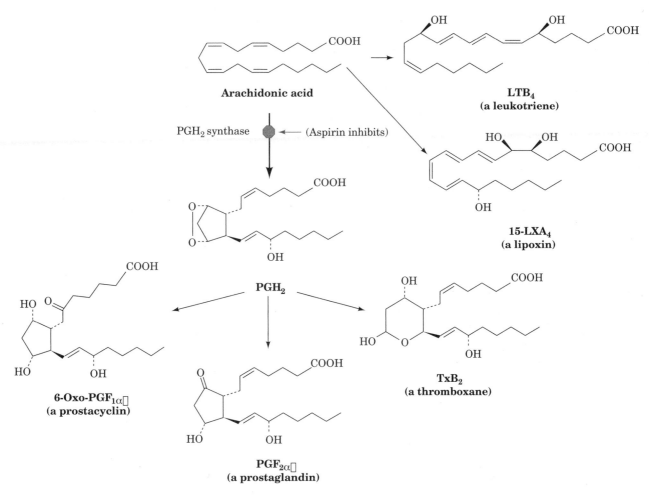

Arachidonic acid

PGH₂ synthase

(Aspirin inhibits)

LTB₄
(a leukotriene)

15-LXA₄
(a lipoxin)

PGH₂

6-Oxo-PGF₁α
(a prostacyclin)

PGF₂α
(a prostaglandin)

TxB₂
(a thromboxane)

Figure 9-12 Eicosanoids.

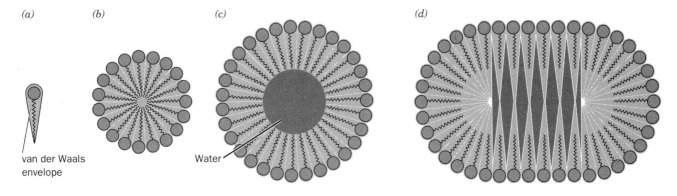

(a)

van der Waals
envelope

(b)

(c)

Water

(d)

Figure 9-13 Aggregates of single-tailed lipids.

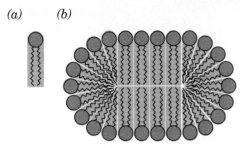

Figure 9-14 Bilayer formation by phospholipids.

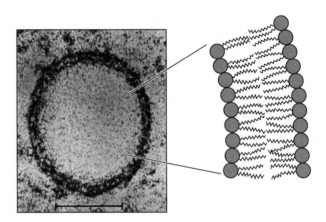

Figure 9-15 Electron micrograph of a liposome.

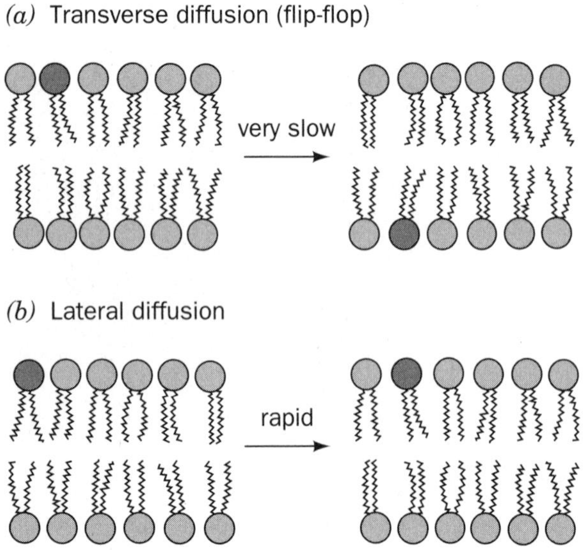

(a) Transverse diffusion (flip-flop)

very slow

(b) Lateral diffusion

rapid

Figure 9-16 Phospholipid diffusion in a lipid bilayer.

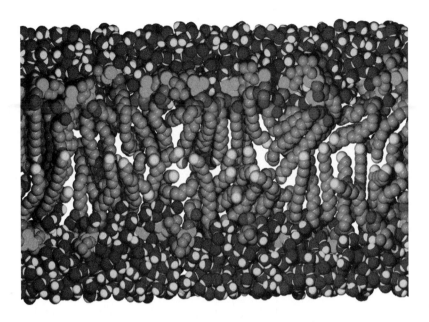

Figure 9-17 Model (snapshot) of a lipid bilayer at a particular instant in time.

(a) Above transition temperature

(b) Below transition temperature

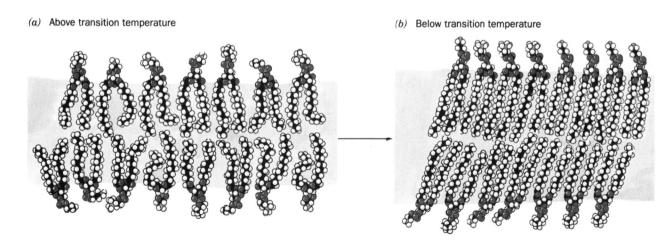

Figure 9-18 Phase transition in a lipid bilayer.

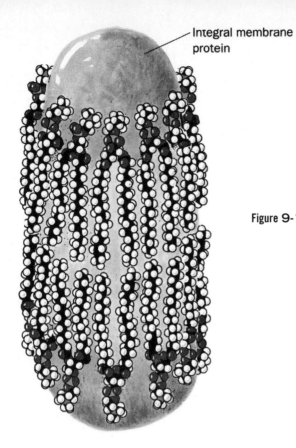

Integral membrane protein

Figure 9-19 Model of an integral membrane protein.

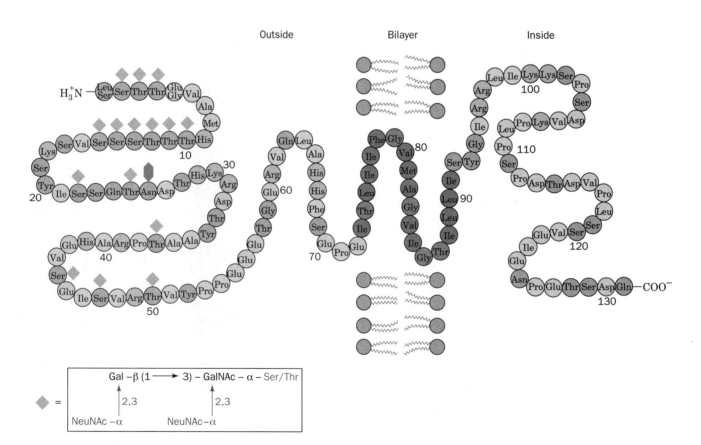

Figure 9-20 Human erythrocyte glycophorin A.

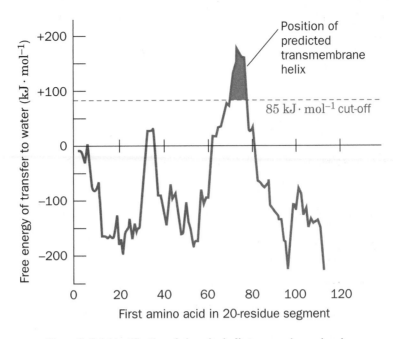

Figure 9-21 Identification of glycophorin A's transmembrane domain.

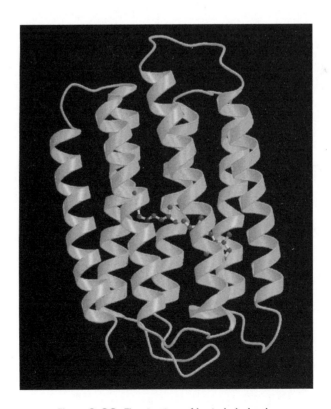

Figure 9-22 The structure of bacteriorhodopsin.

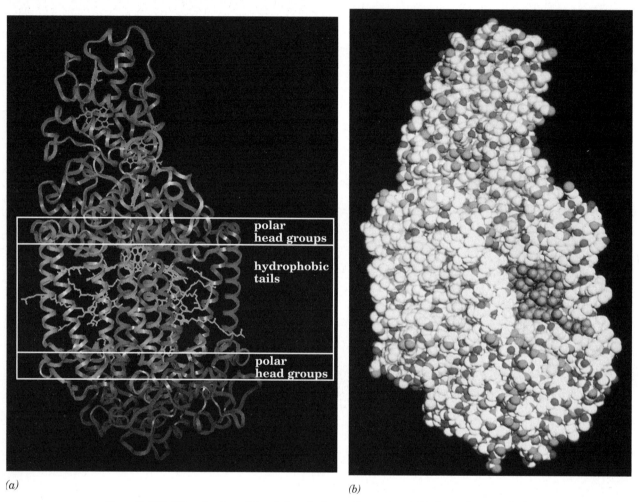

polar
head groups

hydrophobic
tails

polar
head groups

(a)

(b)

Figure 9-23 X-Ray structure of the photosynthetic reaction center of *Rhodopseudomonas viridis.*

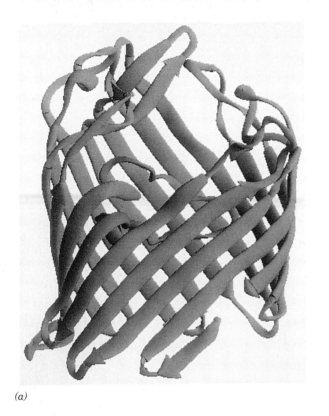

(a)

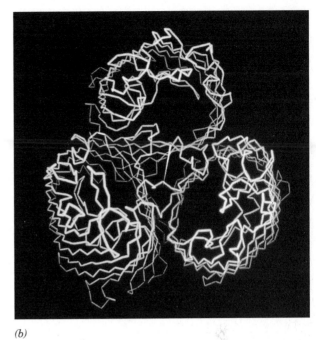

(b)

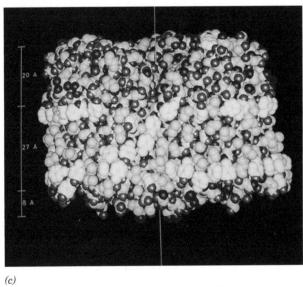

(c)

Figure 9-24 X-Ray structure of the *E. coli* OmpF porin.

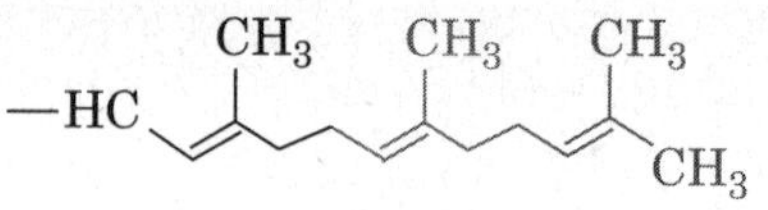

Farnesyl residue

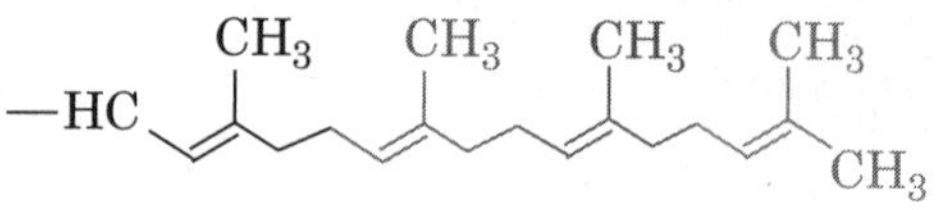

Geranylgeranyl residue

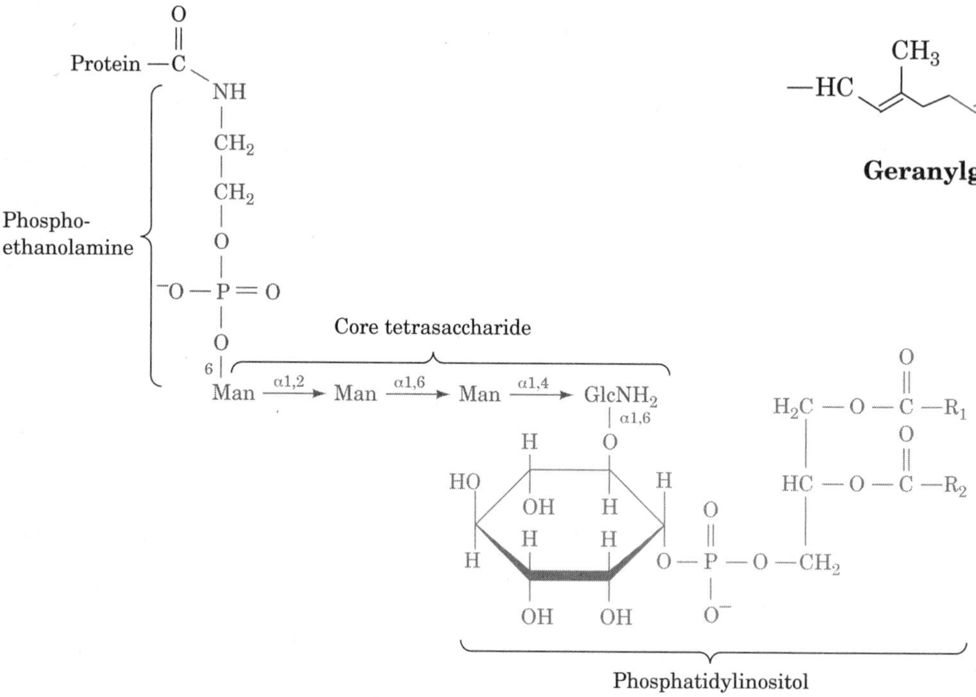

Protein

Phospho-
ethanolamine

Core tetrasaccharide

$$\text{Man} \xrightarrow{\alpha 1,2} \text{Man} \xrightarrow{\alpha 1,6} \text{Man} \xrightarrow{\alpha 1,4} \text{GlcNH}_2$$

Phosphatidylinositol

Figure 9-25 The core structure of the GPI anchors of proteins.

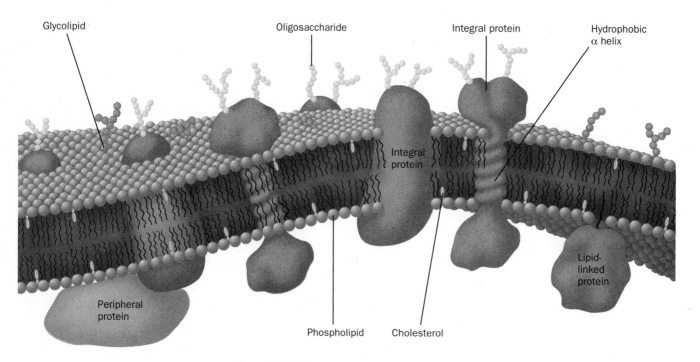

Glycolipid

Oligosaccharide

Integral protein

Hydrophobic
α helix

Integral
protein

Lipid-
linked
protein

Peripheral
protein

Phospholipid

Cholesterol

Figure 9-26 Diagram of a plasma membrane.

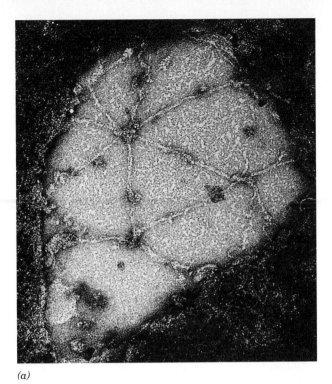

(a)

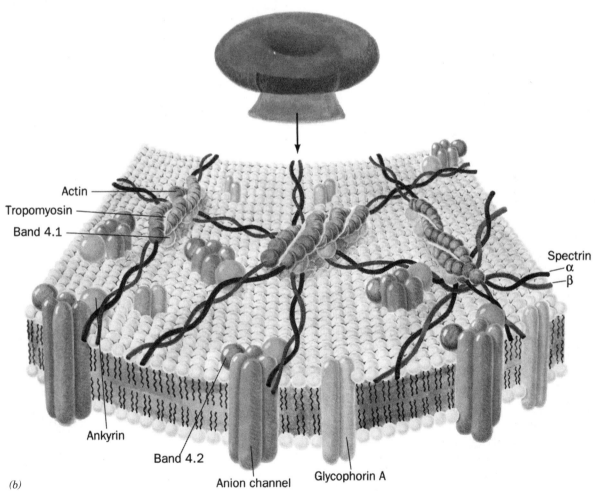

(b)

Figure 9-30 The human erythrocyte membrane skeleton.

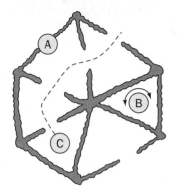

Figure 9-32 Model rationalizing the various mobilities of membrane proteins.

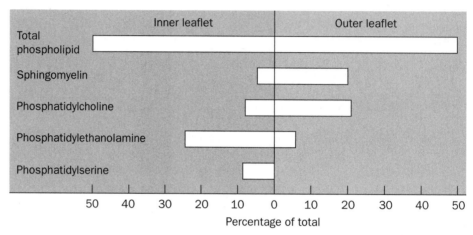

Figure 9-33 Asymmetric distribution of membrane phospholipids in the human erythrocyte membrane.

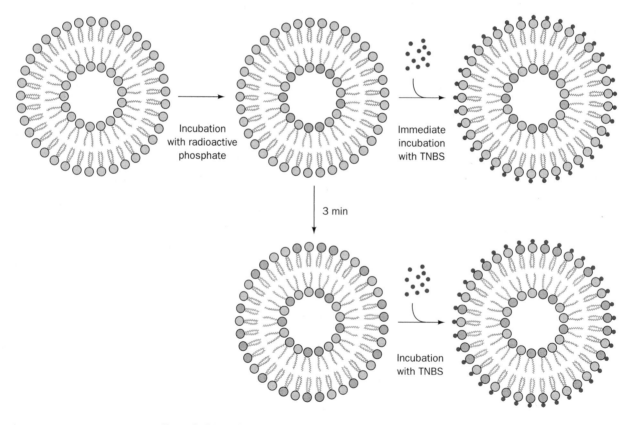

Figure 9-35 The location of lipid synthesis in a bacterial membrane.

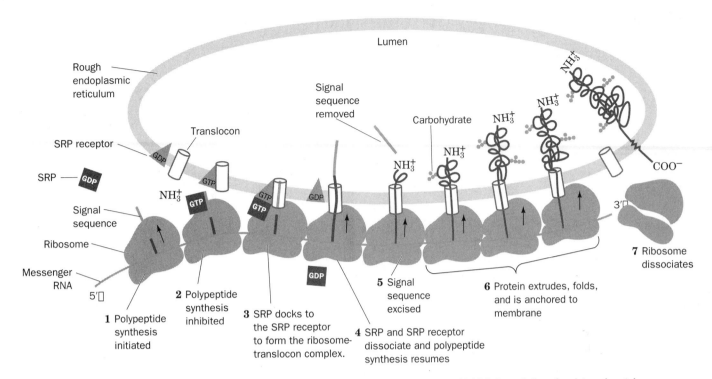

Figure 9-36 Key to Function. The ribosomal synthesis, membrane insertion, and initial glycosylation of an integral protein via the secretory pathway:

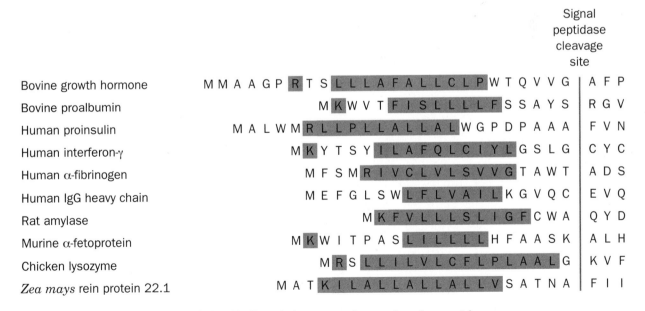

Figure 9-37 The N-terminal sequences of some eukaryotic preproteins.

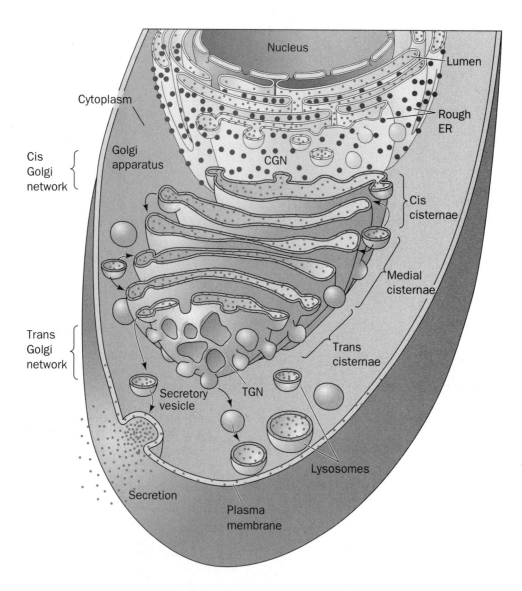

Figure 9-41 The posttranslational processing of proteins.

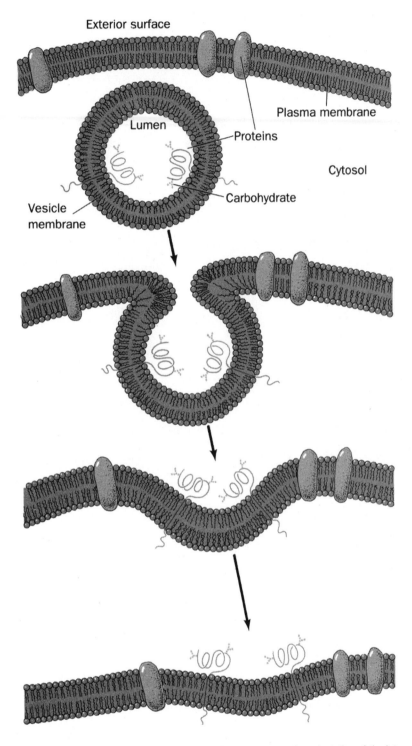

Figure 9-43 The fusion of a vesicle with the plasma membrane preserves the orientation of the integral proteins embedded in the vesicle bilayer.

(a)

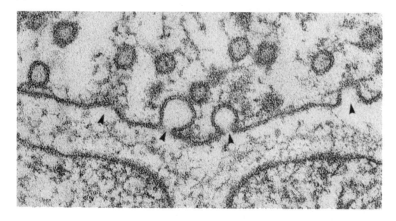

(b)

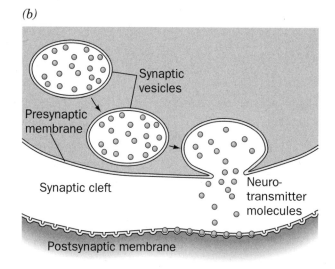

Figure 9-47 Vesicle fusion at a synapse.

Membrane Transport

$$\Delta \overline{G}_A = RT \ln\left(\frac{[A]_{in}}{[A]_{out}}\right) + Z_A \mathscr{F} \Delta\Psi \qquad [10\text{-}3]$$

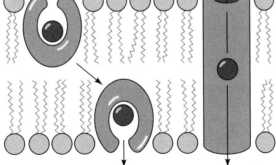

(a) Carrier ionophore   *(b)* Channel-forming ionophore

Figure 10-1 Ionophore action.

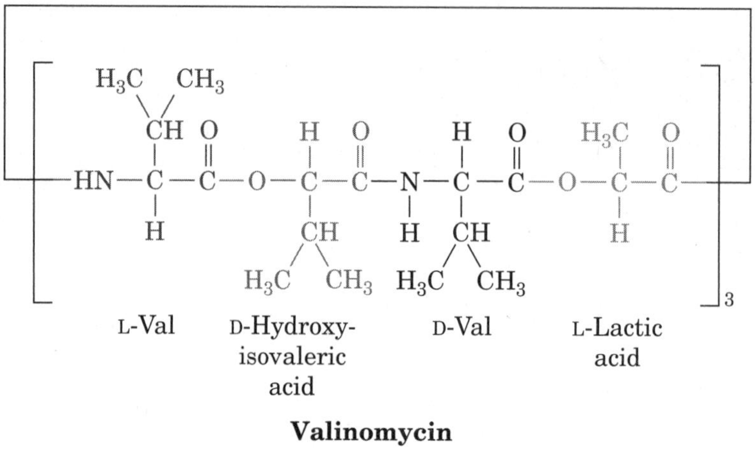

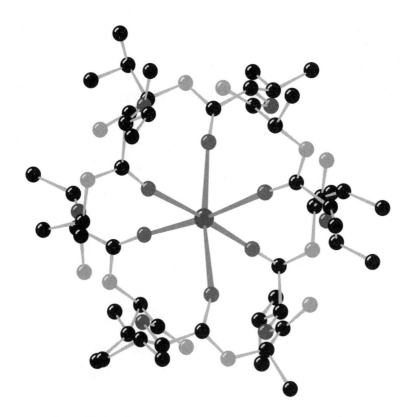

H₃C CH₃

CH O   H O   H O   H₃C O
│   ║   │ ║   │ ║   │   ║
—HN—C—C—O—C—C—N—C—C—O—C—C—
│   │   │   │
H   CH   H CH   H
│   │   │
H₃C CH₃   H₃C CH₃

    L-Val   D-Hydroxy-   D-Val   L-Lactic
               isovaleric   acid
                  acid

Valinomycin

Figure 10-2 Valinomycin.

Figure 10-3 X-Ray structure of valinomycin in complex with a K⁺ ion.

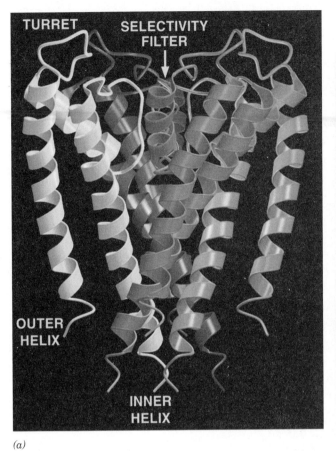

(a)

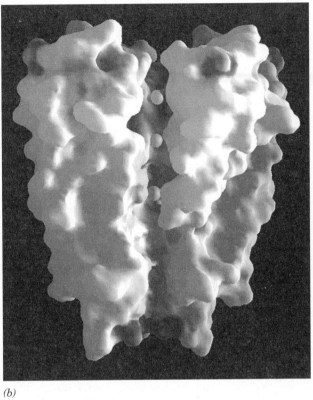

(b)

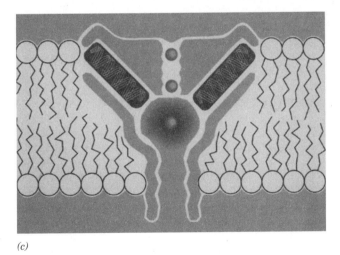

(c)

Figure 10-7 X-Ray structure of the KscA K$^+$ channel.

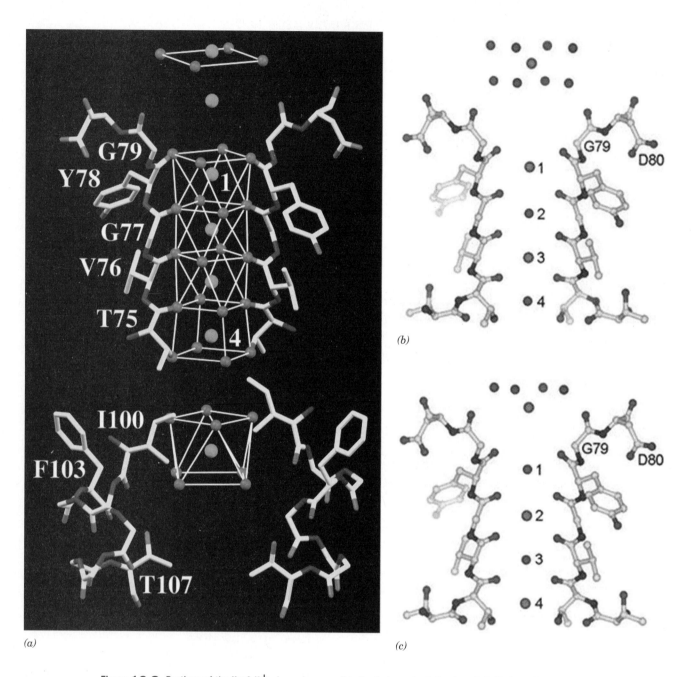

(a)

(b)

(c)

Figure 10-8 Portions of the KcsA K$^+$ channel responsible for its ion selectivity viewed similarly to Fig. 10-7.

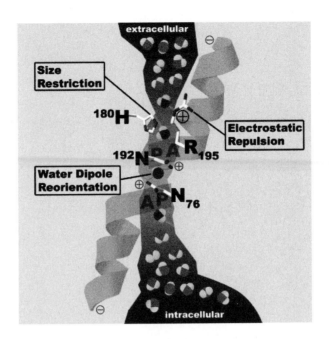

Figure 10-16 Schematic model of the water-conducting pore of aquaporin AQP1 viewed from within the membrane with the extracellular surface above.

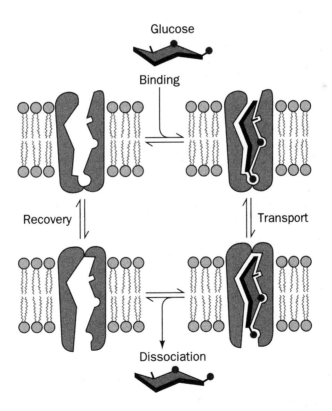

Figure 10-17 *Key to Function.* Model for glucose transport.

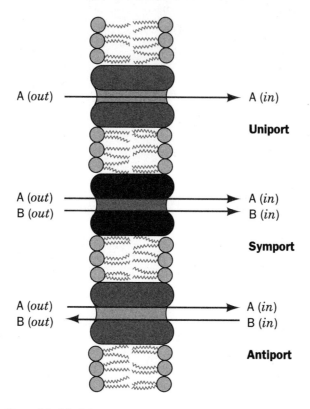

Figure 10-19 Uniport, symport, and antiport translocation systems.

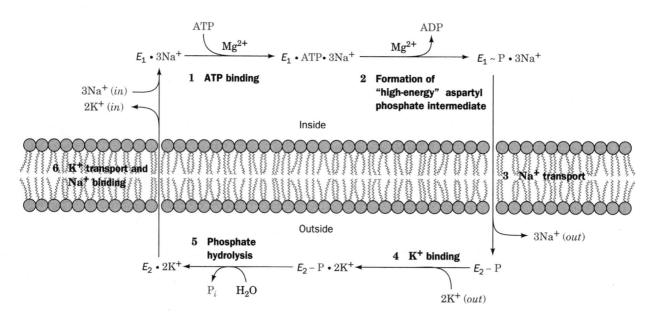

Figure 10-21 *Key to Function.* Scheme for the active transport of Na$^+$ and K$^+$ by the (Na$^+$–K$^+$)–ATPase.

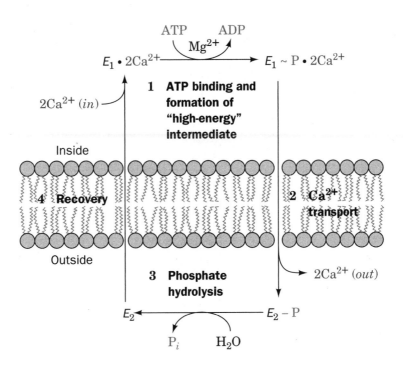

Figure 10-22 Scheme for the active transport of Ca^{2+} by the Ca^{2+}–ATPase.

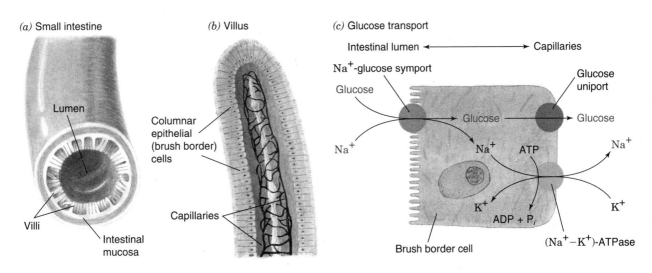

Figure 10-24 Glucose transport in the intestinal epithelium.

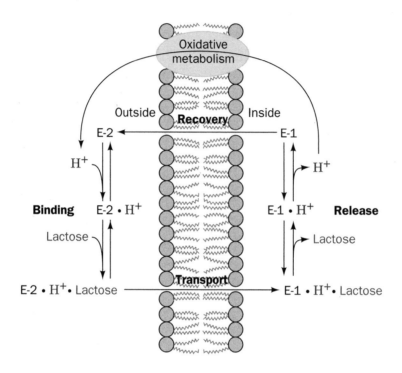

Figure 10-25 Scheme for the cotransport of H^+ and lactose by lactose permease in *E. coli*.

CHAPTER 11

Enzymatic Catalysis

Table 11-1 Catalytic Power of Some Enzymes

Enzyme	Nonenzymatic Reaction Rate (s^{-1})	Enzymatic Reaction Rate (s^{-1})	Rate Enhancement
Carbonic anhydrase	1.3×10^{-1}	1×10^{6}	7.7×10^{6}
Chorismate mutase	2.6×10^{-5}	50	1.9×10^{6}
Triose phosphate isomerase	4.3×10^{-6}	4300	1.0×10^{9}
Carboxypeptidase A	3.0×10^{-9}	578	1.9×10^{11}
AMP nucleosidase	1.0×10^{-11}	60	6.0×10^{12}
Staphylococcal nuclease	1.7×10^{-13}	95	5.6×10^{14}

Source: Radzicka, A. and Wolfenden, R., *Science* **267,** 91 (1995).

Table 11-2 Enzyme Classification According to Reaction Type

Classification	Type of Reaction Catalyzed
1. Oxidoreductases	Oxidation–reduction reactions
2. Transferases	Transfer of functional groups
3. Hydrolases	Hydrolysis reactions
4. Lyases	Group elimination to form double bonds
5. Isomerases	Isomerization
6. Ligases	Bond formation coupled with ATP hydrolysis

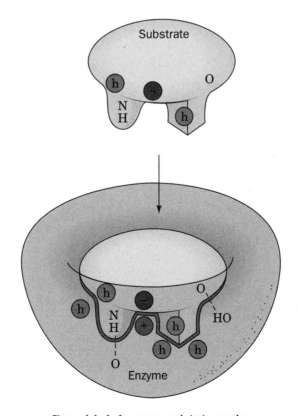

Figure 11-1 An enzyme–substrate complex.

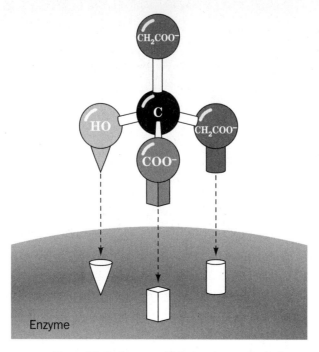

Figure 11-2 Stereospecificity in substrate binding.

$$\begin{array}{c} \text{COO}^- \\ | \\ \text{CH}_2 \\ | \\ \text{HO} - \text{C} - \text{COO}^- \\ | \\ \text{CH}_2 \\ | \\ \text{COO}^- \end{array} \xrightleftharpoons[\text{aconitase}]{} \begin{array}{c} \text{COO}^- \\ | \\ \text{CH}_2 \\ | \\ \text{H} - \text{C} - \text{COO}^- \\ | \\ \text{HO} - \text{C} - \text{H} \\ | \\ \text{COO}^- \end{array}$$

Citrate   **Isocitrate**

Table 11-3 Characteristics of Common Coenzymes

Coenzyme	Reaction Mediated	Vitamin Source	Human Deficiency Disease
Biocytin	Carboxylation	Biotin	a
Coenzyme A	Acyl transfer	Pantothenate	a
Cobalamin coenzymes	Alkylation	Cobalamin (B$_{12}$)	Pernicious anemia
Flavin coenzymes	Oxidation–reduction	Riboflavin (B$_2$)	a
Lipoic acid	Acyl transfer	—	a
Nicotinamide coenzymes	Oxidation–reduction	Nicotinamide (niacin)	Pellagra
Pyridoxal phosphate	Amino group transfer	Pyridoxine (B$_6$)	a
Tetrahydrofolate	One-carbon group transfer	Folic acid	Megaloblastic anemia
Thiamine pyrophosphate	Aldehyde transfer	Thiamine (B$_1$)	Beriberi

[a]No specific name; deficiency in humans is rare or unobserved.

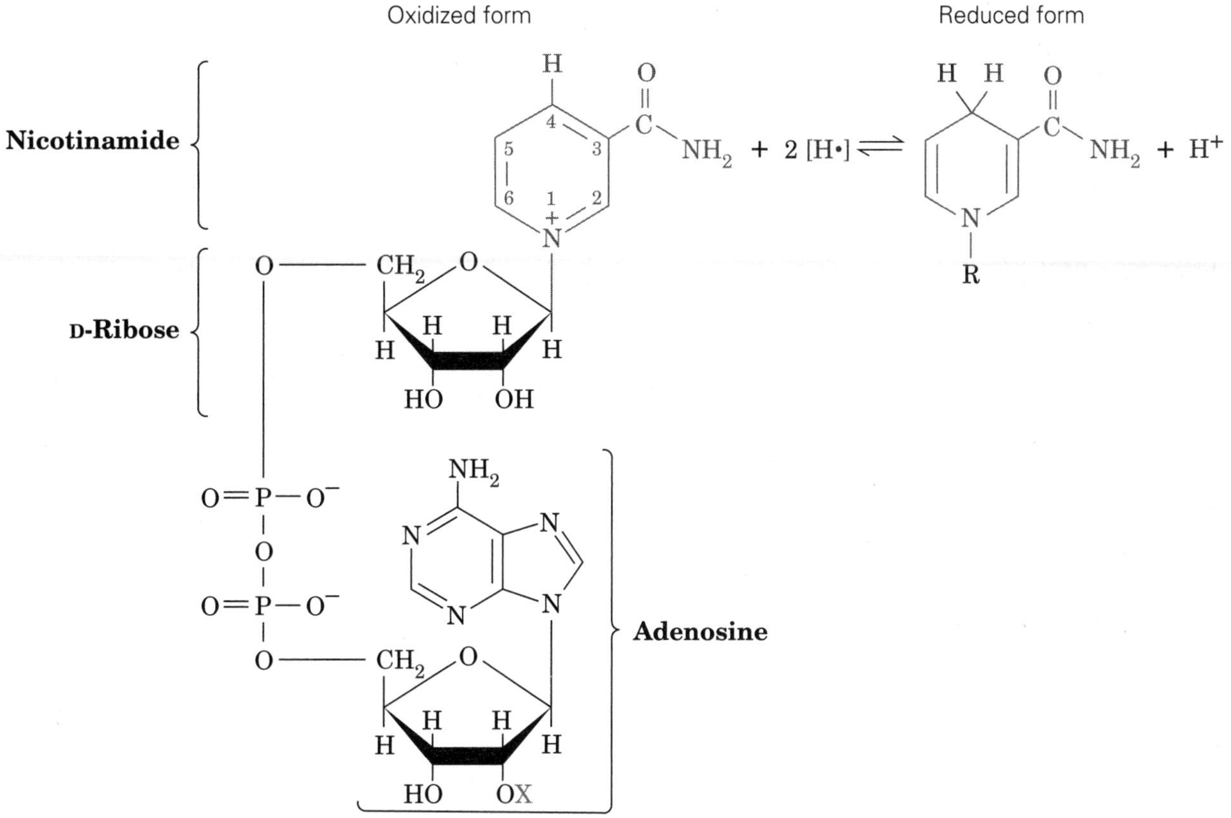

Nicotinamide

D-Ribose

Adenosine

Oxidized form

Reduced form

$+ 2\,[\text{H}\cdot] \rightleftharpoons$

$+\ \text{H}^+$

X = H   **Nicotinamide adenine dinucleotide (NAD$^+$)**
X = PO$_3^{2-}$ **Nicotinamide adenine dinucleotide phosphate (NADP$^+$)**

Figure 11-3 The structures and reaction of nicotinamide adenine dinucleotide (NAD$^+$) and nicotinamide adenine dinucleotide phosphate (NADP$^+$).

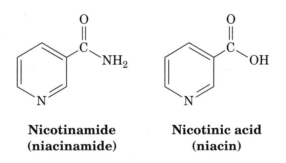

Nicotinamide
(niacinamide)

Nicotinic acid
(niacin)

Figure 11-4 The structures of nicotinamide and nicotinic acid.

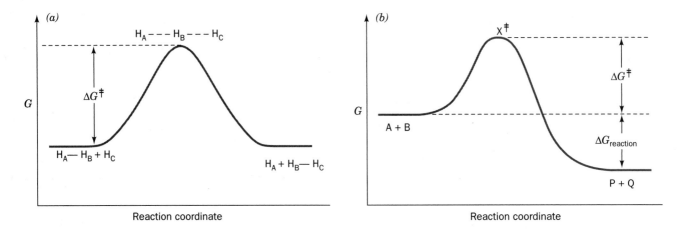

Figure 11-5 Transition state diagrams.

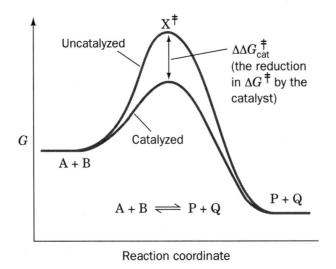

Figure 11-7 Effect of a catalyst on the transition state diagram of a reaction.

TYPES OF CATALYTIC MECHANISMS

1. Acid–base catalysis

2. Covalent catalysis

3. Metal ion catalysis

4. Proximity and orientation effects

5. Preferential binding of the transition state complex

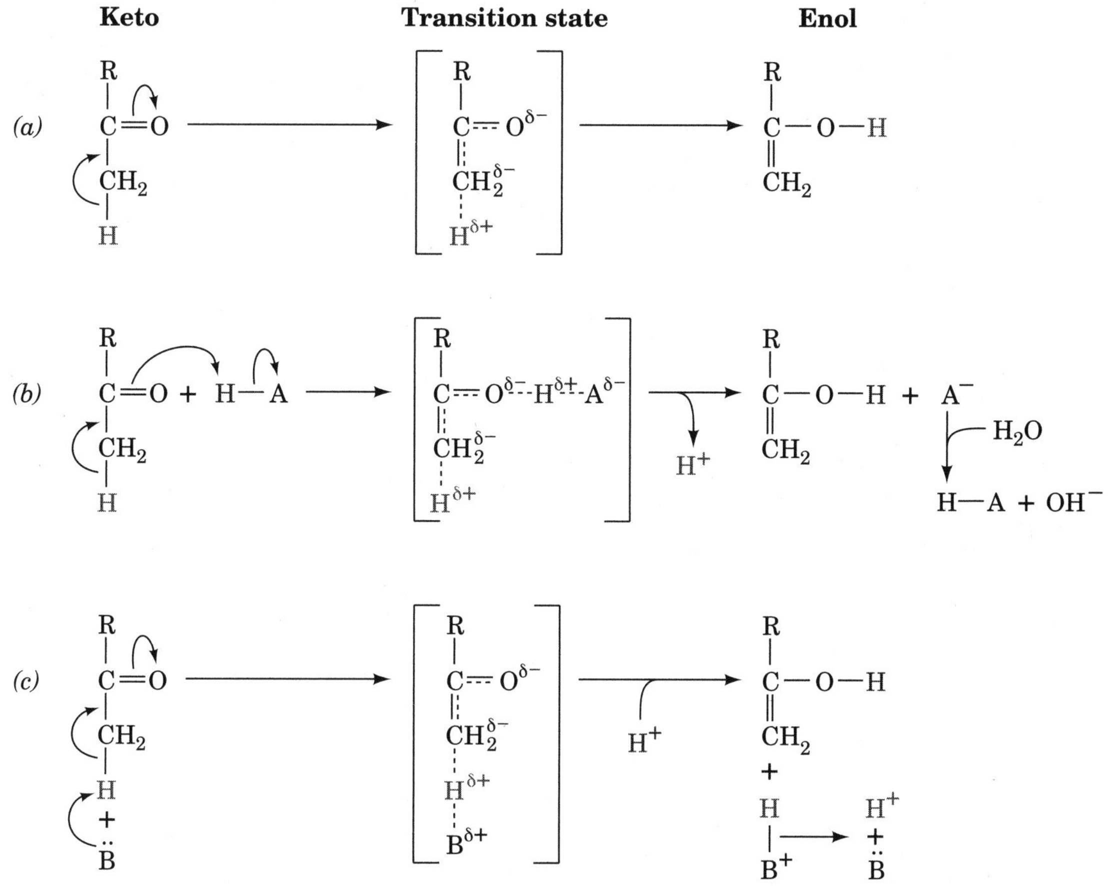

Figure 11-8 Mechanisms of keto–enol tautomerization.

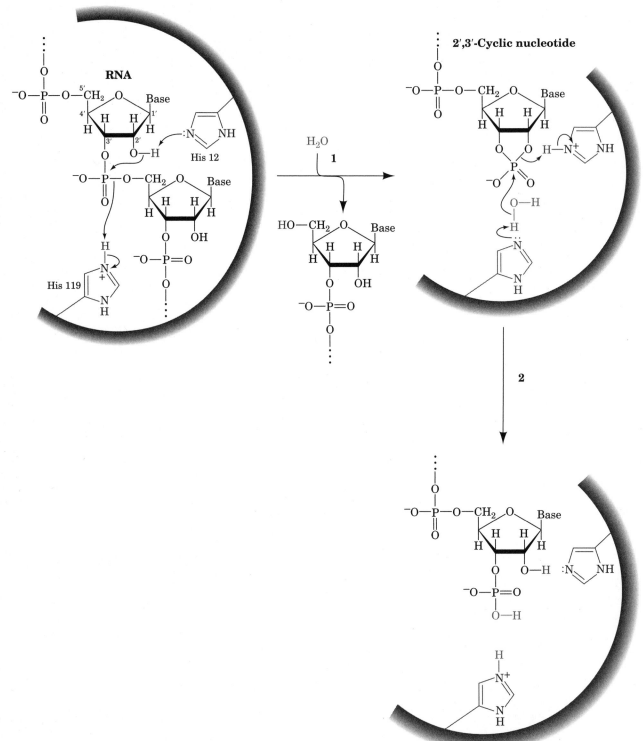

Figure 11-10 The RNase A mechanism.

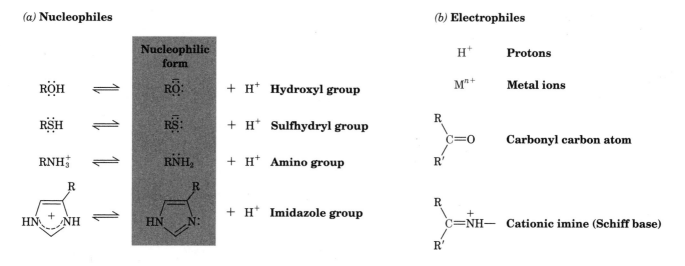

Figure 11-11 The decarboxylation of acetoacetate.

(a) **Nucleophiles**

		Nucleophilic form			
R$\ddot{O}$H	⇌	R$\ddot{\ddot{O}}$:	+ H$^+$	**Hydroxyl group**
R$\ddot{S}$H	⇌	R$\ddot{\ddot{S}}$:	+ H$^+$	**Sulfhydryl group**
R$\overset{+}{N}$H$_3$	⇌	R$\ddot{N}$H$_2$	+ H$^+$	**Amino group**
HN$\overset{+}{}$NH (imidazolium)	⇌	HN N: (imidazole)	+ H$^+$	**Imidazole group**

(b) **Electrophiles**

H$^+$	**Protons**
M^{n+}	**Metal ions**
$\underset{R'}{\overset{R}{C}}$=O	**Carbonyl carbon atom**
$\underset{R'}{\overset{R}{C}}$=$\overset{+}{N}$H—	**Cationic imine (Schiff base)**

Figure 11-12 Biologically important nucleophilic and electrophilic groups.

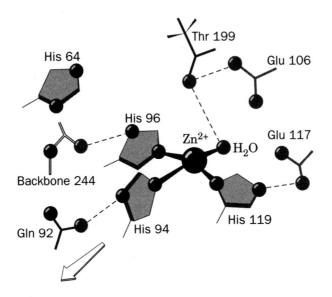

Figure 11-13 The active site of human carbonic anhydrase.

$$Im-Zn^{2+}-\overset{Im}{\underset{H}{\overset{|}{O^-}}} + \overset{O}{\underset{O}{C}}$$

$$\Updownarrow$$

$$Im-Zn^{2+}----\overset{Im}{\underset{H}{\overset{|}{O}}}-\overset{O}{C}-O^-$$

$$\Updownarrow -H_2O$$

$$Im-Zn^{2+}-\overset{Im}{\underset{H}{\overset{|}{O^-}}} + H^+ + H-O-\overset{O}{C}-O^-$$

Im = imidazole

p-Nitrophenylacetate → **N-Acetylimidazolium** + **p-Nitrophenolate**

Imidazole

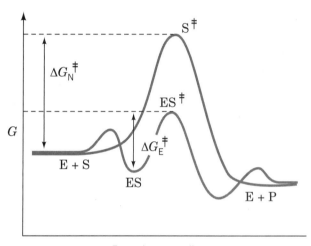

Figure 11-15 *Key to Function.* Effect of preferential transition state binding.

L-Proline　　**Planar transition state**　　**D-Proline**

Pyrrole-2-carboxylate　　**Δ-1-Pyrroline-2-carboxylate**

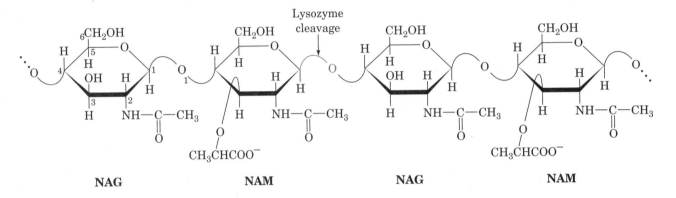

Figure 11-16 The lysozyme cleavage site.

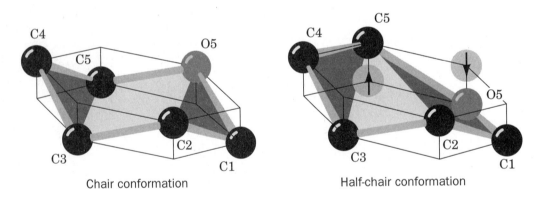

Figure 11-18 Chair and half-chair conformations.

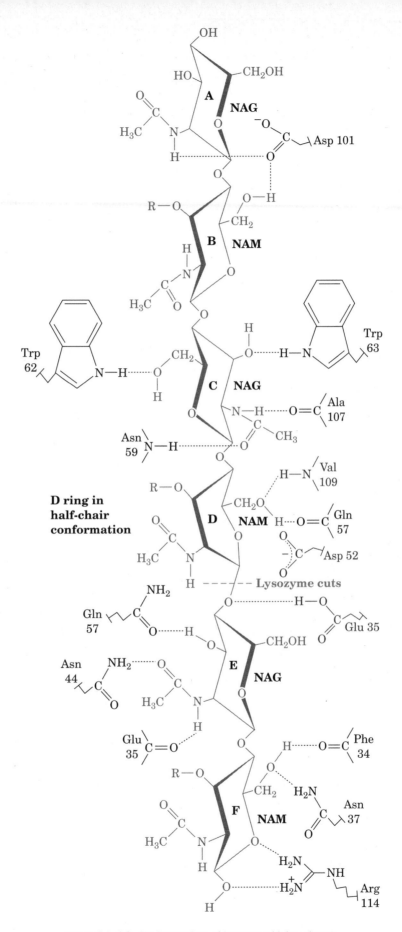

Figure 11-19 The interactions of lysozyme with its substrate.

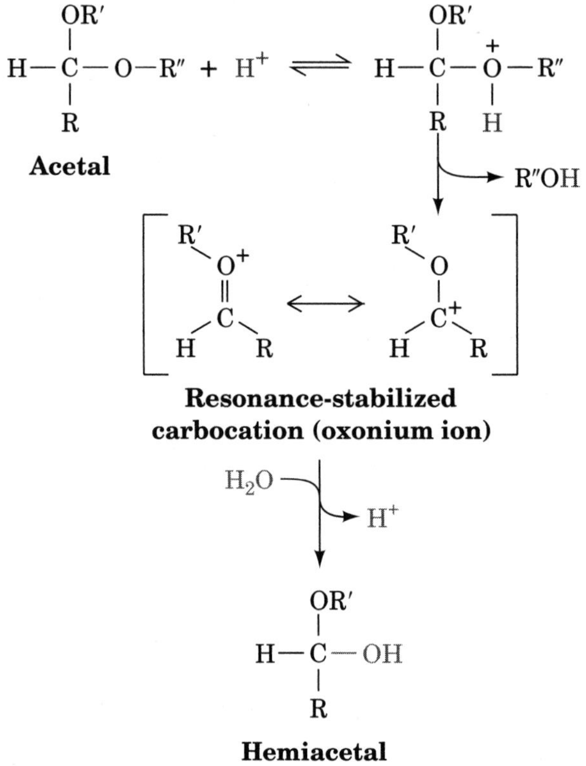

Figure 11-20 The mechanism of the nonenzymatic acid-catalyzed hydrolysis of an acetal to a hemiacetal.

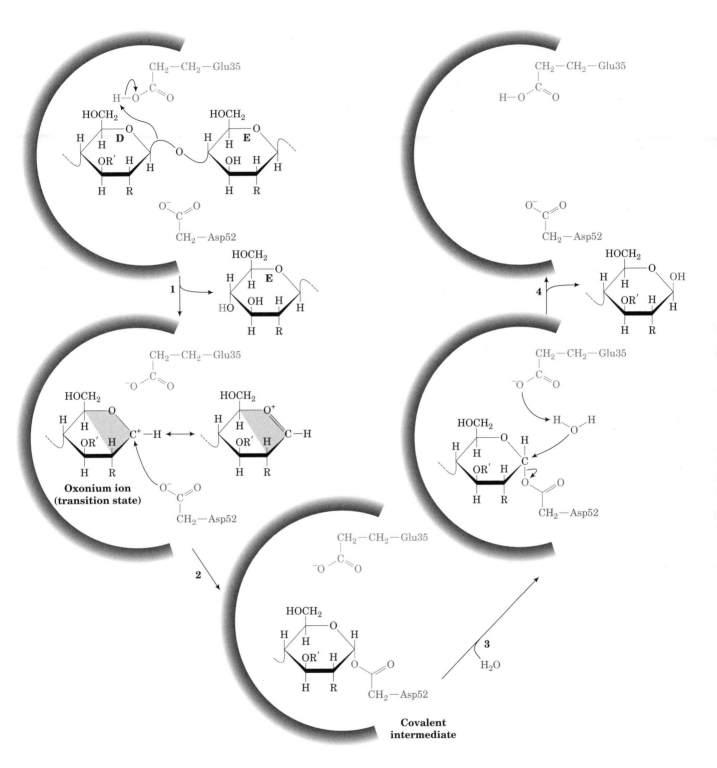

Figure 11-21 The lysozyme reaction mechanism.

Table 11-4 Binding Free Energies of HEW Lysozyme Subsites

Site	Bound Saccharide	Binding Free Energy $(kJ \cdot mol^{-1})$
A	NAG	-7.5
B	NAM	-12.3
C	NAG	-23.8
D	**NAM**	**+12.1**
E	NAG	-7.1
F	NAM	-7.1

Source: Chipman, D.M. and Sharon, N., *Science* **165,** 459 (1969).

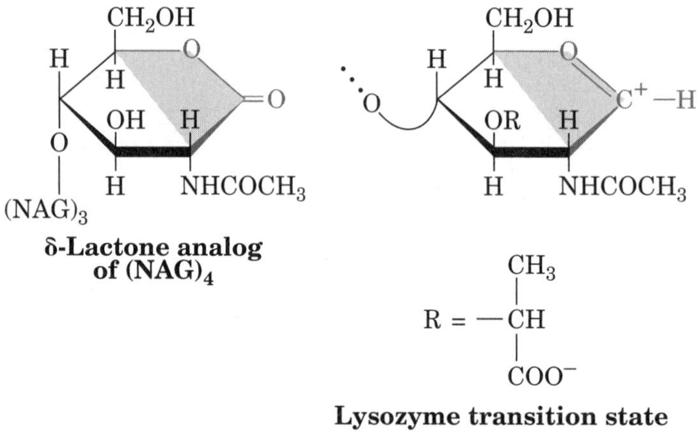

δ-Lactone analog of (NAG)₄

Lysozyme transition state

$$R = -CH \begin{matrix} CH_3 \\ | \\ COO^- \end{matrix}$$

Figure 11-22 Transition state analog inhibition of lysozyme.

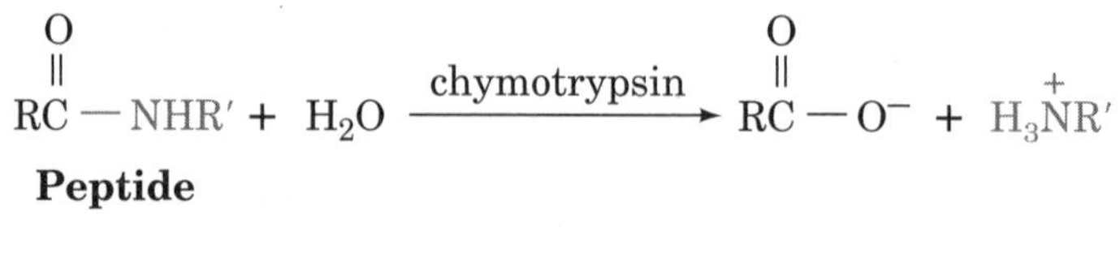

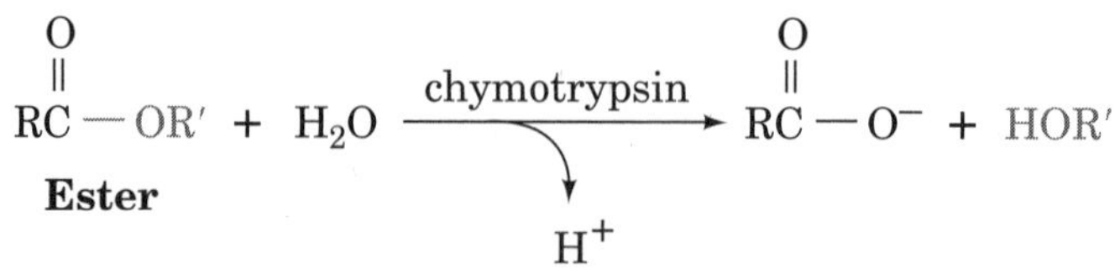

$$\overset{O}{\underset{\|}{RC}} - NHR' + H_2O \xrightarrow{\text{chymotrypsin}} \overset{O}{\underset{\|}{RC}} - O^- + H_3\overset{+}{N}R'$$

Peptide

$$\overset{O}{\underset{\|}{RC}} - OR' + H_2O \xrightarrow{\text{chymotrypsin}} \overset{O}{\underset{\|}{RC}} - O^- + HOR'$$

Ester

H^+

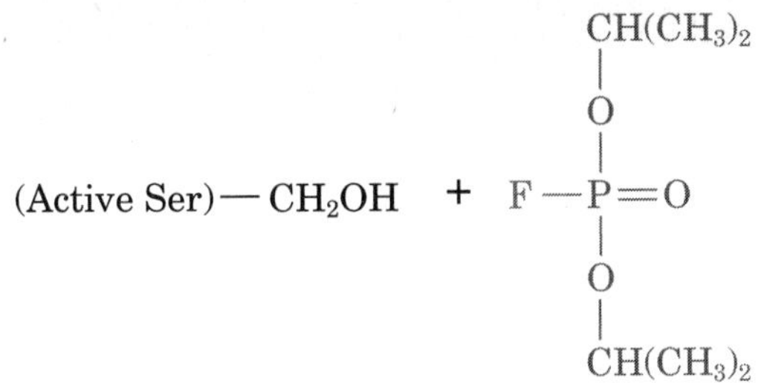

(Active Ser) — CH$_2$OH   +   F — P=O

CH(CH$_3$)$_2$
|
O
|
F — P=O
|
O
|
CH(CH$_3$)$_2$

**Diisopropylphospho-
fluoridate (DIPF)**

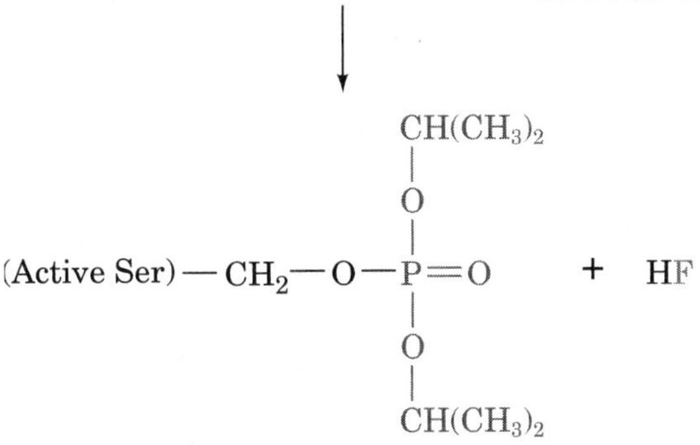

(Active Ser) — CH$_2$ — O — P=O   +   HF

CH(CH$_3$)$_2$
|
O
|
CH$_2$ — O — P=O
|
O
|
CH(CH$_3$)$_2$

DIP–Enzyme

Chymotrypsin

His 57   TPCK   HCl   Chymotrypsin

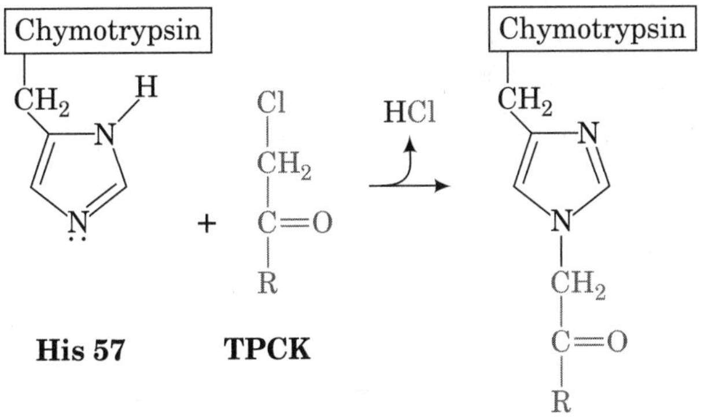

Figure 11-24 Reaction of TPCK with His 57 of chymotrypsin.

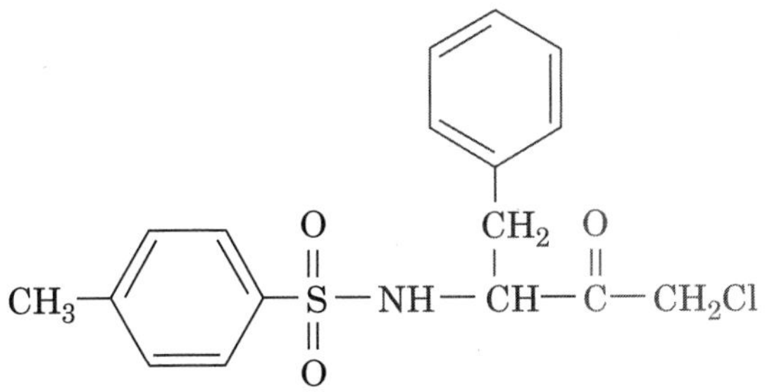

CH$_3$ —〔benzene ring〕— S(=O)(=O) — NH — CH(CH$_2$—phenyl) — C(=O) — CH$_2$Cl

Tosyl-L-phenylalanine chloromethylketone

CH$_3$ —〔benzene ring〕— S(=O)(=O) — NH — CH — C(=O) — CH$_2$Cl

$\overset{+}{N}H_3$
|
CH$_2$
|
CH$_2$
|
CH$_2$
|
CH$_2$

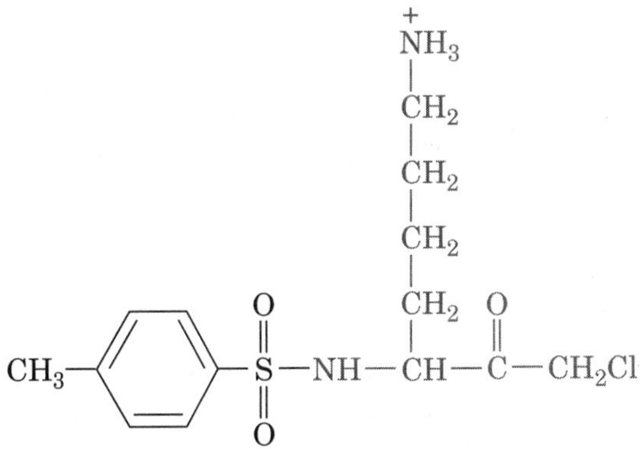

Tosyl-L-lysine chloromethylketone

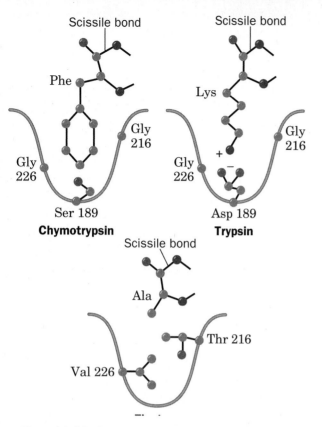

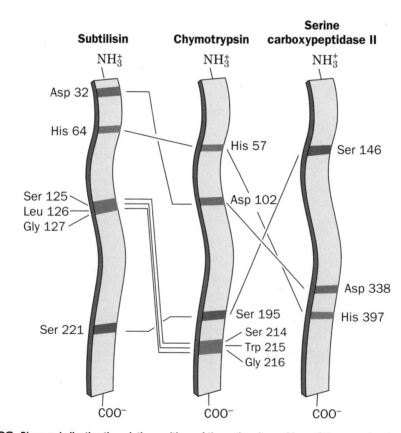

Figure 11-27 Specificity pockets of three serine proteases.

Figure 11-28 Diagram indicating the relative positions of the active site residues of three unrelated serine proteases.

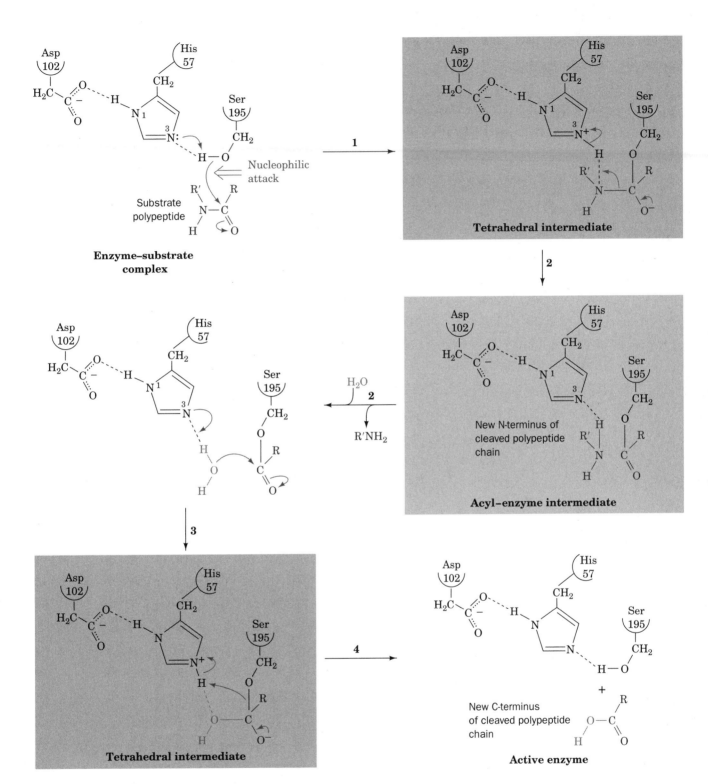

Figure 11-29 *Key to Function.* The catalytic mechanism of the serine proteases.

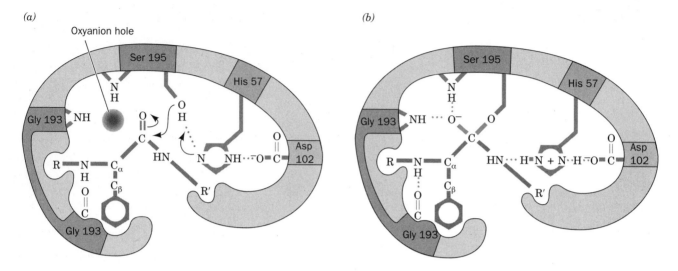

(a)
Oxyanion hole
Ser 195
His 57
Gly 193
Asp 102
Gly 193
(b)
Ser 195
His 57
Gly 193
Asp 102
Gly 193

Figure 11-30 Transition state stabilization in the serine proteases.

$$\overset{+}{H_3N}\text{—}\overset{10}{\text{Val}}\text{—}(\text{Asp})_4\text{—}\overset{15}{\text{Lys}}\text{—}\overset{16}{\text{Ile}}\text{—Val—}\cdots$$

Trypsinogen

↓ enteropeptidase or trypsin

$$\overset{+}{H_3N}\text{—Val—}(\text{Asp})_4\text{—Lys} \quad + \quad \text{Ile—Val—}\cdots$$

Trypsin

Figure 11-33 The activation of trypsinogen to trypsin.

CHAPTER 12

Enzyme Kinetics, Inhibition, and Regulation

$$A \longrightarrow P$$

$$v = \frac{d[P]}{dt} = -\frac{d[A]}{dt} = k[A] \qquad [12\text{-}1]$$

$$\frac{d[A]}{[A]} = d\ln[A] = -k\,dt \qquad [12\text{-}4]$$

$$\int_{[A]_O}^{[A]} d\ln[A] = -k\int_0^t dt \qquad [12\text{-}5]$$

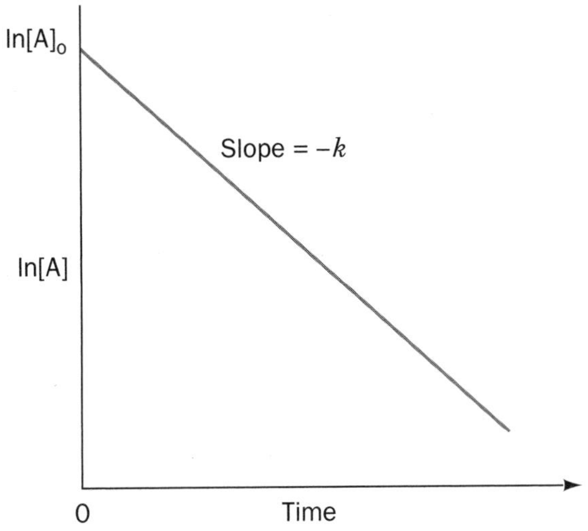

$$\boxed{\ln[A] = \ln[A]_O - kt} \qquad [12\text{-}6]$$

Figure 12-1 A plot of a first-order rate equation.

$$\ln\left(\frac{[A]_O/2}{[A]_O}\right) = -kt_{1/2} \qquad [12\text{-}8]$$

$$t_{1/2} = \frac{\ln 2}{k} = \frac{0.693}{k} \qquad [12\text{-}9]$$

$$\boxed{\frac{1}{[A]} = \frac{1}{[A]_O} + kt} \qquad [12\text{-}11]$$

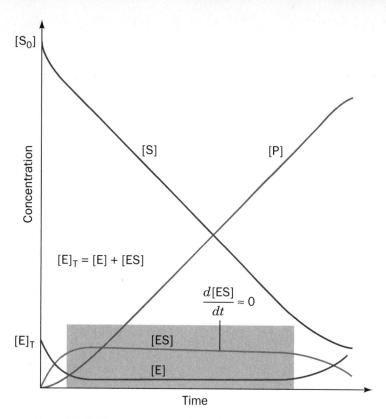

Figure 12-2 The progress curves for a simple enzyme-catalyzed reaction.

$$E + S \underset{k_{-1}}{\overset{k_1}{\rightleftharpoons}} ES \xrightarrow{k_2} P + E \qquad [12\text{-}12]$$

$$v = \frac{d[P]}{dt} = k_2[ES] \qquad [12\text{-}13]$$

$$\frac{d[ES]}{dt} = k_1[E][S] - k_{-1}[ES] - k_2[ES] \qquad [12\text{-}14]$$

$$K_S = \frac{k_{-1}}{k_1} = \frac{[E][S]}{[ES]} \qquad [12\text{-}15]$$

$$\frac{d[ES]}{dt} = 0 \qquad [12\text{-}16]$$

$$[E]_T = [E] + [ES] \qquad\qquad [12\text{-}17]$$

$$k_1[E][S] = k_{-1}[ES] + k_2[ES] \qquad\qquad [12\text{-}18]$$

$$\frac{([E]_T - [ES])[S]}{[ES]} = \frac{k_{-1} + k_2}{k_1} \qquad\qquad [12\text{-}19]$$

$$K_M = \frac{k_{-1} + k_2}{k_1} \qquad\qquad [12\text{-}20]$$

$$K_M[ES] = ([E]_T - [ES])[S] \qquad\qquad [12\text{-}21]$$

$$[ES] = \frac{[E]_T[S]}{K_M + [S]} \qquad\qquad [12\text{-}22]$$

$$v_0 = \left(\frac{d[P]}{dt}\right)_{t=0} = k_2[ES] = \frac{k_2[E]_T[S]}{K_M + [S]} \qquad\qquad [12\text{-}23]$$

$$V_{max} = k_2[E]_T \qquad\qquad [12\text{-}24]$$

$$\boxed{v_0 = \frac{V_{max}[S]}{K_M + [S]}} \qquad\qquad [12\text{-}25]$$

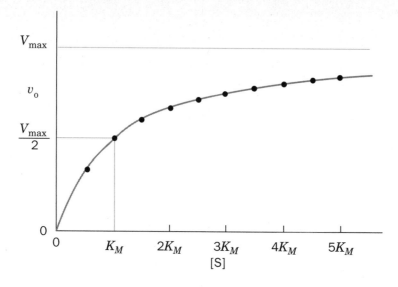

Figure 12-3 *Key to Function.* A plot of the initial velocity v_0 of a simple enzymatic reaction versus the substrate concentration [S].

$$K_M = \frac{k_{-1}}{k_1} + \frac{k_2}{k_1} = K_S + \frac{k_2}{k_1} \qquad [12\text{-}26]$$

$$k_{cat} = \frac{V_{max}}{[E]_T} \qquad [12\text{-}27]$$

$$v_O \approx \left(\frac{k_2}{K_M}\right)[E]_T[S] \approx \left(\frac{k_{cat}}{K_M}\right)[E][S] \qquad [12\text{-}28]$$

Table 12-1 The Values of K_M, k_{cat}, and k_{cat}/K_M for Some Enzymes and Substrates

Enzyme	Substrate	K_M (M)	k_{cat} (s^{-1})	k_{cat}/K_M (M$^{-1} \cdot$ s^{-1})
Acetylcholinesterase	Acetylcholine	9.5×10^{-5}	1.4×10^4	1.5×10^8
Carbonic anhydrase	CO_2	1.2×10^{-2}	1.0×10^6	8.3×10^7
	HCO_3^-	2.6×10^{-2}	4.0×10^5	1.5×10^7
Catalase	H_2O_2	2.5×10^{-2}	1.0×10^7	4.0×10^8
Chymotrypsin	*N*-Accetylglycine ethyl ester	4.4×10^{-1}	5.1×10^{-2}	1.2×10^{-1}
	N-Acetylvaline ethyl ester	8.8×10^{-2}	1.7×10^{-1}	1.9
	N-Acetyltyrosine ethyl ester	6.6×10^{-4}	1.9×10^2	2.9×10^5
Fumarase	Fumarate	5.0×10^{-6}	8.0×10^2	1.6×10^8
	Malate	2.5×10^{-5}	9.0×10^2	3.6×10^7
Urease	Urea	2.5×10^{-2}	1.0×10^4	4.0×10^5

$$\frac{1}{v_\text{o}} = \left(\frac{K_M}{V_\text{max}}\right)\frac{1}{[\text{S}]} + \frac{1}{V_\text{max}}$$

[12-29]

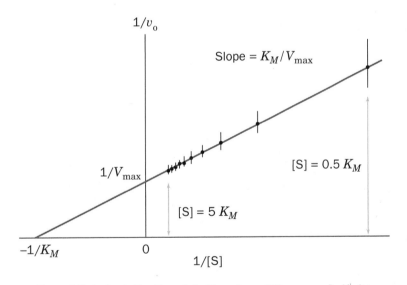

Figure 12-4 *Key to Function.* A double-reciprocal (Lineweaver–Burk) plot.

$$R_1 - \overset{\overset{\displaystyle O}{\|}}{C} - NH - R_2 \ + \ H_2O \ \xrightarrow{\text{trypsin}} \ R_1 - \overset{\overset{\displaystyle O}{\|}}{C} - O^- \ + \ H_3\overset{+}{N} - R_2$$

Polypeptide

(b)

$$CH_3 - \overset{\overset{\displaystyle H}{|}}{\underset{\underset{\displaystyle H}{|}}{C}} - OH \ + \ NAD^+ \xrightarrow[\substack{\downarrow \\ H^+}]{\substack{\text{alcohol} \\ \text{dehydrogenase}}} CH_3 - \overset{\overset{\displaystyle O}{\|}}{C}H \ + \ NADH$$

Figure 12-5 Some bisubstrate reactions.

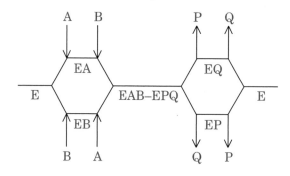

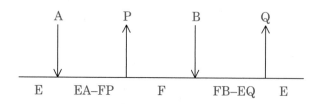

$$\text{E} + \text{S} \underset{k_{-1}}{\overset{k_1}{\rightleftharpoons}} \text{ES} \overset{k_2}{\longrightarrow} \text{P} + \text{E}$$

$$+$$

$$\text{I}$$

$$K_{\text{I}} \Updownarrow$$

$$\text{EI} + \text{S} \longrightarrow \text{NO REACTION}$$

$$K_{\text{I}} = \frac{[\text{E}][\text{I}]}{[\text{EI}]} \qquad [12\text{-}30]$$

$$v_{\text{O}} = \frac{V_{\text{max}}[\text{S}]}{\alpha K_M + [\text{S}]} \qquad [12\text{-}31]$$

$$\alpha = 1 + \frac{[\text{I}]}{K_{\text{I}}} \qquad [12\text{-}32]$$

$$\frac{1}{v_{\text{O}}} = \left(\frac{\alpha K_M}{V_{\text{max}}}\right)\frac{1}{[\text{S}]} + \frac{1}{V_{\text{max}}} \qquad [12\text{-}33]$$

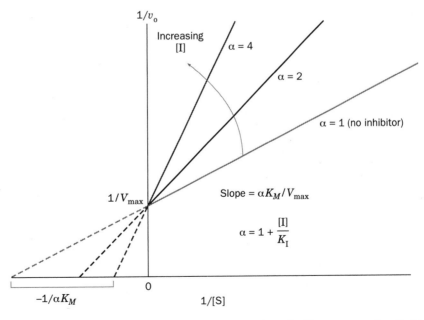

Figure 12-7 A Lineweaver–Burk plot of the competitively inhibited Michaelis–Menten enzyme described by Fig. 12-6.

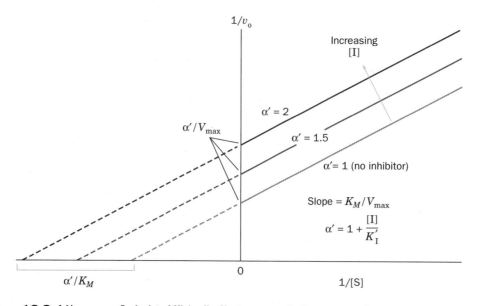

—Phe—Pro—
HIV protease substrate

bond to be
cleaved

Saquinavir

$K_I = 0.40$ nM

Ritonavir

$K_I = 0.015$ nM

$$E + S \underset{k_{-1}}{\overset{k_1}{\rightleftharpoons}} ES \xrightarrow{k_2} P + E$$

$$+$$

$$I$$

$$K'_I \Updownarrow$$

$$ESI \longrightarrow \text{NO REACTION}$$

$$K'_I = \frac{[ES][I]}{[ESI]} \qquad [12\text{-}34]$$

$1/v_o$

Increasing
[I]

$\alpha' = 2$

α'/V_{max}

$\alpha' = 1.5$

$\alpha' = 1$ (no inhibitor)

Slope $= K_M/V_{max}$

$$\alpha' = 1 + \frac{[I]}{K'_I}$$

α'/K_M

0

$1/[S]$

Figure 12-8 A Lineweaver–Burk plot of Michaelis–Menten enzyme in the presence of an uncompetitive inhibitor.

$$
E + S \underset{k_{-1}}{\overset{k_1}{\rightleftharpoons}} ES \xrightarrow{k_2} P + E
$$

$$
\begin{array}{ccc}
+ & & + \\
I & & I
\end{array}
$$

$$
K_{\mathrm{I}} \Updownarrow \qquad\qquad K_{\mathrm{I}}' \Updownarrow
$$

$$
\mathrm{EI} \qquad\qquad \mathrm{ESI} \longrightarrow \mathrm{NO\ REACTION}
$$

$$
K_{\mathrm{I}} = \frac{[\mathrm{E}][\mathrm{I}]}{[\mathrm{EI}]} \qquad \text{and} \qquad K_{\mathrm{I}}' = \frac{[\mathrm{ES}][\mathrm{I}]}{[\mathrm{ESI}]} \qquad\qquad [12\text{-}35]
$$

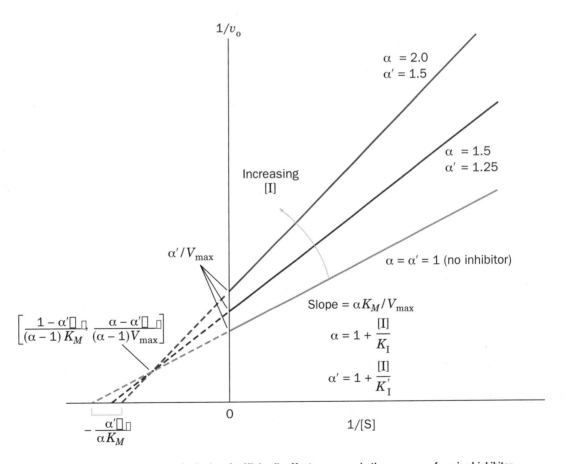

Figure 12-9 A Lineweaver–Burk plot of a Michaelis–Menten enzyme in the presence of a mixed inhibitor.

Table 12-2 Effects of Inhibitors on Michaelis–Menten Reactions[a]

Type of Inhibition	Michaelis–Menten Equation	Lineweaver–Burk Equation	Effect of Inhibitor
None	$v_o = \dfrac{V_{max}[S]}{K_M + [S]}$	$\dfrac{1}{v_o} = \dfrac{K_M}{V_{max}}\dfrac{1}{[S]} + \dfrac{1}{V_{max}}$	None
Competitive	$v_o = \dfrac{V_{max}[S]}{\alpha K_M + [S]}$	$\dfrac{1}{v_o} = \dfrac{\alpha K_M}{V_{max}}\dfrac{1}{[S]} + \dfrac{1}{V_{max}}$	Increases K_M^{app}
Uncompetitive	$v_o = \dfrac{V_{max}[S]}{K_M + \alpha'[S]} = \dfrac{(V_{max}/\alpha')[S]}{K_M/\alpha' + [S]}$	$\dfrac{1}{v_o} = \dfrac{K_M}{V_{max}}\dfrac{1}{[S]} + \dfrac{\alpha'}{V_{max}}$	Decreases K_M^{app} and V_{max}^{app}
Mixed (noncompetitive)	$v_o = \dfrac{V_{max}[S]}{\alpha K_M + \alpha'[S]} = \dfrac{(V_{max}/\alpha')[S]}{(\alpha/\alpha')K_M + [S]}$	$\dfrac{1}{v_o} = \dfrac{\alpha K_M}{V_{max}}\dfrac{1}{[S]} + \dfrac{\alpha'}{V_{max}}$	Decreases V_{max}^{app}; may increase or decrease K_M^{app}

[a] $\alpha = 1 + \dfrac{[1]}{K_1}$ and $\alpha' = 1 + \dfrac{[1]}{K_1'}$

Carbamoyl phosphate   **Aspartate**

aspartate transcarbamoylase

N-Carbamoylaspartate $+ \quad H_2PO_4^-$

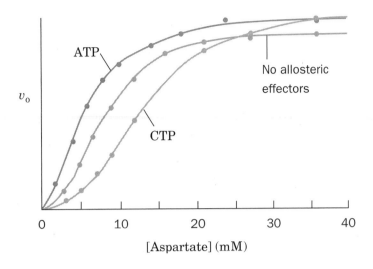

Figure 12-10 Plot of v_0 versus [Aspartate] for the ATCase reaction.

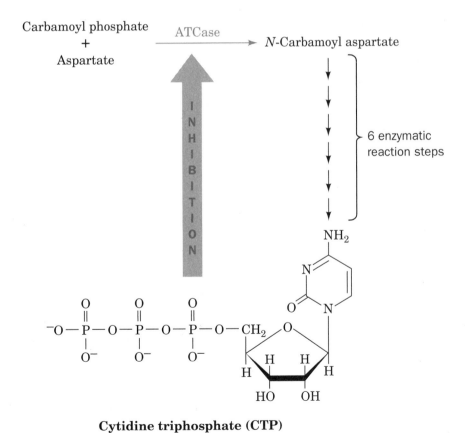

Cytidine triphosphate (CTP)

Figure 12-11 A schematic representation of the pyrimidine biosynthesis pathway.

CHAPTER 13

Introduction to Metabolism

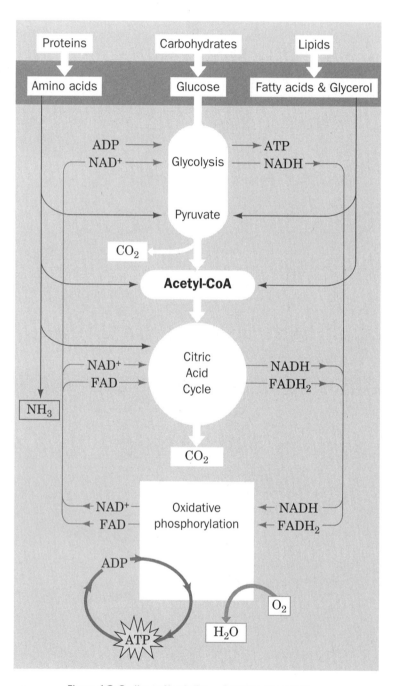

Figure 13-2 *Key to Metabolism.* Overview of catabolism.

Table 13-1 Metabolic Functions of Eukaryotic Organelles

Organelle	Major functions
Mitochondrion	Citric acid cycle, oxidative phosphorylation, fatty acid oxidation, amino acid breakdown
Cytosol	Glycolysis, pentose phosphate pathway, fatty acid biosynthesis, many reactions of gluconeogenesis
Lysosomes	Enzymatic digestion of cell components and ingested matter
Nucleus	DNA replication and transcription, RNA processing
Golgi apparatus	Posttranslational processing of membrane and secretory proteins; formation of plasma membrane and secretory vesicles
Rough endoplasmic reticulum	Synthesis of membrane-bound and secretory proteins
Smooth endoplasmic reticulum	Lipid and steroid biosynthesis
Peroxisomes (glyoxysomes in plants)	Oxidative reactions catalyzed by amino acid oxidases and catalase; glyoxylate cycle reactions in plants

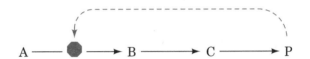

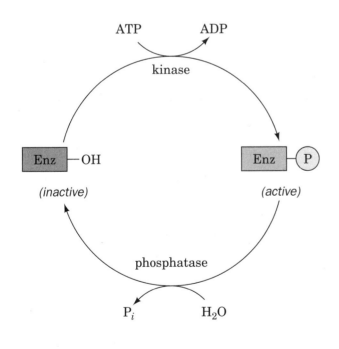

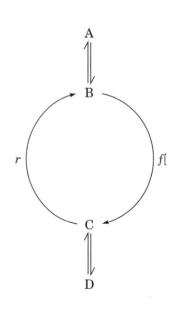

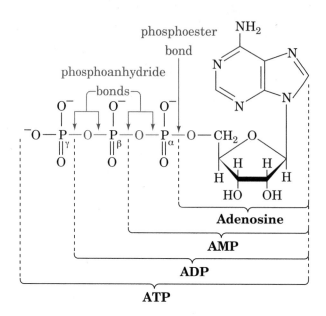

Figure 13-3 *Key to Structure.* The structure of ATP indicating its relationship to ADP, AMP, and adenosine.

Table 13-2 Standard Free Energies of Phosphate Hydrolysis of Some Compounds of Biological Interest

Compound	$\Delta G^{\circ\prime}$ (kJ·mol^{-1})
Phosphoenolpyruvate	−61.9
1,3-Bisphosphoglycerate	−49.4
ATP ($\rightarrow$ AMP + PP$_i$)	−45.6
Acetyl phosphate	−43.1
Phosphocreatine	−43.1
ATP ($\rightarrow$ ADP + P$_i$)	−30.5
Glucose-1-phosphate	−20.9
PP$_i$	−19.2
Fructose-6-phosphate	−13.8
Glucose-6-phosphate	−13.8
Glycerol-3-phosphate	−9.2

Source: Mostly from Jencks, W.P., *in* Fasman, G.D. (Ed.), *Handbook of Biochemistry and Molecular Biology* (3rd ed.), Physical and Chemical Data, Vol. I, pp. 296–304, CRC Press (1976).

Figure 13-4 Resonance and electrostatic stabilization in a phosphoanhydride and its hydrolytic products.

146

(a) $\Delta G^{\circ\prime}$ (kJ • mol^{-1})

Endergonic half-reaction 1	P_i + glucose $\rightleftharpoons$ glucose-6-P + H_2O	+13.8
Exergonic half-reaction 2	ATP + H_2O $\rightleftharpoons$ ADP + P_i	−30.5
Overall coupled reaction	ATP + glucose $\rightleftharpoons$ ADP + glucose-6-P	−16.7

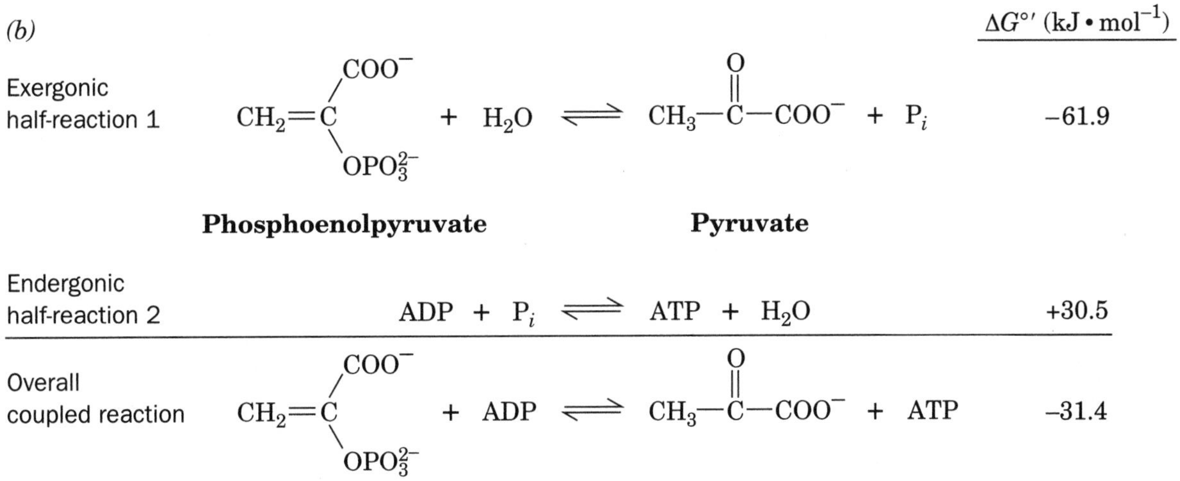

(b) $\Delta G^{\circ\prime}$ (kJ • mol^{-1})

Exergonic half-reaction 1

$$CH_2{=}C\overset{COO^-}{\underset{OPO_3^{2-}}{}} + H_2O \rightleftharpoons CH_3-\overset{O}{\overset{\|}{C}}-COO^- + P_i \qquad -61.9$$

Phosphoenolpyruvate   **Pyruvate**

Endergonic half-reaction 2   ADP + P_i $\rightleftharpoons$ ATP + H_2O +30.5

Overall coupled reaction

$$CH_2{=}C\overset{COO^-}{\underset{OPO_3^{2-}}{}} + ADP \rightleftharpoons CH_3-\overset{O}{\overset{\|}{C}}-COO^- + ATP \qquad -31.4$$

Figure 13-5 Some coupled reactions involving ATP.

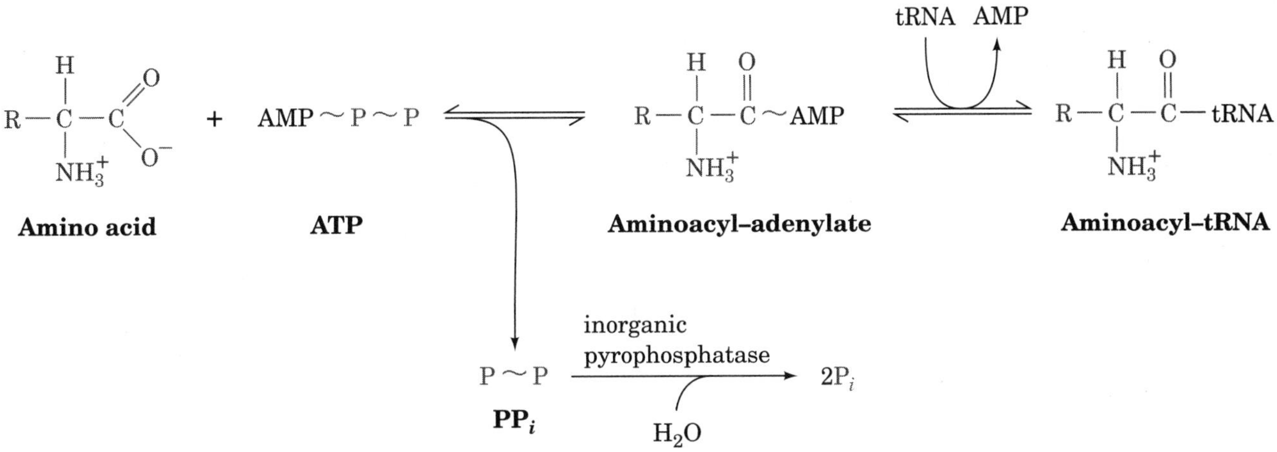

Figure 13-6 Pyrophosphate cleavage in the synthesis of an aminoacyl–tRNA.

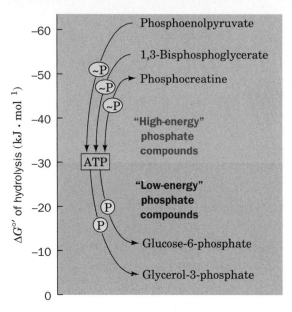

Figure 13-7 Position of ATP relative to "high-energy" and "low-energy" phosphate compounds.

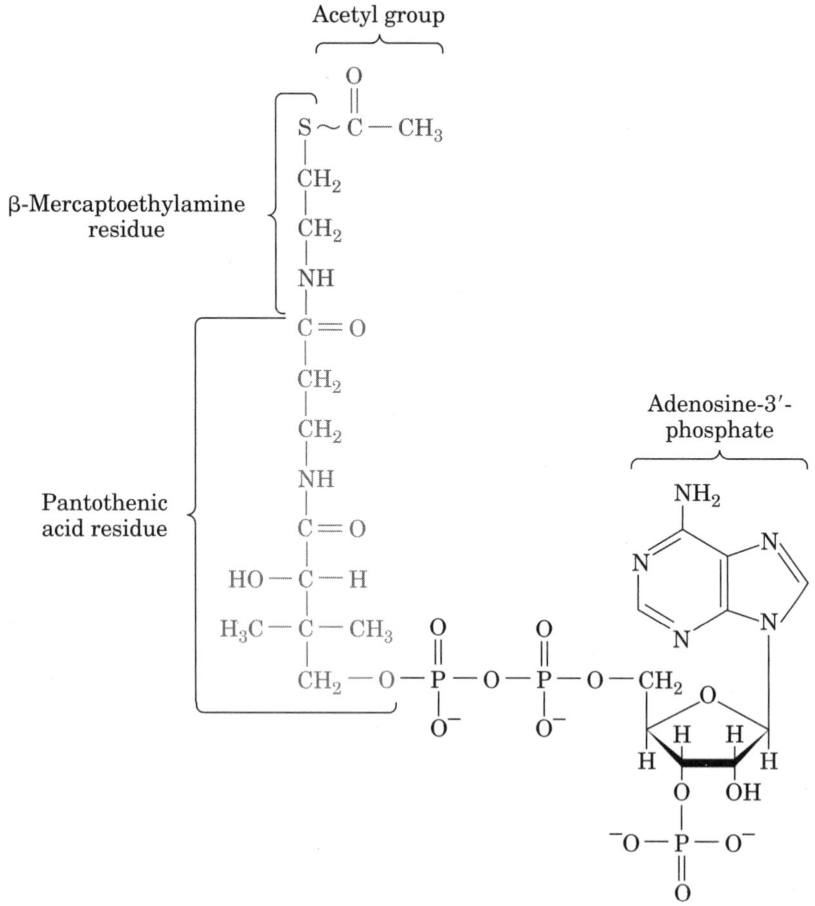

Acetyl group

$$S \sim \overset{\overset{\displaystyle O}{\|}}{C} - CH_3$$

β-Mercaptoethylamine residue

CH_2
CH_2
NH

Pantothenic acid residue

$C = O$
CH_2
CH_2
NH
$C = O$
$HO - C - H$
$H_3C - C - CH_3$
$CH_2 - O - P - O - P - O - CH_2$

Adenosine-3'-phosphate

NH_2

Acetyl-coenzyme A (acetyl-CoA)

Figure 13-9 The chemical structure of acetyl-CoA.

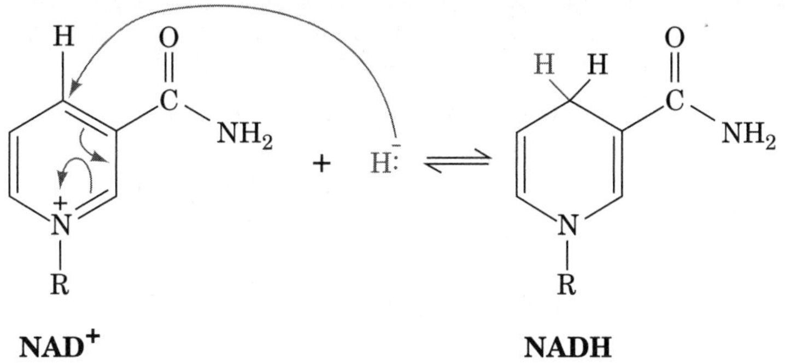

NAD⁺ $\quad$ **+** $\quad$ H:⁻ $\quad\rightleftharpoons\quad$ **NADH**

Figure 13-10 Reduction of NAD⁺ to NADH.

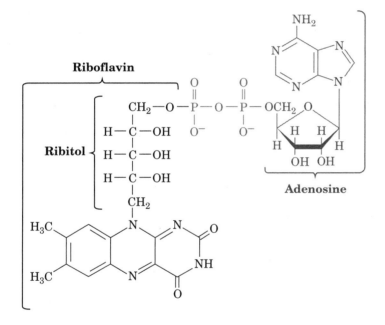

Figure 13-11 Flavin adenine dinucleotide (FAD).

Flavin adenine dinucleotide (FAD)
(oxidized or quinone form)

$\Updownarrow$ H•

FADH• (radical or semiquinone form)

$\Updownarrow$ H•

FADH₂ (reduced or hydroquinone form)

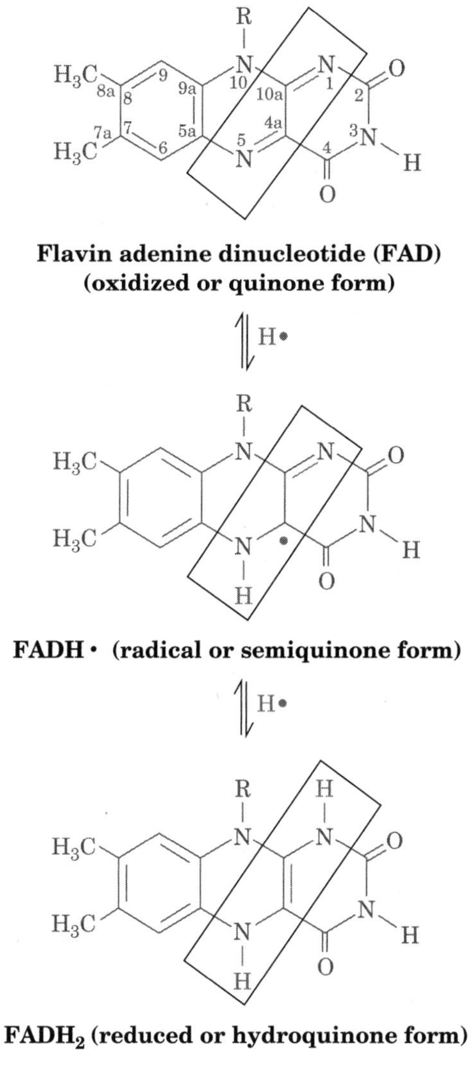

Figure 13-12 Reduction of FAD to FADH₂.

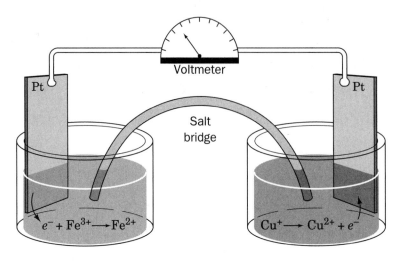

Figure 13-13 An electrochemical cell.

$$\Delta G = \Delta G^\circ + RT \ln \left(\frac{[A_{red}][B_{ox}^{n+}]}{[A_{ox}^{n+}][B_{red}]} \right) \qquad [13\text{-}4]$$

$$\Delta G = -n\mathscr{F}\Delta\mathscr{E} \qquad [13\text{-}7]$$

$$\boxed{\Delta\mathscr{E} = \Delta\mathscr{E}^\circ - \frac{RT}{n\mathscr{F}} \ln \left(\frac{[A_{red}][B_{ox}^{n+}]}{[A_{ox}^{n+}][B_{red}]} \right)} \qquad [13\text{-}8]$$

$$\Delta\mathscr{E}^\circ = \mathscr{E}^\circ(e^-\text{acceptor}) - \mathscr{E}^\circ(e^-\text{donor}) \qquad [13\text{-}11]$$

Table 13-3 Standard Reduction Potentials of Some Biochemically Important Half-Reactions

Half-Reaction	$e^{\circ\prime}$ (V)
$\frac{1}{2} O_2 + 2\,H^+ + 2\,e^- \rightleftharpoons H_2O$	0.815
$SO_4^{2-} + 2\,H^+ + 2\,e^- \rightleftharpoons SO_3^{2-} + H_2O$	0.48
$NO_3^- + 2\,H^+ + 2\,e^- \rightleftharpoons NO_2^- + H_2O$	0.42
Cytochrome a_3 (Fe^{3+}) $+ e^- \rightleftharpoons$ cytochrome a_3 (Fe^{2+})	0.385
$O_2(g) + 2\,H^+ + 2\,e^- \rightleftharpoons H_2O_2$	0.295
Cytochrome a (Fe^{3+}) $+ e^- \rightleftharpoons$ cytochrome a (Fe^{2+})	0.29
Cytochrome c (Fe^{3+}) $+ e^- \rightleftharpoons$ cytochrome c (Fe^{2+})	0.235
Cytochrome c_1 (Fe^{3+}) $+ e^- \rightleftharpoons$ cytochrome c_1 (Fe^{2+})	0.22
Cytochrome b (Fe^{3+}) $+ e^- \rightleftharpoons$ cytochrome b (Fe^{2+}) (*mitochondrial*)	0.077
Ubiquinone $+ 2\,H^+ + 2\,e^- \rightleftharpoons$ ubiquinol	0.045
Fumarate$^- + 2\,H^+ + 2\,e^- \rightleftharpoons$ succinate$^-$	0.031
FAD $+ 2\,H^+ + 2\,e^- \rightleftharpoons$ FADH$_2$ (*in flavoproteins*)	~0.
Oxaloacetate$^- + 2\,H^+ + 2\,e^- \rightleftharpoons$ malate$^-$	−0.166
Pyruvate$^- + 2\,H^+ + 2\,e^- \rightleftharpoons$ lactate$^-$	−0.185
Acetaldehyde $+ 2\,H^+ + 2\,e^- \rightleftharpoons$ ethanol	−0.197
FAD $+ 2\,H^+ + 2\,e^- \rightleftharpoons$ FADH$_2$ (*free coenzyme*)	−0.219
$S + 2\,H^+ + 2\,e^- \rightleftharpoons H_2S$	−0.23
Lipoic acid $+ 2\,H^+ + 2\,e^- \rightleftharpoons$ dihydrolipoic acid	−0.29
$NAD^+ + H^+ + 2\,e^- \rightleftharpoons$ NADH	−0.315
$NADP^+ + H^+ + 2\,e^- \rightleftharpoons$ NADPH	−0.320
Cystine $+ 2\,H^+ + 2\,e^- \rightleftharpoons$ 2 cysteine	−0.340
Acetoacetate$^- + 2\,H^+ + 2\,e^- \rightleftharpoons$ β-hydroxybutyrate$^-$	−0.346
$H^+ + e^- \rightleftharpoons \frac{1}{2} H_2$	−0.421
Acetate$^- + 3\,H^+ + 2\,e^- \rightleftharpoons$ acetaldehyde $+ H_2O$	−0.581

Source: Mostly from Loach, P.A., *In* Fasman, G.D. (Ed.), *Handbook of Biochemistry and Molecular Biology* (3rd ed.), Physical and Chemical Data, Vol. I, pp. 123–130, CRC Press (1976).

CHAPTER 14

Glucose Catabolism

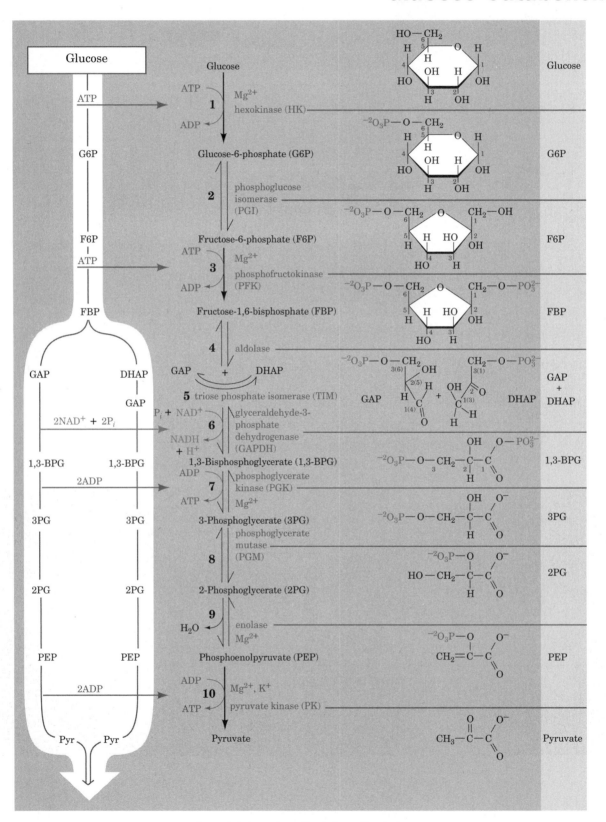

Figure 14-1 *Key to Metabolism.* Glycolysis.

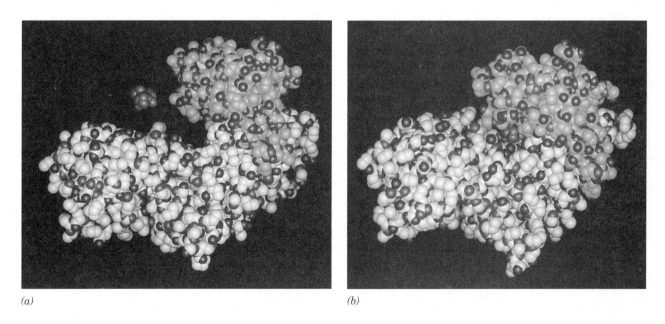

CH$_2$OH

H   O   H

H

OH   H

HO   OH

H   OH

Glucose   + ATP

hexokinase
Mg^{2+}

CH$_2$OPO$_3^{2-}$

H   O   H

H

OH   H

HO   OH

H   OH

Glucose-6-phosphate   + ADP + H$^+$
(G6P)

Mg^{2+}

O$^-$ O$^-$ O$^-$ H
                                        O—CH$_2$

Adenosine—O—P—O—P—O—P—O$^-$ O   H

‖   ‖   ‖   H
O   O   O   OH   H

HO   OH

H   OH

ATP   **Glucose**

Figure 14-2 Substrate-induced conformational changes in yeast hexokinase.

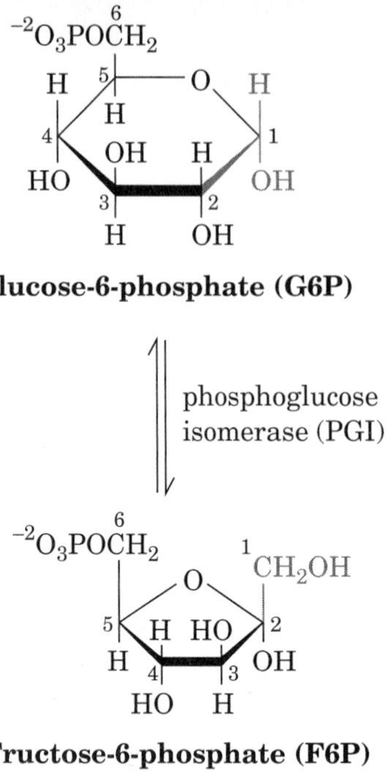

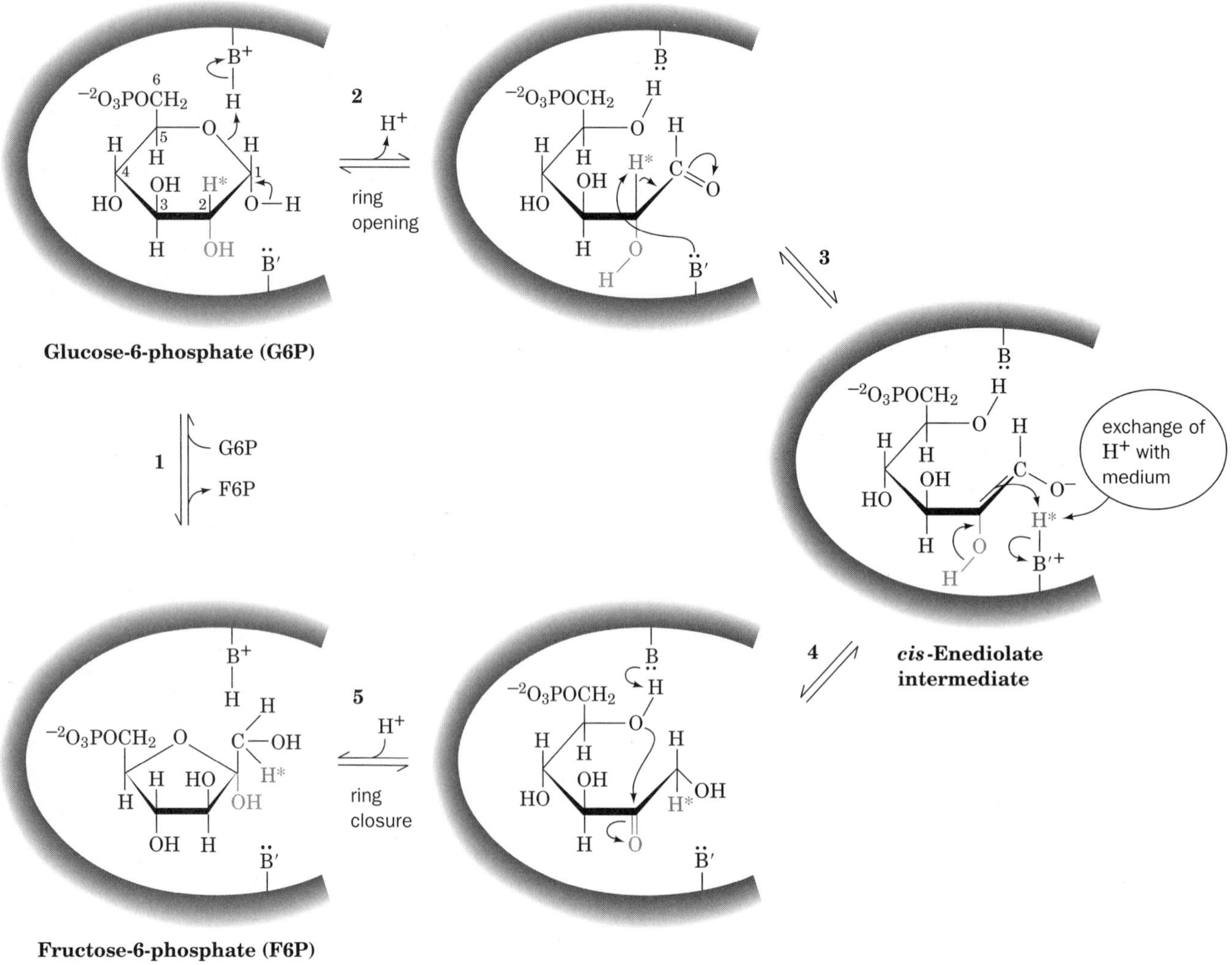

Figure 14-3 The reaction mechanism of phosphoglucose isomerase.

154

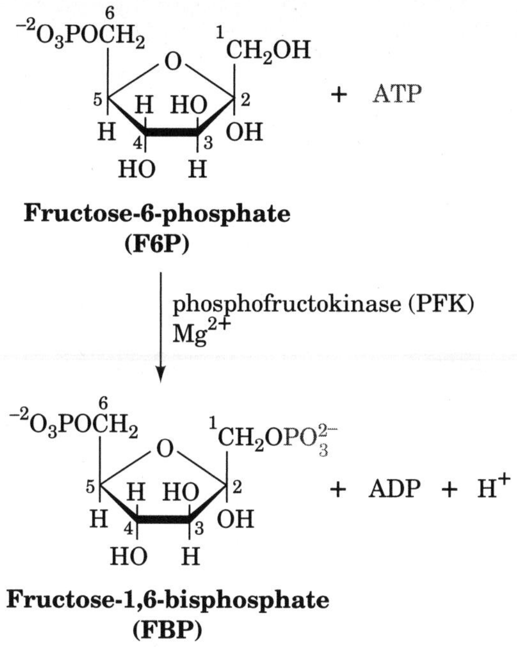

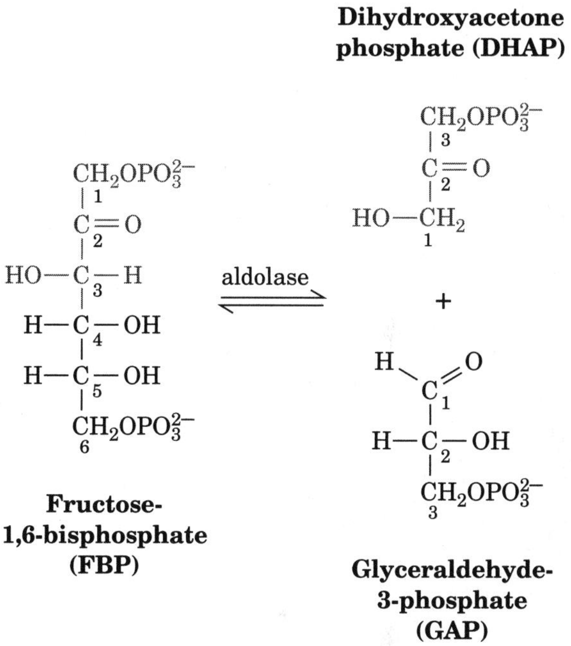

Fructose-6-phosphate
(F6P)

$+$ ATP

phosphofructokinase (PFK)
Mg^{2+}

Fructose-1,6-bisphosphate
(FBP)

$+$ ADP $+$ H$^+$

Dihydroxyacetone
phosphate (DHAP)

aldolase

Fructose-
1,6-bisphosphate
(FBP)

Glyceraldehyde-
3-phosphate
(GAP)

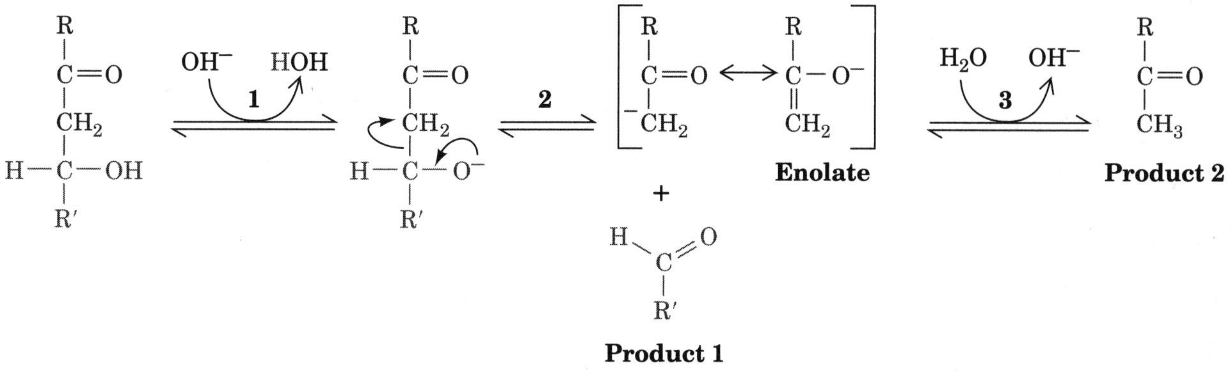

Enolate

Product 2

Product 1

Figure 14-4 The mechanism of base-catalyzed aldol cleavage.

155

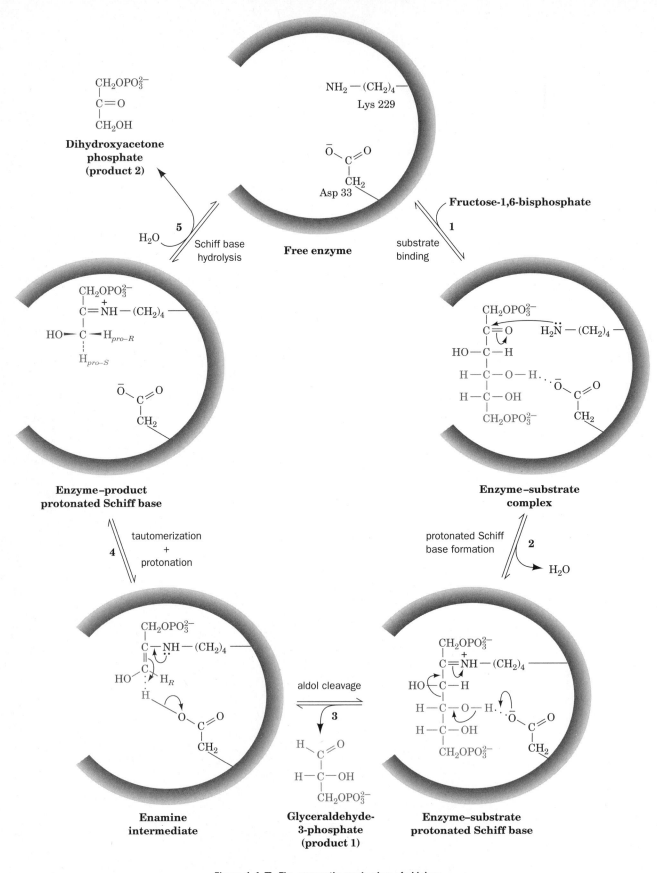

Figure 14-5 The enzymatic mechanism of aldolase.

156

Glyceraldehyde-
3-phosphate
(an aldose)

Dihydroxyacetone
phosphate
(a ketose)

Enediol
intermediate

Phosphoglyco-
hydroxamate

Proposed enediolate
intermediate

2-Phosphoglycolate

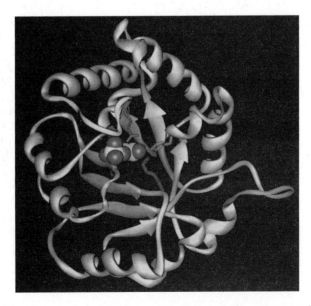

Figure 14-6 A ribbon diagram of yeast TIM in complex with its transition state analog 2-phosphoglycolate.

157

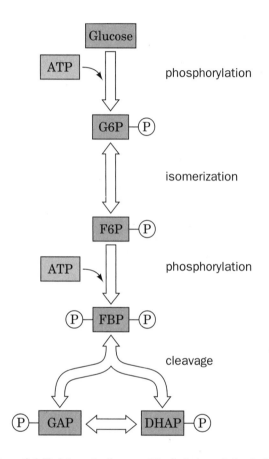

Figure 14-7 Schematic diagram of the first stage of glycolysis.

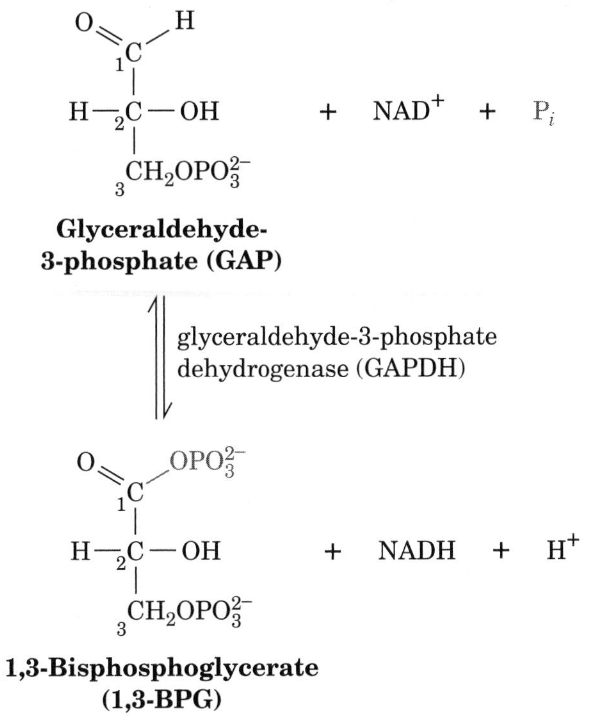

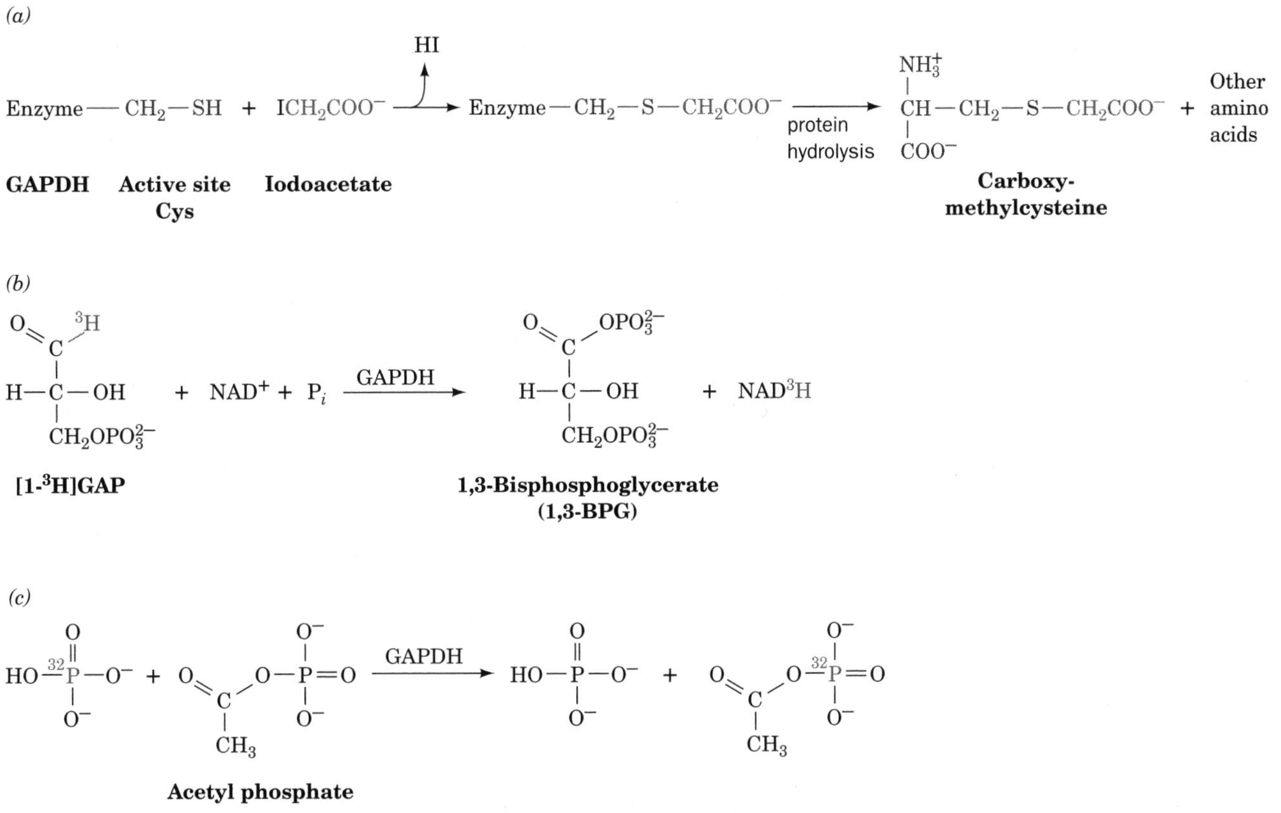

Figure 14-8 Reactions that were used to elucidate the enzymatic mechanism of GAPDH.

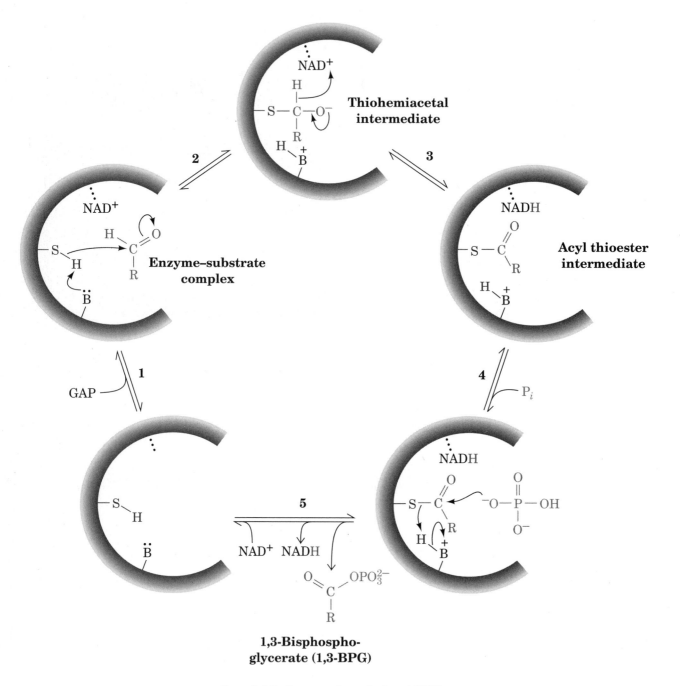

Figure 14-9 The enzymatic mechanism of GAPDH.

$$\underset{\substack{3}}{\underset{\substack{2}}{\underset{\substack{1}}{O}}}$$

$$\begin{array}{c} O=\overset{\displaystyle}{\underset{1}{C}}-OPO_3^{2-} \\ \\ H-\overset{\displaystyle}{\underset{2}{C}}-OH \qquad + \ ADP \\ \\ \underset{3}{CH_2}OPO_3^{2-} \end{array}$$

**1,3-Bisphosphoglycerate
(1,3-BPG)**

$$\text{Mg}^{2+} \ \Big\updownarrow \ \begin{array}{l}\text{phosphoglycerate}\\ \text{kinase (PGK)}\end{array}$$

$$\begin{array}{c} {}^-O\diagdown \overset{\displaystyle}{\underset{1}{C}} \diagup O \\ \\ H-\overset{\displaystyle}{\underset{2}{C}}-OH \qquad + \ ATP \\ \\ \underset{3}{CH_2}OPO_3^{2-} \end{array}$$

**3-Phosphoglycerate
(3PG)**

Figure 14-1O A space-filling model of yeast phosphoglycerate kinase.

$$\text{GAP} + P_i + \text{NAD}^+ \longrightarrow \text{1,3-BPG} + \text{NADH} \qquad \Delta G^{\circ\prime} = +6.7 \ \text{kJ}\cdot\text{mol}^{-1}$$

$$\underline{\text{1,3-BPG} + \text{ADP} \longrightarrow \text{3PG} + \text{ATP} \qquad\qquad\qquad \Delta G^{\circ\prime} = -18.8 \ \text{kJ}\cdot\text{mol}^{-1}}$$

$$\text{GAP} + P_i + \text{NAD}^+ + \text{ADP} \longrightarrow \text{3PG} + \text{NADH} + \text{ATP}$$
$$\Delta G^{\circ\prime} = -12.1 \ \text{kJ}\cdot\text{mol}^{-1}$$

$$\begin{array}{c} O=\overset{\displaystyle}{\underset{1}{C}}\diagup O^- \\ \\ H-\overset{\displaystyle}{\underset{2}{C}}-OH \\ \\ H-\overset{\displaystyle}{\underset{3}{C}}-OPO_3^{2-} \\ \\ H \end{array} \qquad \underset{\substack{\text{phosphoglycerate}\\ \text{mutase (PGM)}}}{\rightleftharpoons} \qquad \begin{array}{c} O=\overset{\displaystyle}{\underset{1}{C}}\diagup O^- \\ \\ H-\overset{\displaystyle}{\underset{2}{C}}-OPO_3^{2-} \\ \\ H-\overset{\displaystyle}{\underset{3}{C}}-OH \\ \\ H \end{array}$$

**3-Phosphoglycerate
(3PG)**   **2-Phosphoglycerate
(2PG)**

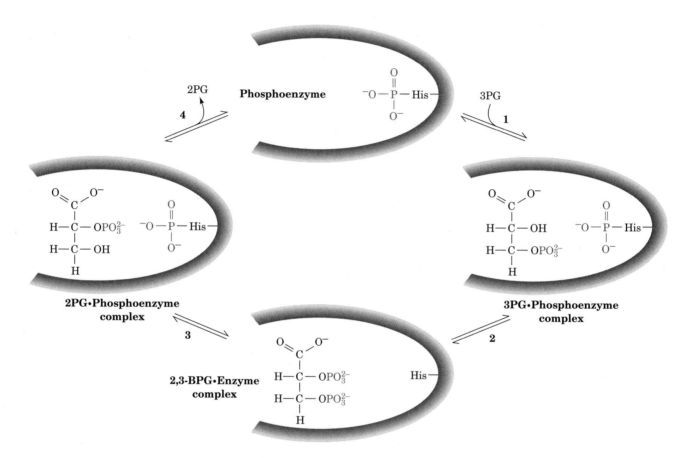

Figure 14-12 A proposed reaction mechanism for phosphoglycerate mutase.

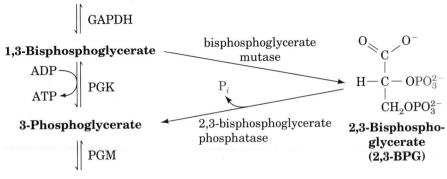

Glyceraldehyde 3-phosphate

GAPDH

1,3-Bisphosphoglycerate

bisphosphoglycerate mutase

ADP

ATP

PGK

P_i

3-Phosphoglycerate

2,3-bisphosphoglycerate phosphatase

PGM

2-Phosphoglycerate

2,3-Bisphospho-glycerate (2,3-BPG)

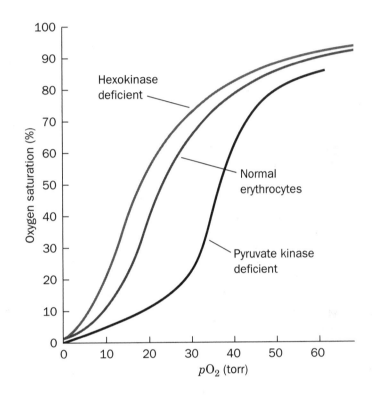

Hexokinase deficient

Normal erythrocytes

Pyruvate kinase deficient

Oxygen saturation (%)

pO_2 (torr)

2-Phosphoglycerate (2PG)

enolase

Phosphoenolpyruvate (PEP)

$+ H_2O$

Phosphoenolpyruvate (PEP)

$+ ADP + H^+$

pyruvate kinase (PK)

$+ ATP$

Pyruvate

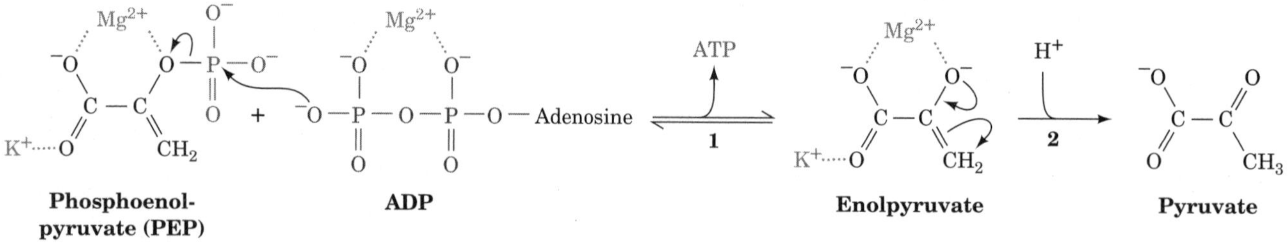

Figure 14-13 The mechanism of the reaction catalyzed by pyruvate kinase.

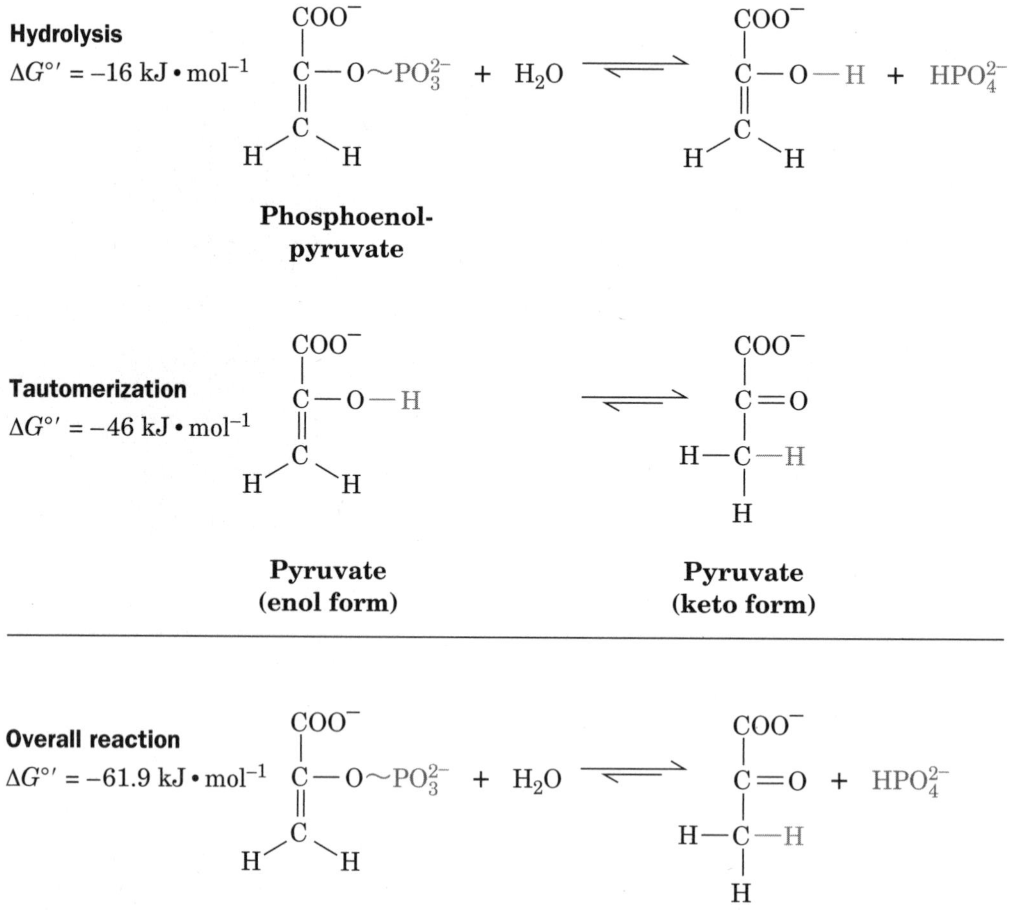

Figure 14-14 The hydrolysis of PEP.

164

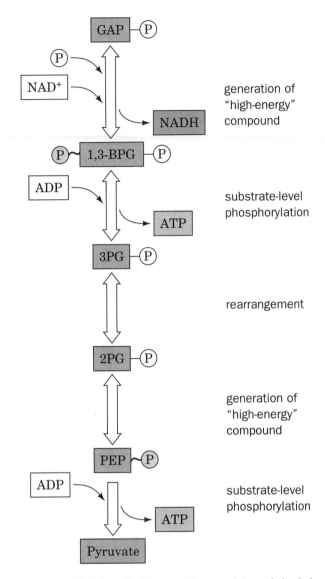

Figure 14-15 Schematic diagram of the second stage of glycolysis.

$$\text{Glucose} + 2\,\text{NAD}^+ + 2\,\text{ADP} + 2\,\text{P}_i \longrightarrow$$
$$2\,\text{pyruvate} + 2\,\text{NADH} + 2\,\text{ATP} + 2\,\text{H}_2\text{O} + 4\,\text{H}^+$$

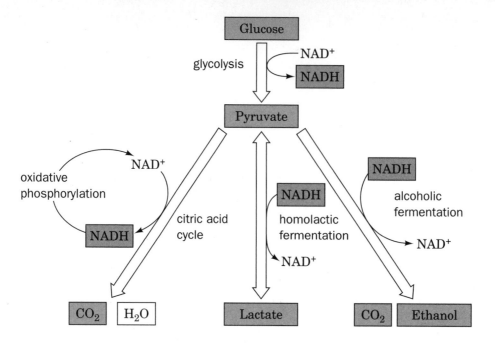

Figure 14-16 Metabolic fate of pyruvate.

Pyruvate   **NADH**

lactate dehydrogenase (LDH)

L-Lactate   **NAD⁺**

NADH   **Pyruvate**

Arg 109

CH₂—His 195

Arg 171

L-Lactate

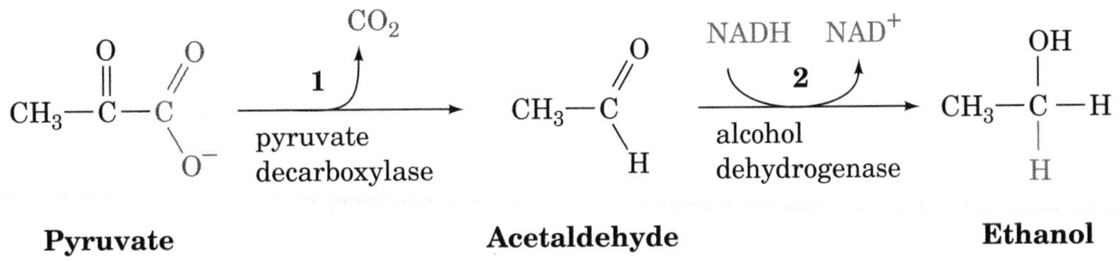

Figure 14-18 The two reactions of alcoholic fermentation.

Thiamine Pyrophosphate (TPP)

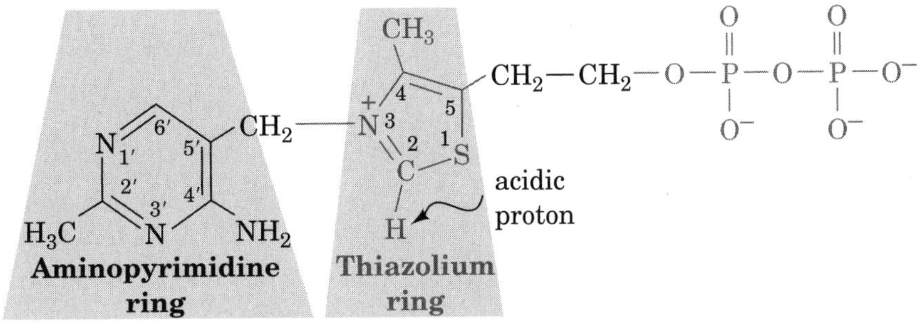

Thiamine pyrophosphate (TPP)

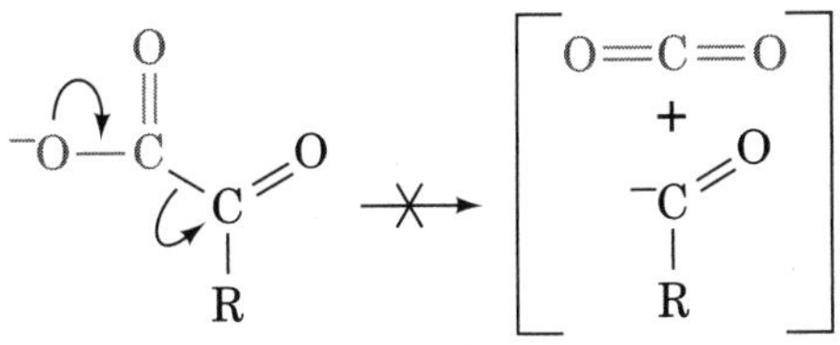

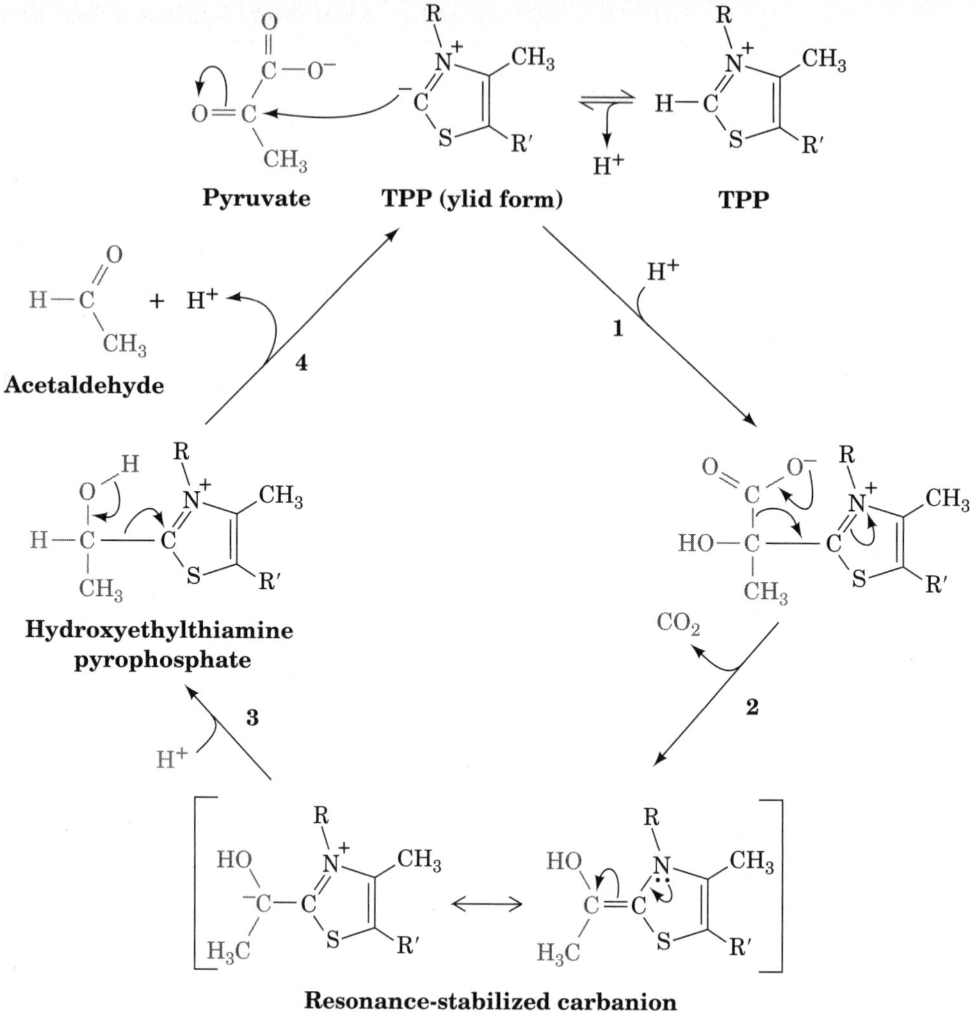

Figure 14-20 The reaction mechanism of pyruvate decarboxylase.

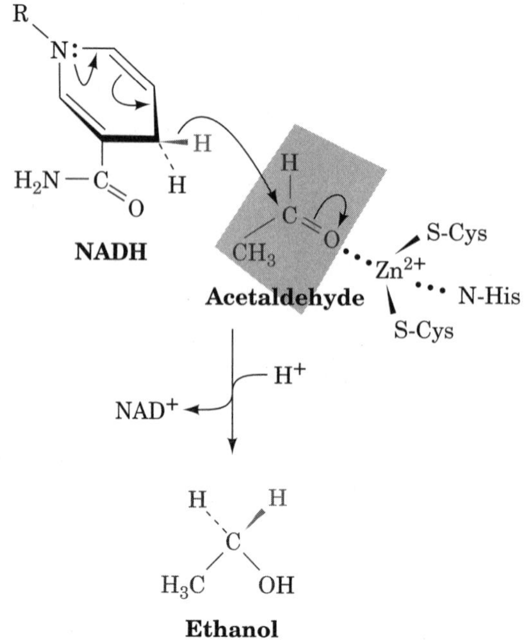

Table 14-1 $\Delta G^{\circ\prime}$ and ΔG for the Reactions of Glycolysis in Heart Muscle[a]

Reaction	Enzyme	$\Delta G^{\circ\prime}$ $(kJ \cdot mol^{-1})$	ΔG $(kJ \cdot mol^{-1})$
1	Hexokinase	−20.9	−27.2
2	PGI	+2.2	−1.4
3	PFK	−17.2	−25.9
4	Aldolase	+22.8	−5.9
5	TIM	+7.9	~0
6 + 7	GAPDH + PGK	−16.7	−1.1
8	PGM	+4.7	−0.6
9	Enolase	−3.2	−2.4
10	PK	−23.0	−13.9

[a]Calculated from data in Newsholme, E.A. and Start, C., *Regulation in Metabolism*, p. 97, Wiley (1973).

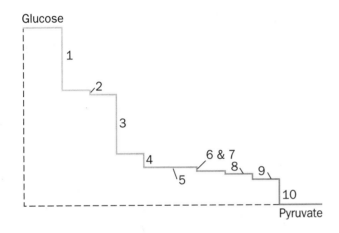

Figure 14-21 Diagram of free energy changes in glycolysis.

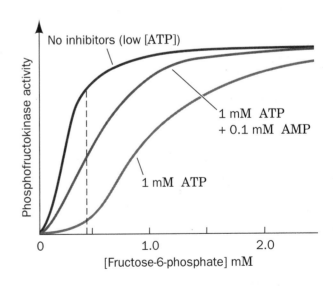

Figure 14-23 PFK activity versus F6P concentration.

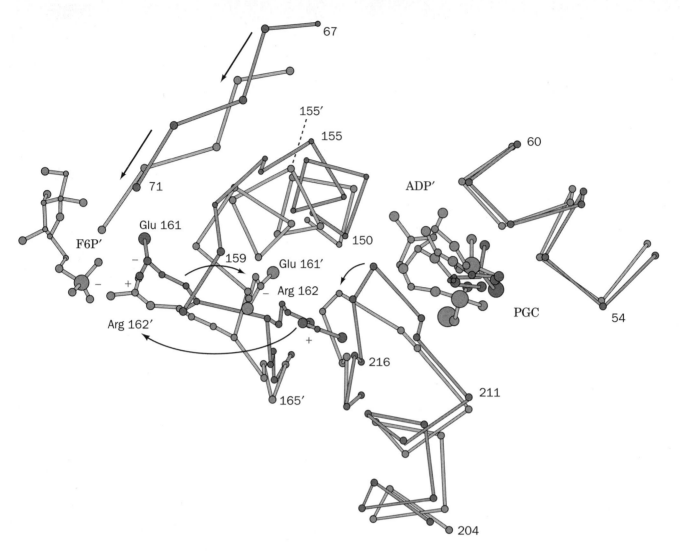

Figure 14-24 Allosteric changes in PFK from

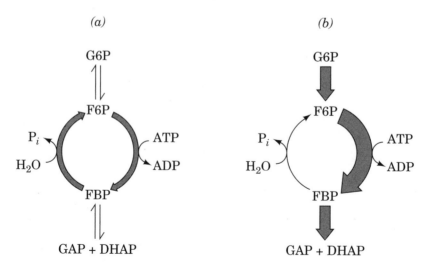

Figure 14-25 Substrate cycling in the regulation of PFK.

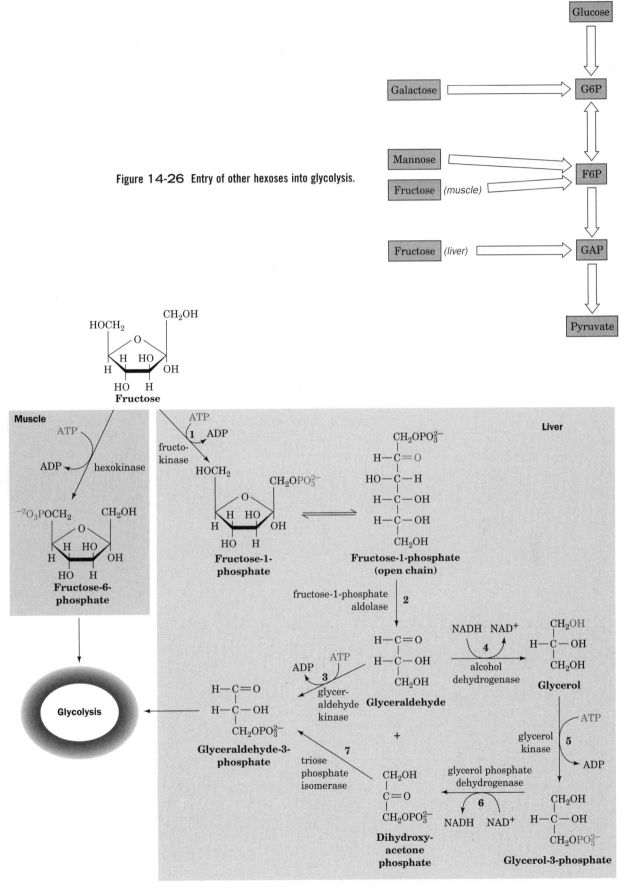

Figure 14-26 Entry of other hexoses into glycolysis.

Figure 14-27 The metabolism of fructose.

171

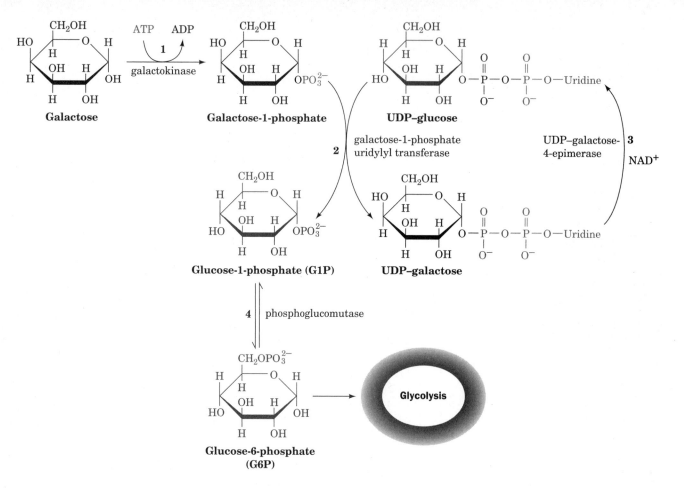

Figure 14-28 The metabolism of galactose.

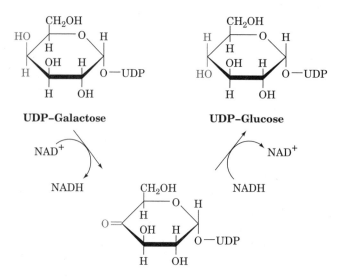

α-D-**Glucose** α-D-**Galactose**

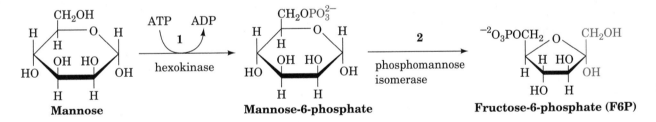

Figure 14-29 The metabolism of mannose.

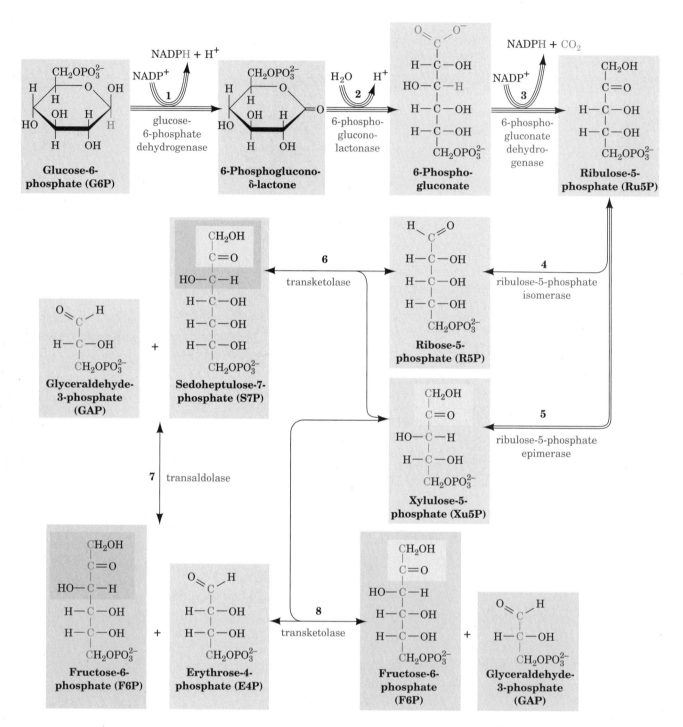

Figure 14-30 *Key to Metabolism.* The pentose phosphate pathway.

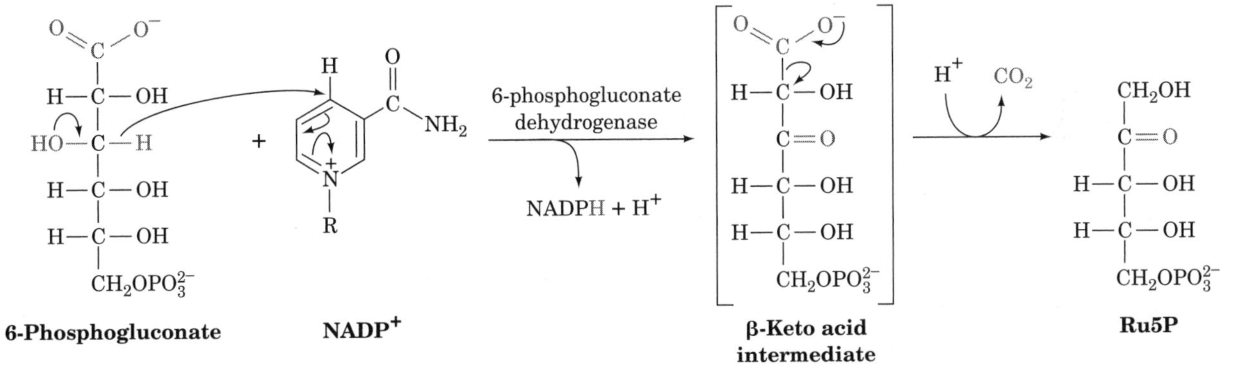

Figure 14-31 The 6-phosphogluconate dehydrogenase reaction.

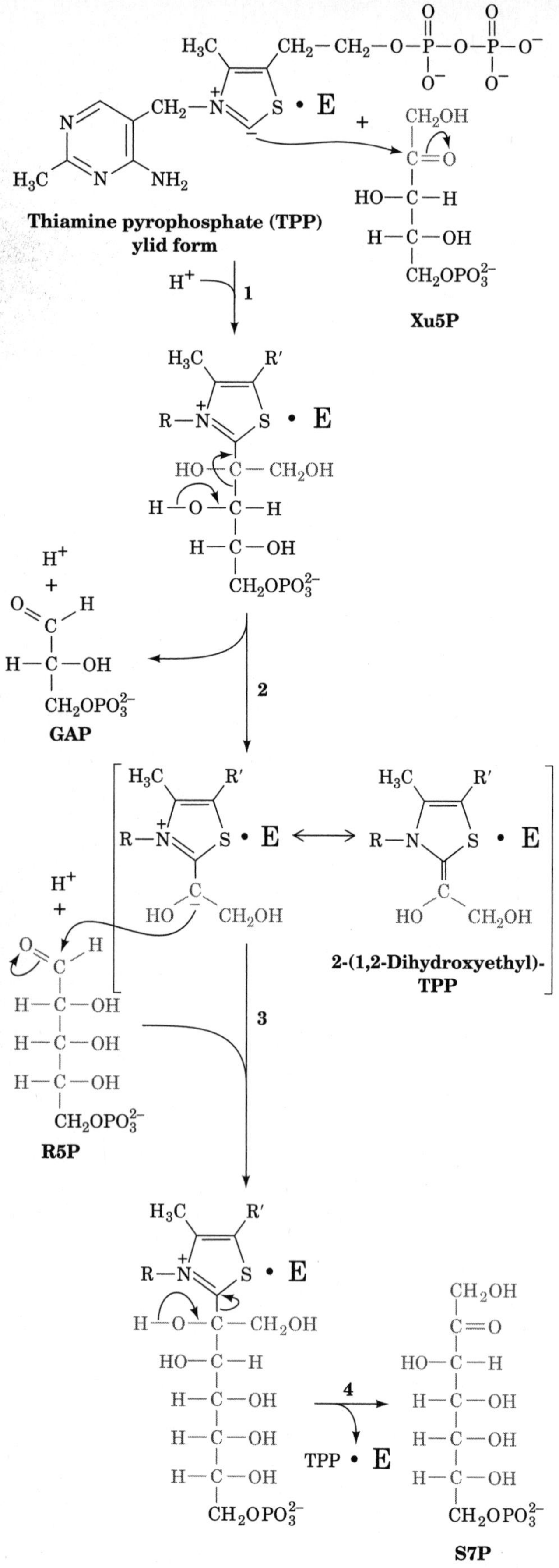

Figure 14-32 Mechanism of transketolase.

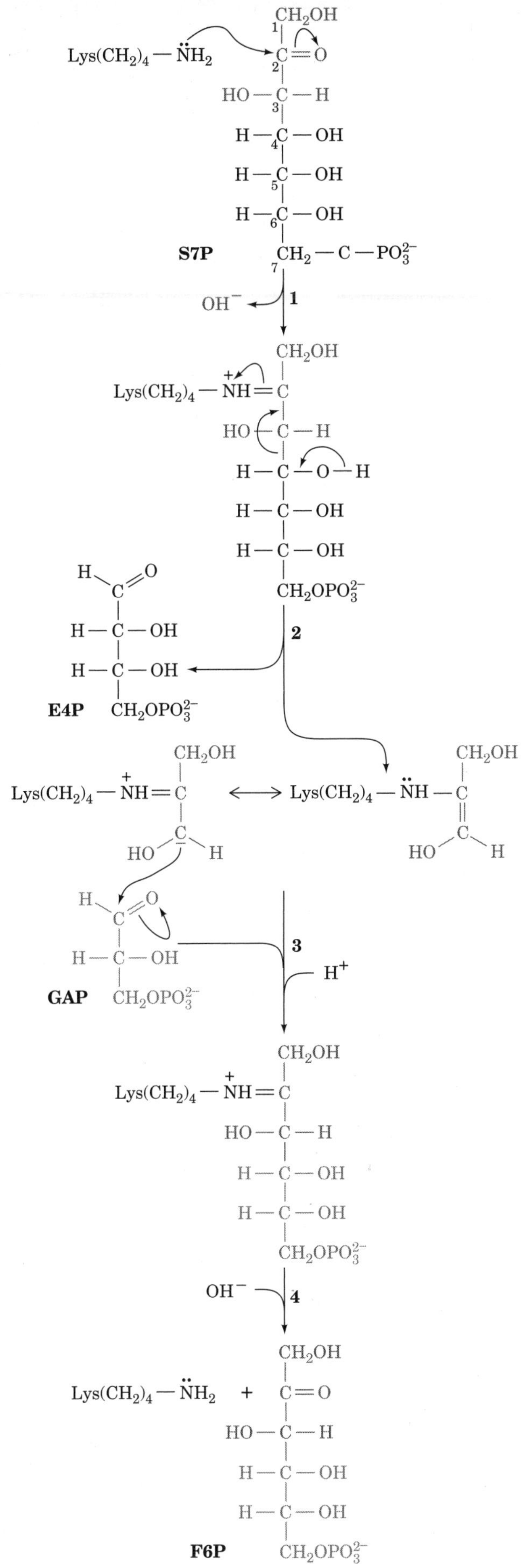

Figure 14-33 Mechanism of transaldolase.

$$\textbf{(6)} \quad C_5 + C_5 \rightleftharpoons C_7 + C_3$$

$$\textbf{(7)} \quad C_7 + C_3 \rightleftharpoons C_6 + C_4$$

$$\textbf{(8)} \quad \underline{C_5 + C_4 \rightleftharpoons C_6 + C_3}$$

$$\text{(Sum)} \quad 3\,C_5 \rightleftharpoons 2\,C_6 + C_3$$

Figure 14-34 Summary of carbon skeleton rearrangements in the pentose phosphate pathway.

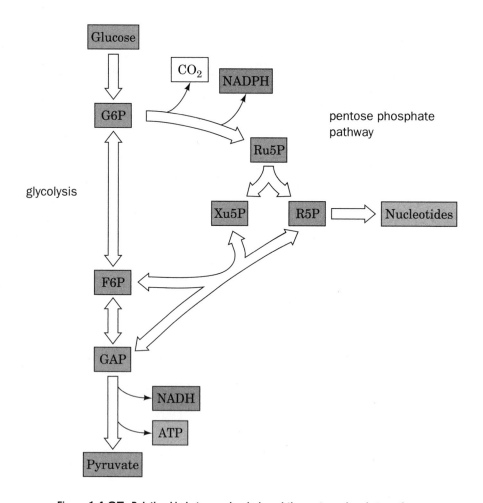

Figure 14-35 Relationship between glycolysis and the pentose phosphate pathway.

CHAPTER 15

Glycogen Metabolism and Gluconeogenesis

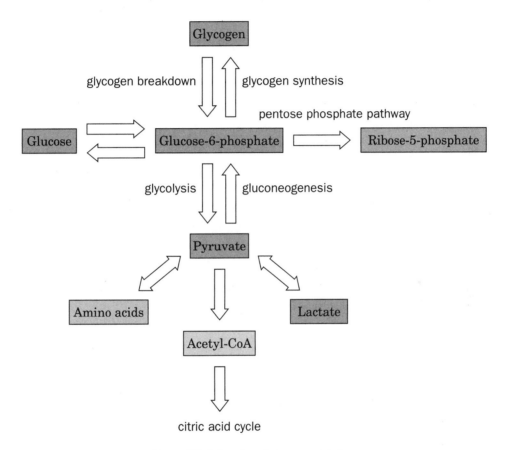

Figure 15-1 Overview of glucose metabolism.

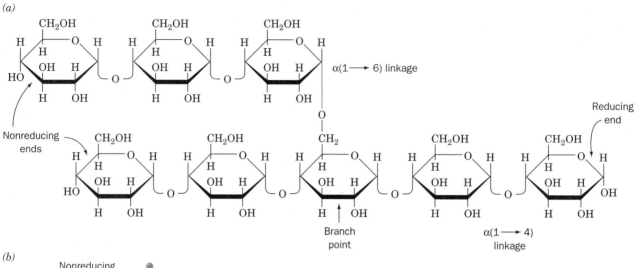

(a)

α(1⟶6) linkage

Reducing
end

Nonreducing
ends

α(1⟶4)
linkage

Branch
point

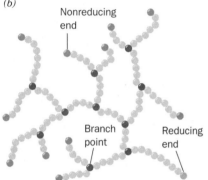

(b)

Nonreducing
end

Branch
point

Reducing
end

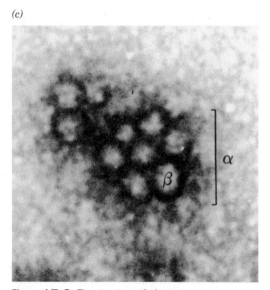

(c)

α

β

Figure 15-2 The structure of glycogen.

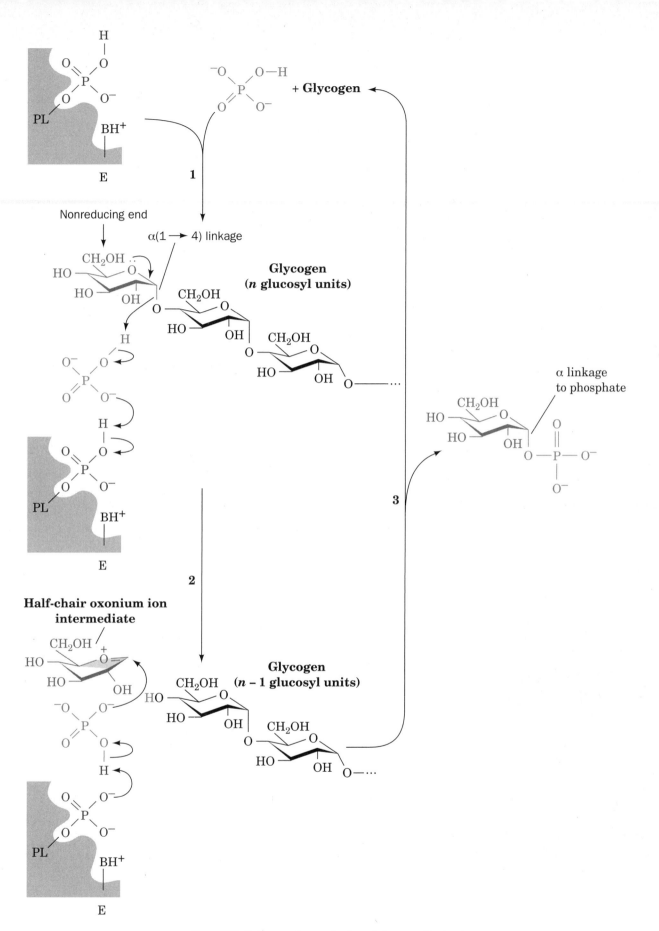

Figure 15-4 The reaction mechanism of glycogen phosphorylase.

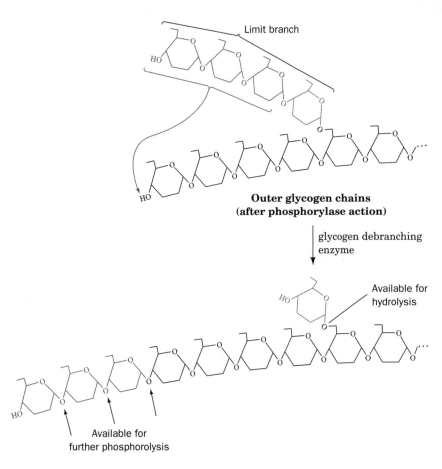

**Outer glycogen chains
(after phosphorylase action)**

glycogen debranching
enzyme

Available for
hydrolysis

Available for
further phosphorolysis

Figure 15-6 The reactions catalyzed by debranching enzyme.

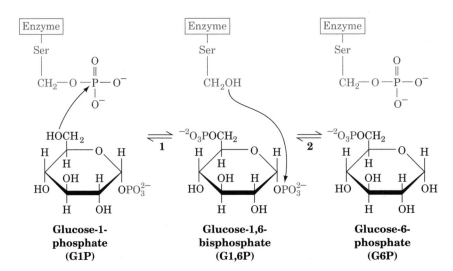

Figure 15-7 The mechanism of phosphoglucomutase.

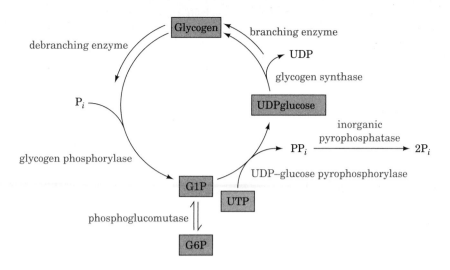

Figure 15-8 Opposing pathways of glycogen synthesis and degradation.

$$
\begin{array}{lc}
 & \Delta G^{\circ\prime} \ (\mathrm{kJ \cdot mol^{-1}}) \\
\hline
\mathrm{G1P + UTP \rightleftharpoons UDPG + PP_i} & \sim 0 \\
\mathrm{H_2O + PP_i \rightarrow 2\,P_i} & -19.2 \\
\hline
\mathrm{Overall\ G1P + UTP \rightarrow UDPG + 2\,P_i} & -19.2 \\
\end{array}
$$

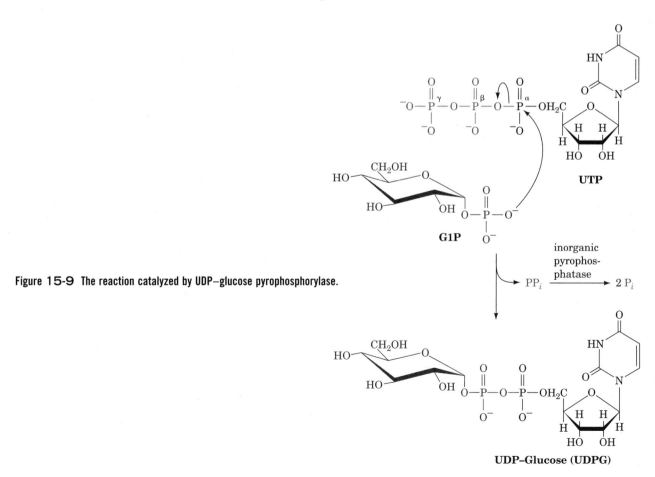

Figure 15-9 The reaction catalyzed by UDP–glucose pyrophosphorylase.

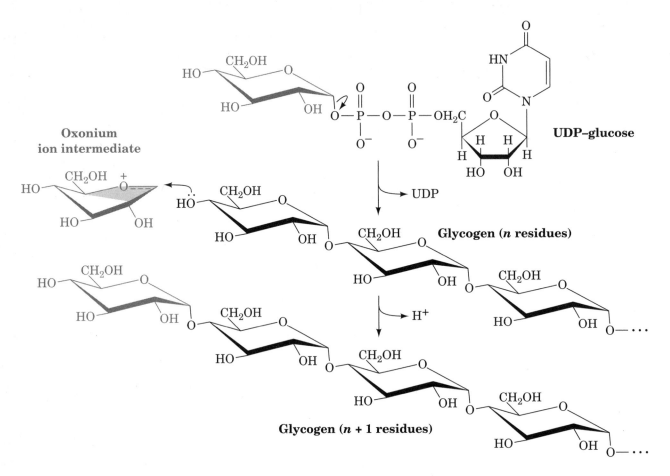

Figure 15-10 The reaction catalyzed by glycogen synthase.

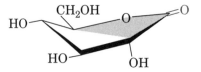

1,5-Gluconolactone

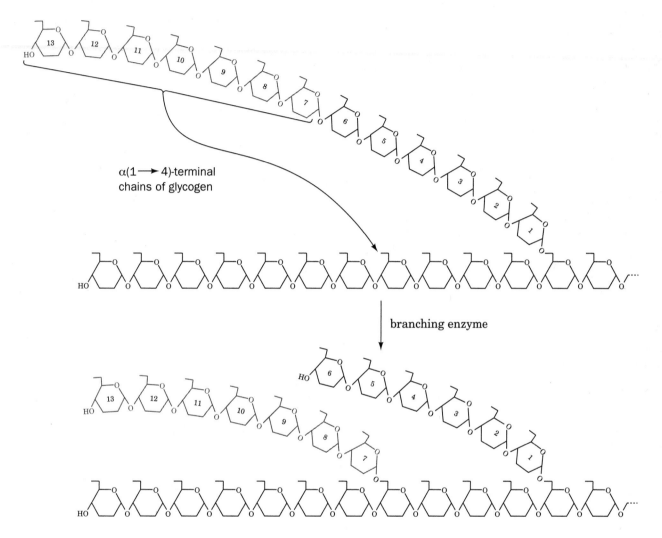

α(1 ⟶ 4)-terminal
chains of glycogen

branching enzyme

Figure 15-11 The branching of glycogen.

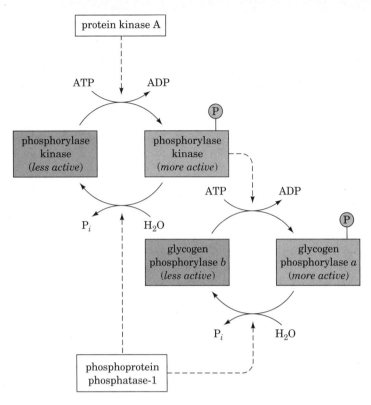

Figure 15-12 The glycogen phosphorylase interconvertible enzyme system.

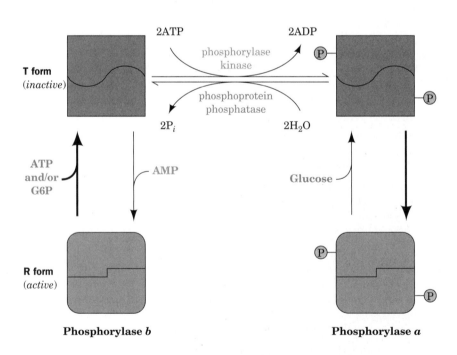

Figure 15-13 The control of glycogen phosphorylase activity.

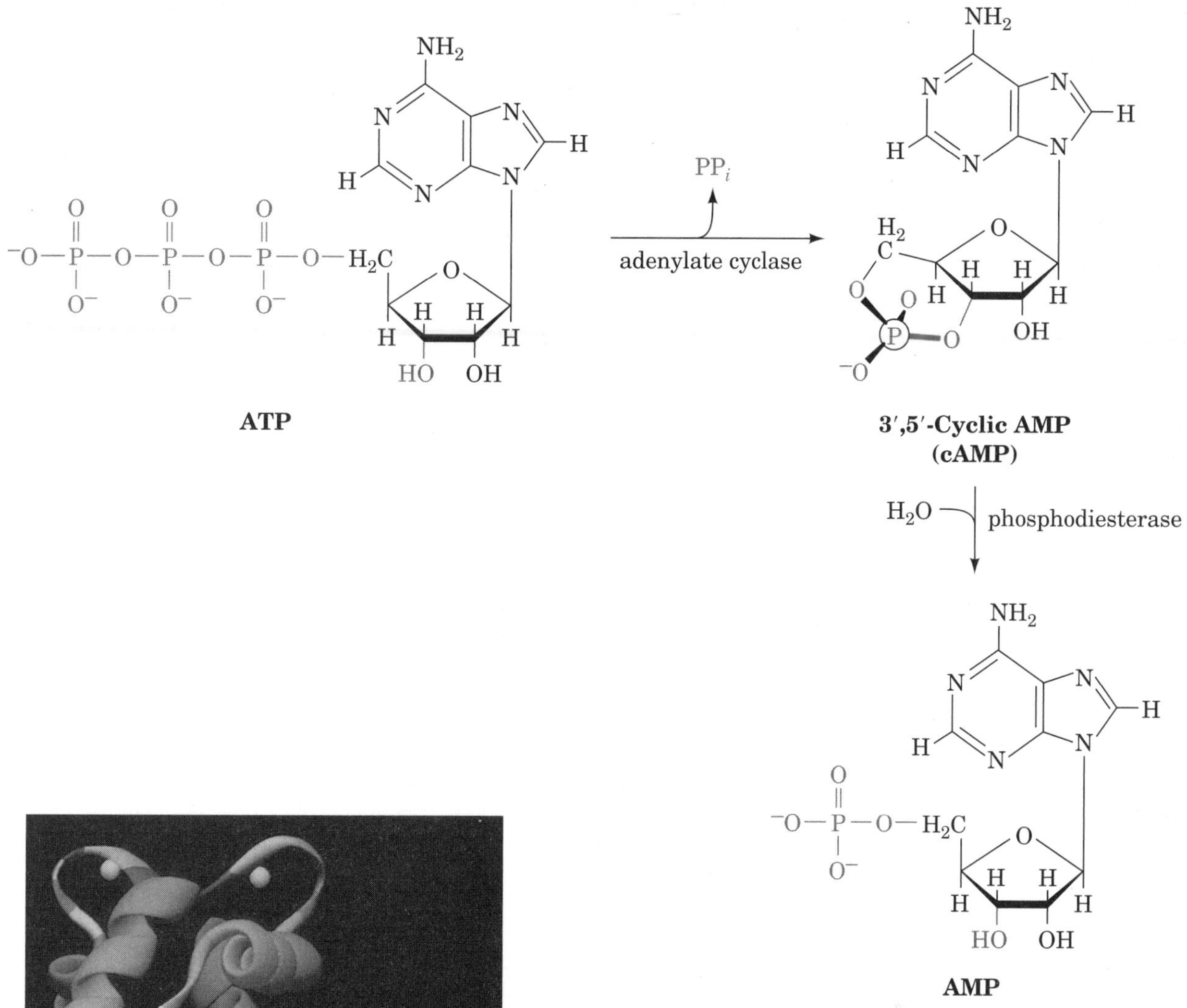

ATP

PP$_i$

adenylate cyclase

3′,5′-Cyclic AMP
(cAMP)

H$_2$O

phosphodiesterase

AMP

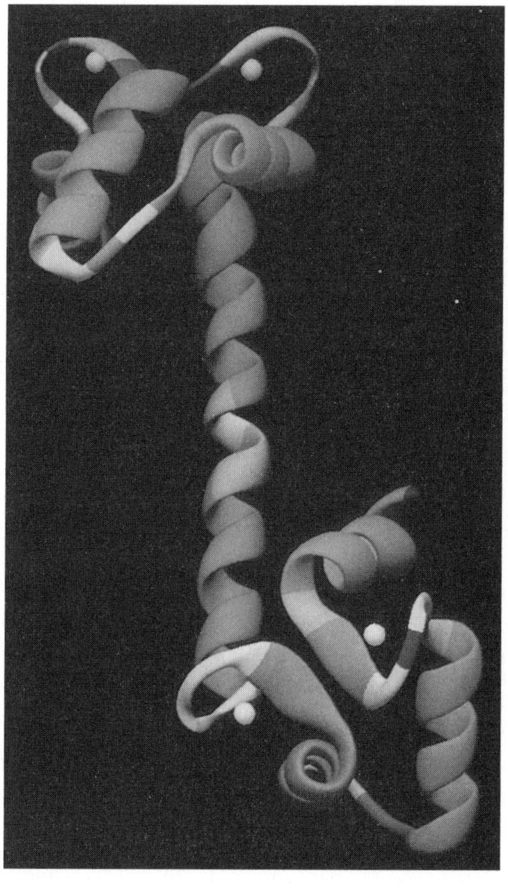

Figure 15-17 The X-ray structure of rat testis calmodulin.

187

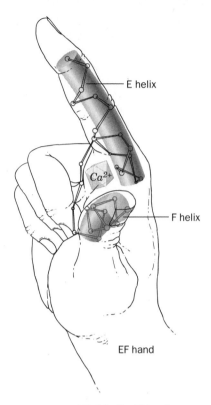

Figure 15-18 The EF hand.

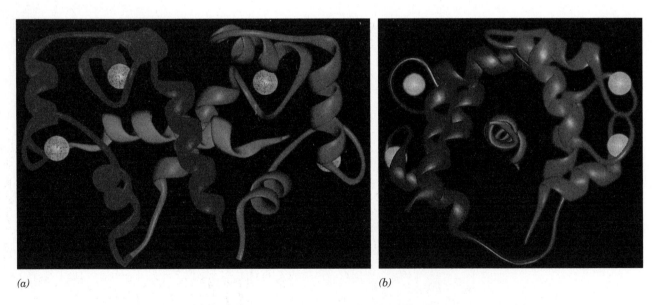

(a)   (b)

Figure 15-19 NMR structure of calmodulin in complex with a 26-residue target polypeptide.

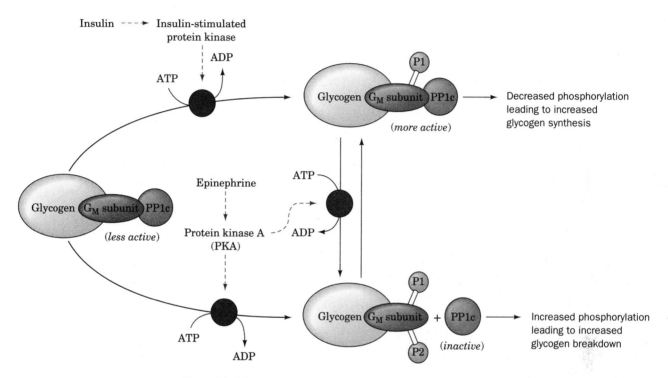

Figure 15-20 Regulation of phosphoprotein phosphatase-1 in muscle.

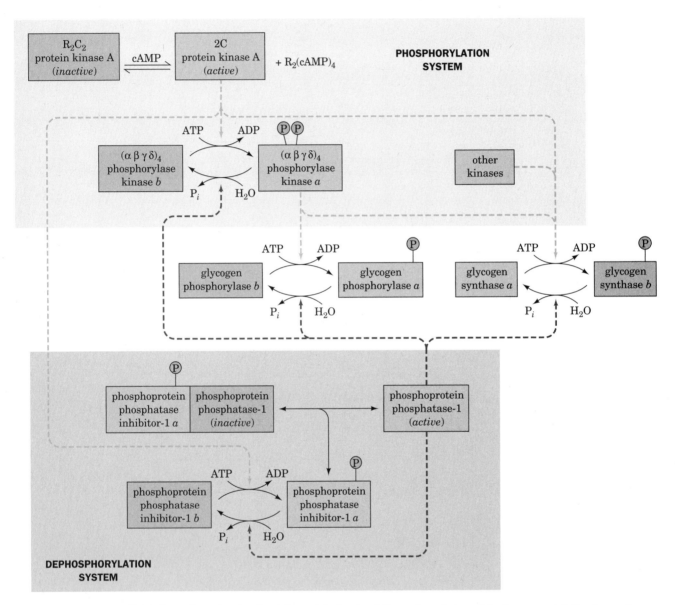

Figure 15-21 *Key to Metabolism.* The major phosphorylation and dephosphorylation systems that regulate glycogen metabolism in muscle.

$$H_3\overset{+}{N}-His-Ser-Glu-Gly-Thr-Phe-Thr-Ser-Asp-Tyr-\;10$$

$$Ser-Lys-Tyr-Leu-Asp-Ser-Arg-Arg-Ala-Gln-\;20$$

$$Asp-Phe-Val-Gln-Trp-Leu-Met-Asn-Thr-COO^-\;29$$

Glucagon

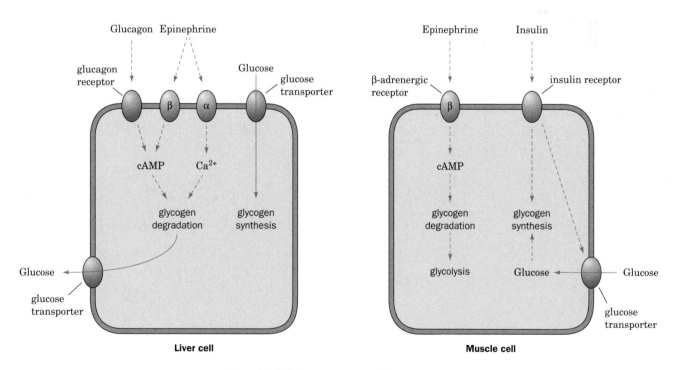

X = CH₃ **Epinephrine**
X = H **Norepinephrine**

Figure 15-22 Hormonal control of glycogen metabolism.

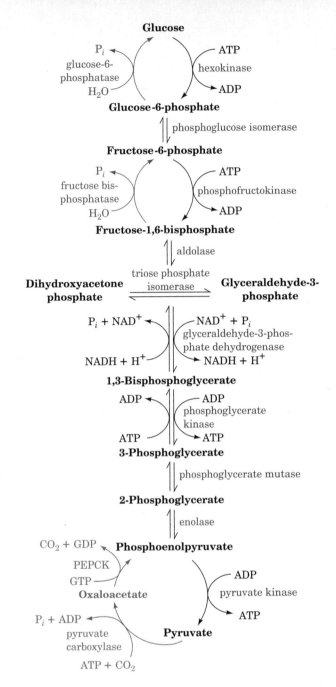

Figure 15-23 *Key to Metabolism.* Comparison of the pathways of gluconeogenesis and glycolysis.

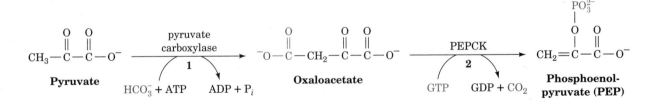

Figure 15-24 The conversion of pyruvate to phosphoenolpyruvate (PEP).

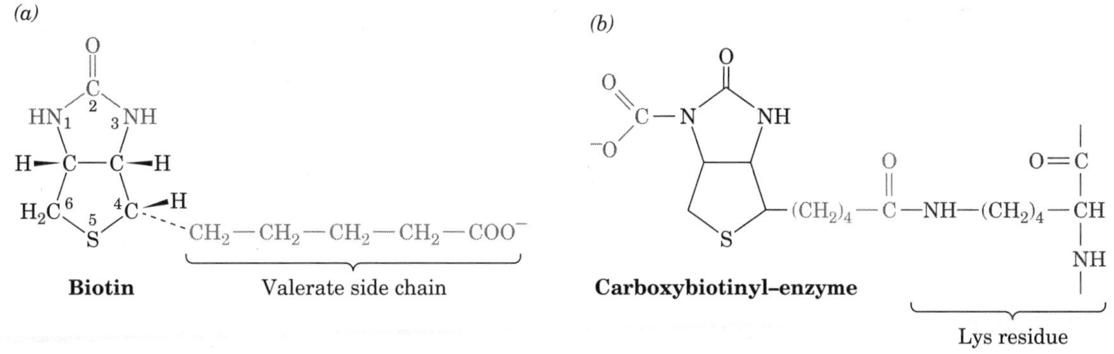

Figure 15-25 Biotin and carboxybiotinyl–enzyme.

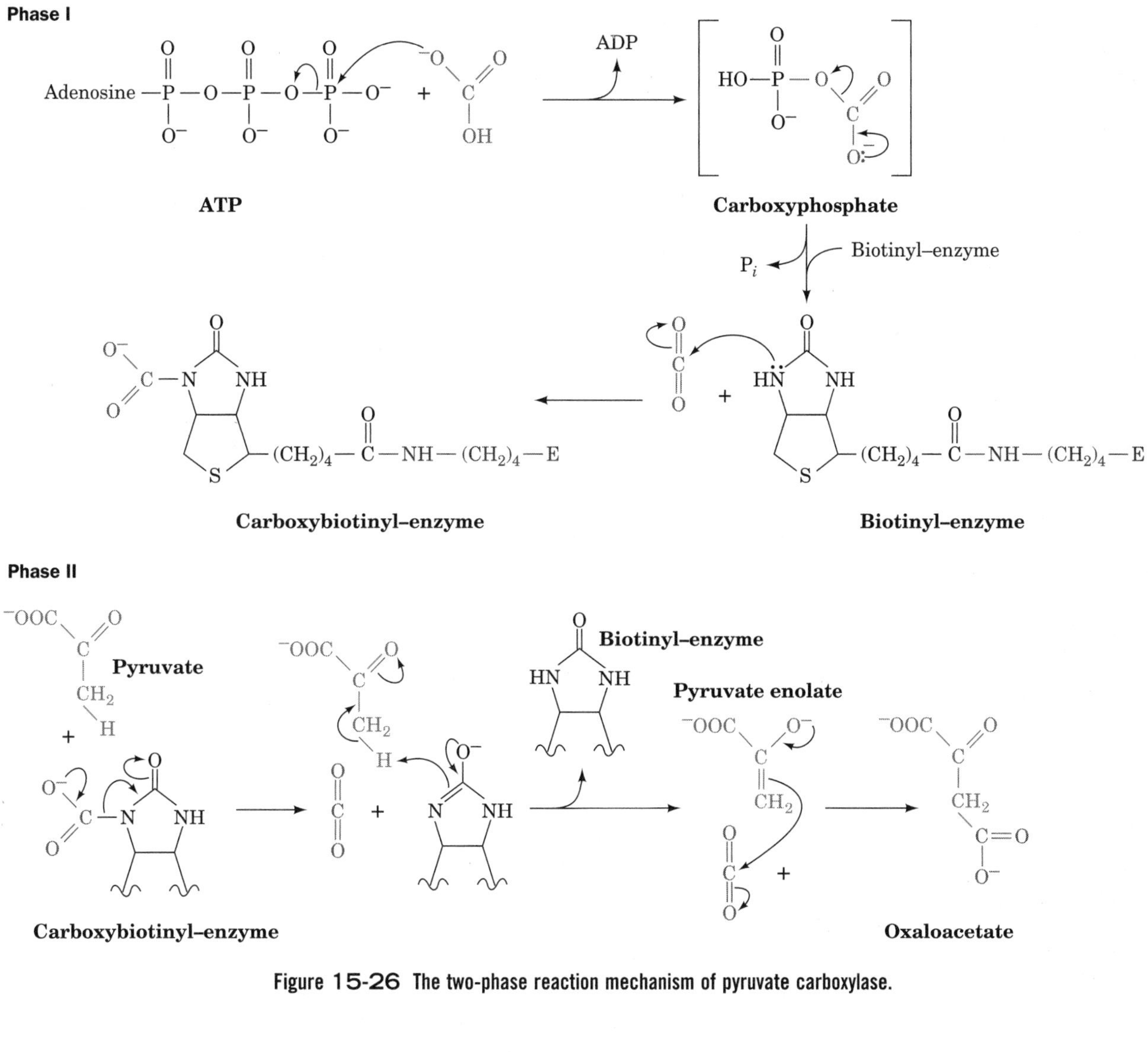

Figure 15-26 The two-phase reaction mechanism of pyruvate carboxylase.

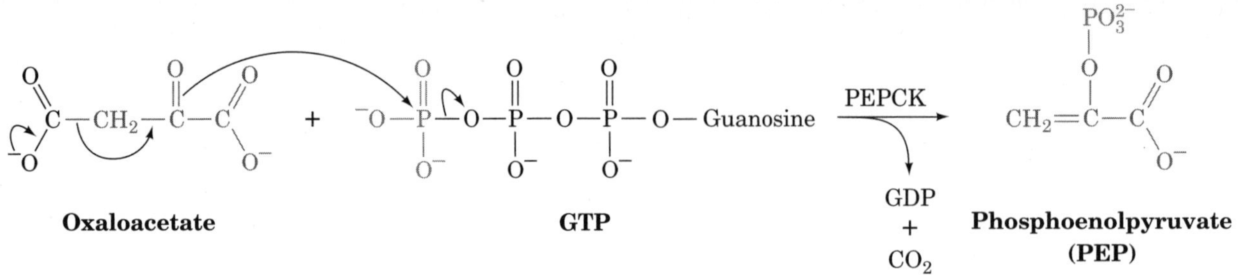

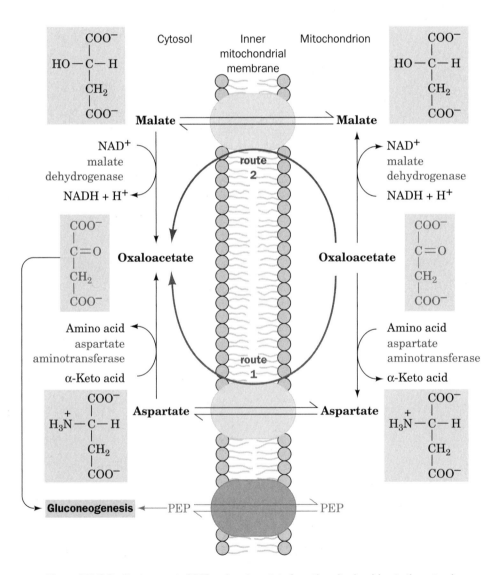

Figure 15-27 The PEPCK mechanism.

Figure 15-28 The transport of PEP and oxaloacetate from the mitochondrion to the cytosol.

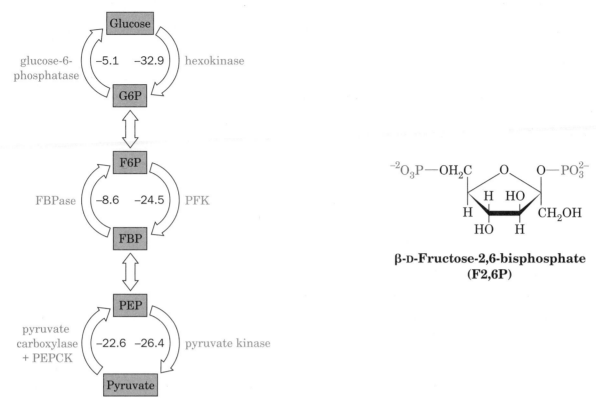

Figure 15-29 Substrate cycles in glucose metabolism.

β-D-Fructose-2,6-bisphosphate
(F2,6P)

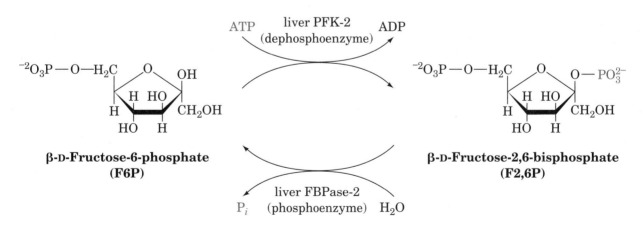

Figure 15-30 The formation and degradation of β-D-fructose-2,6-bisphosphate (F2,6P).

Low blood [glucose]

↓

Increased glucagon secretion

↓

Increased [cAMP]

↓

Increased enzyme phosphorylation

↓

Activation of FBPase-2 and inactivation of PFK-2

↓

Decreased [F2,6P]

↓

Inhibition of PFK and activation of FBPase

↓

Increased gluconeogenesis

Figure 15-31 Sequence of metabolic events linking low blood [glucose] to gluconeogenesis in liver.

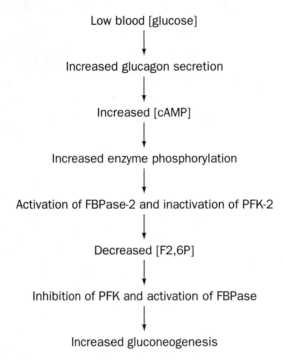

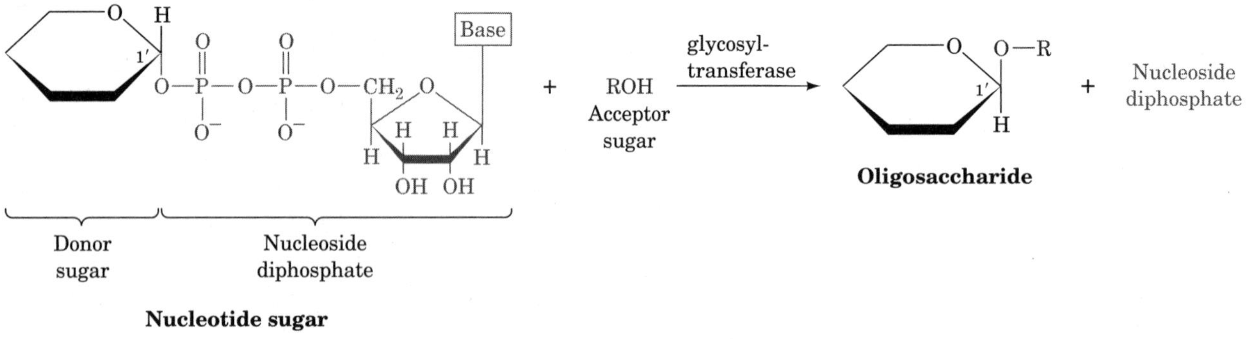

Figure 15-32 Role of nucleotide sugars.

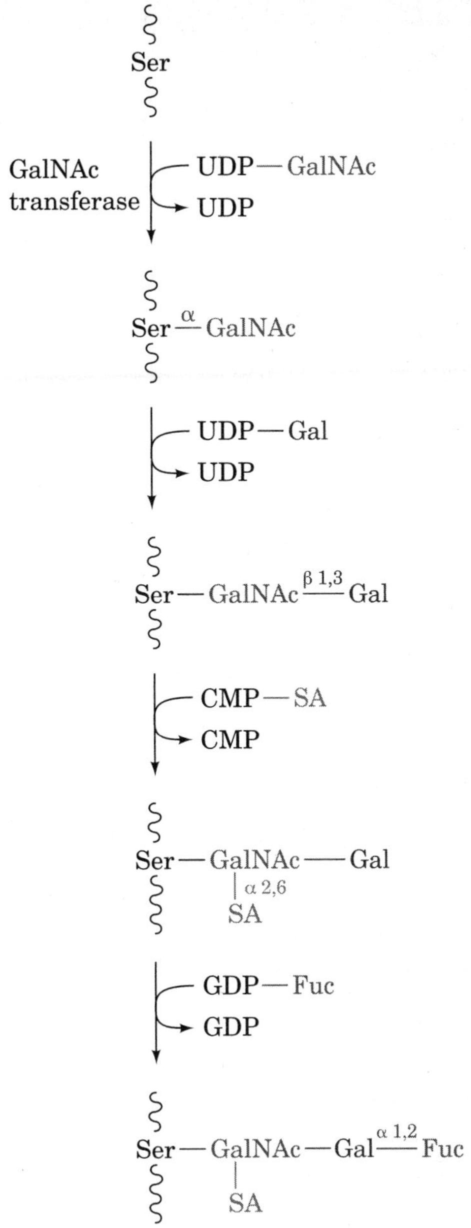

Figure 15-33 Synthesis of an *O*-linked oligosaccharide chain.

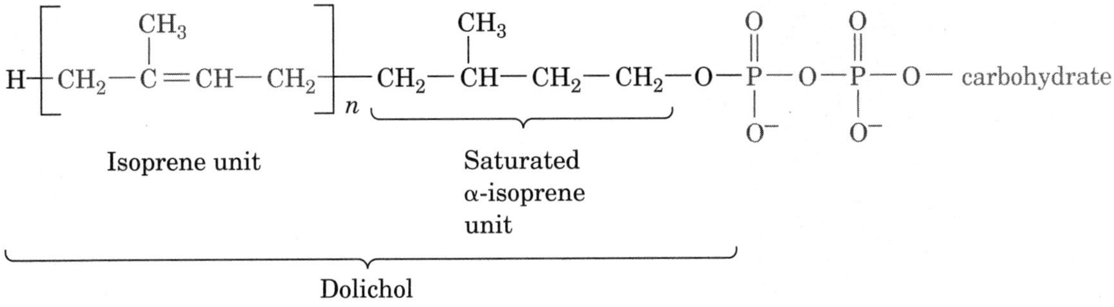

Figure 15-34 Dolichol pyrophosphate glycoside.

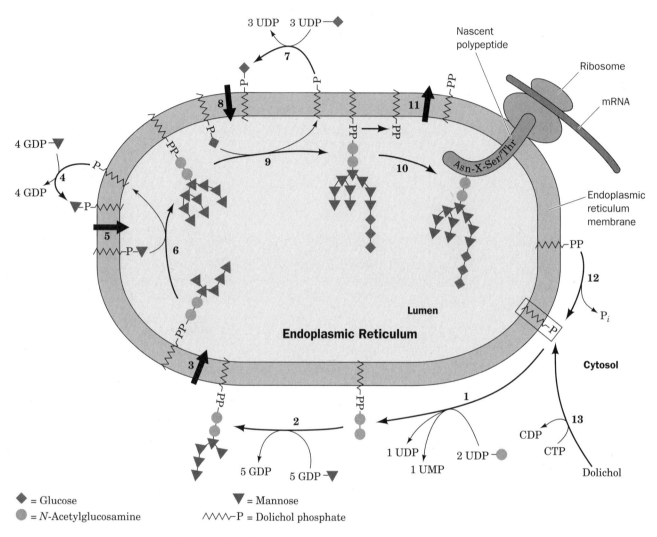

= Glucose

= N-Acetylglucosamine

= Mannose

^^^^–P = Dolichol phosphate

Figure 15-35 The pathway of dolichol-PP-oligosaccharide synthesis.

CHAPTER **16**

Citric Acid Cycle

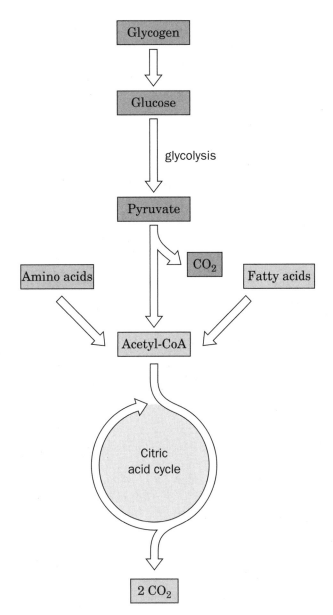

Figure 16-1 Overview of oxidative fuel metabolism.

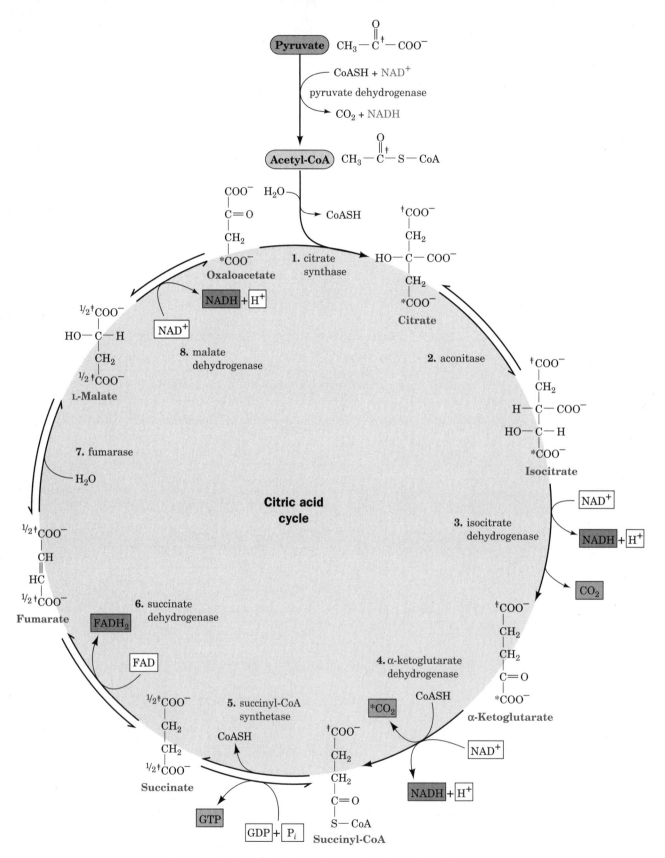

Figure 16-2 *Key to Metabolism.* The reactions of the citric acid cycle.

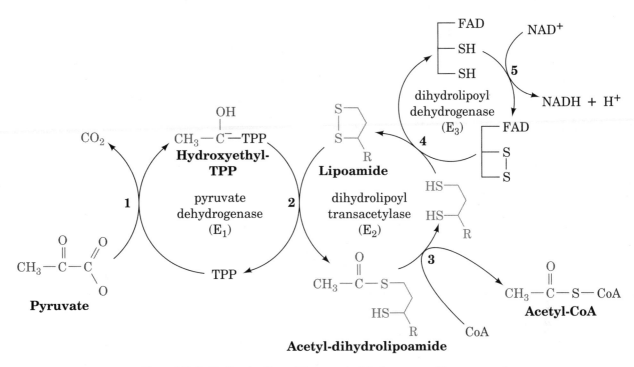

Figure 16-6 The five reactions of the pyruvate dehydrogenase multienzyme complex.

Table 16-1 The Coenzymes and Prosthetic Groups of Pyruvate Dehydrogenase

Cofactor	Location	Function
Thiamine pyrophosphate (TPP)	Bound to E_1	Decarboxylates pyruvate yielding a hydroxyethyl-TPP carbanion
Lipoic acid	Covalently linked to a Lys on E_2 (lipoamide)	Accepts the hydroxyethyl carbanion from TPP as an acetyl group
Coenzyme A (CoA)	Substrate for E_2	Accepts the acetyl group from lipoamide
Flavin adenine dinucleotide (FAD)	Bound to E_3	Reduced by lipoamide
Nicotinamide adenine dinucleotide (NAD^+)	Substrate for E_3	Reduced by $FADH_2$

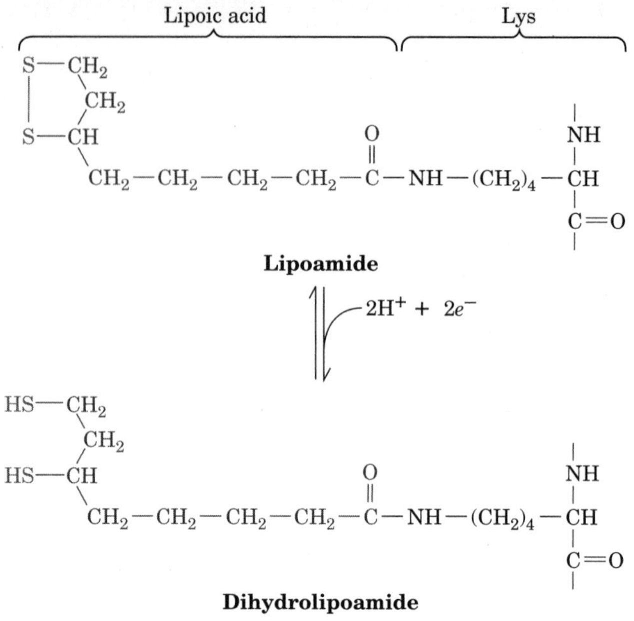

Figure 16-7 Interconversion of lipoamide and dihydrolipoamide.

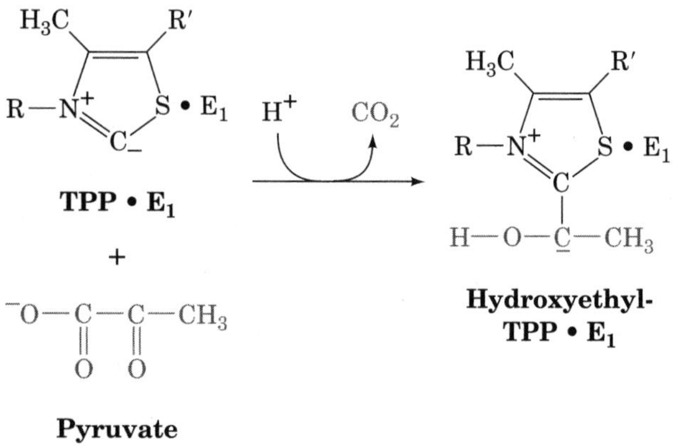

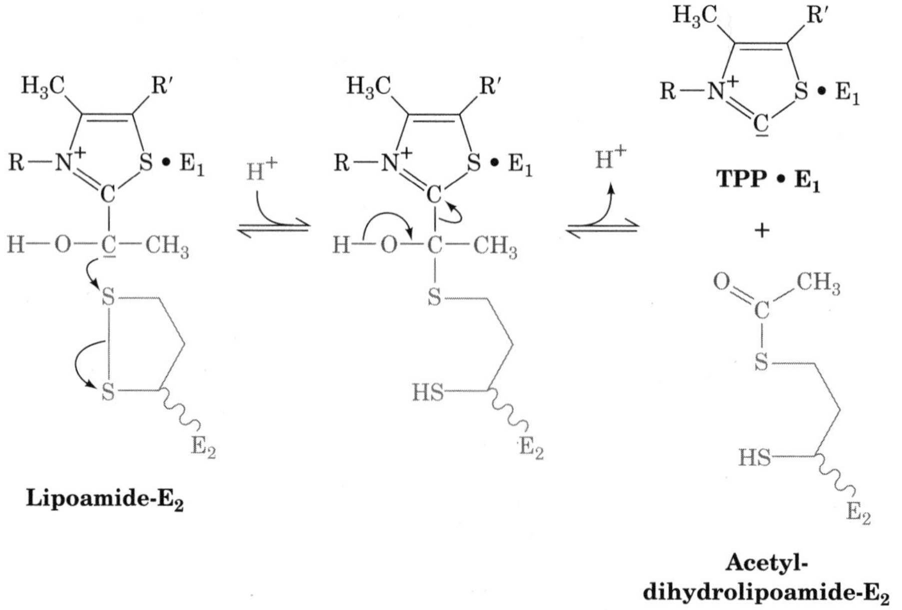

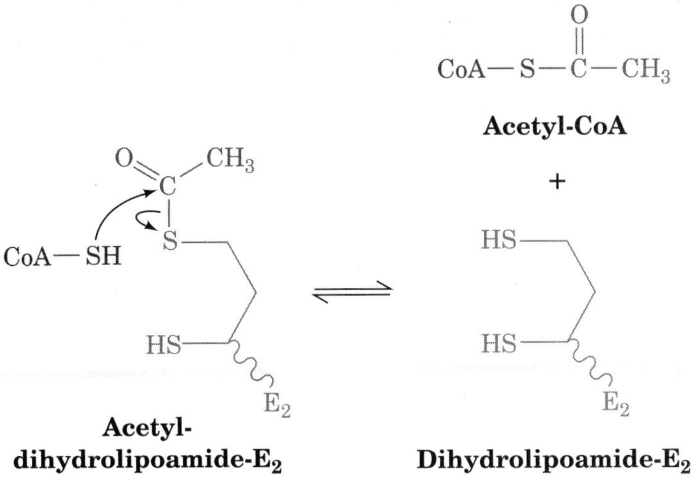

Acetyl-CoA

+

**Acetyl-
dihydrolipoamide-E₂**

Dihydrolipoamide-E₂

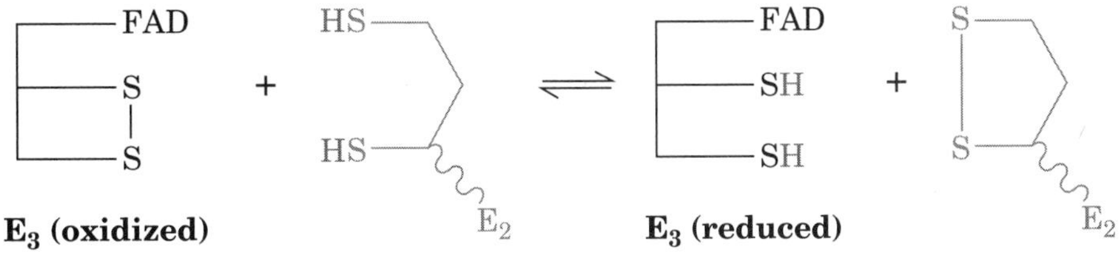

E₃ (oxidized)

E₃ (reduced)

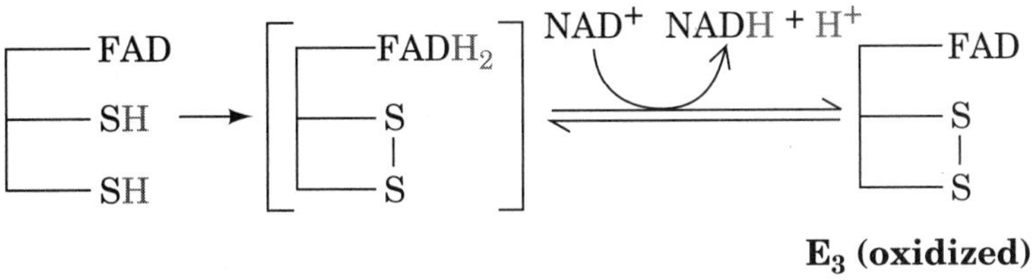

E₃ (oxidized)

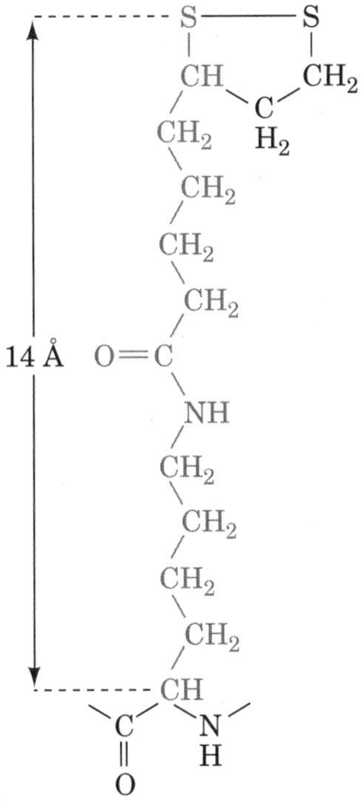

**Lipoyllysyl arm
(fully extended)**

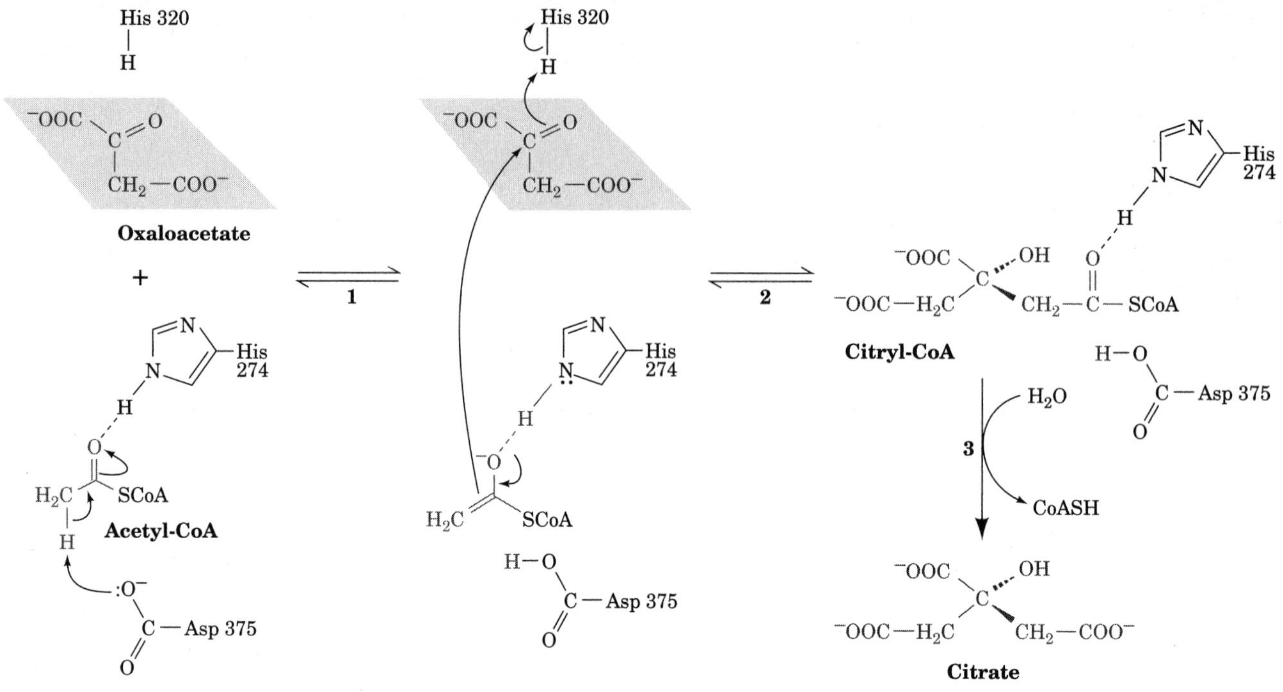

Figure 16-10 The mechanism of the citrate synthase reaction.

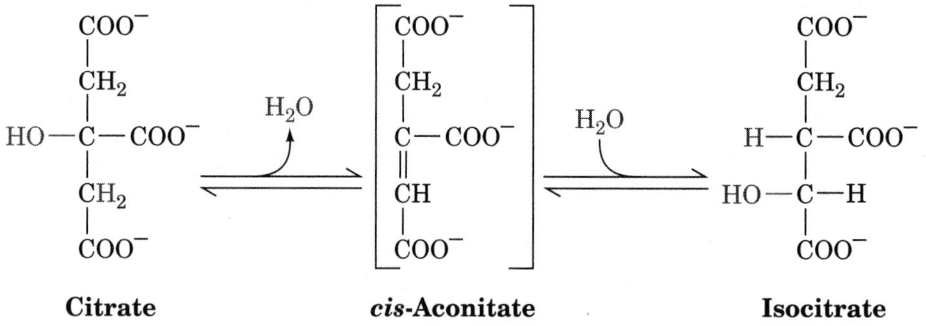

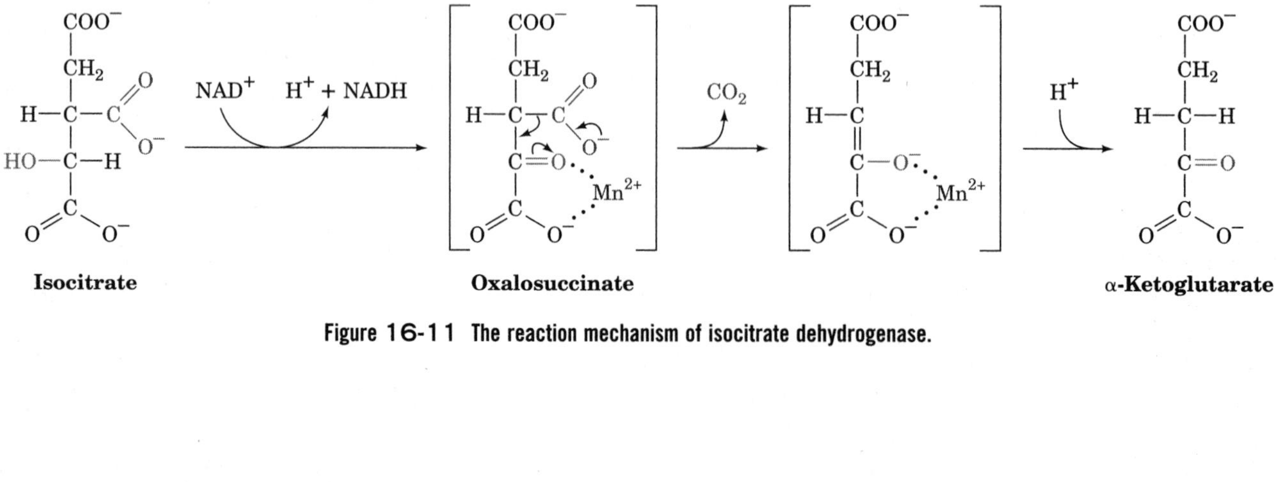

Figure 16-11 The reaction mechanism of isocitrate dehydrogenase.

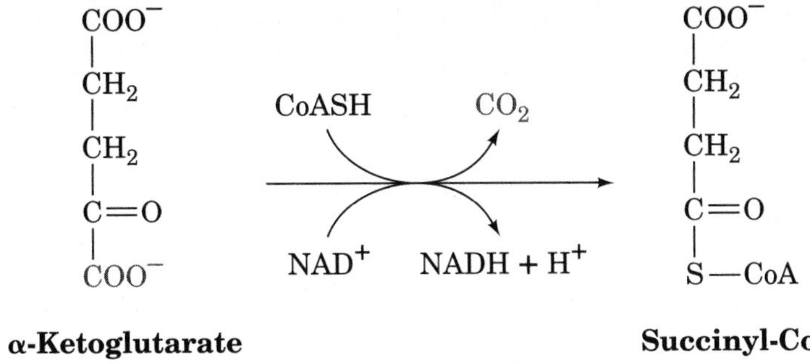

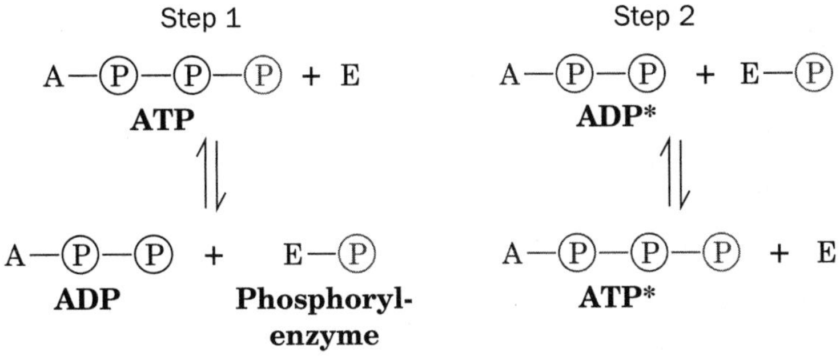

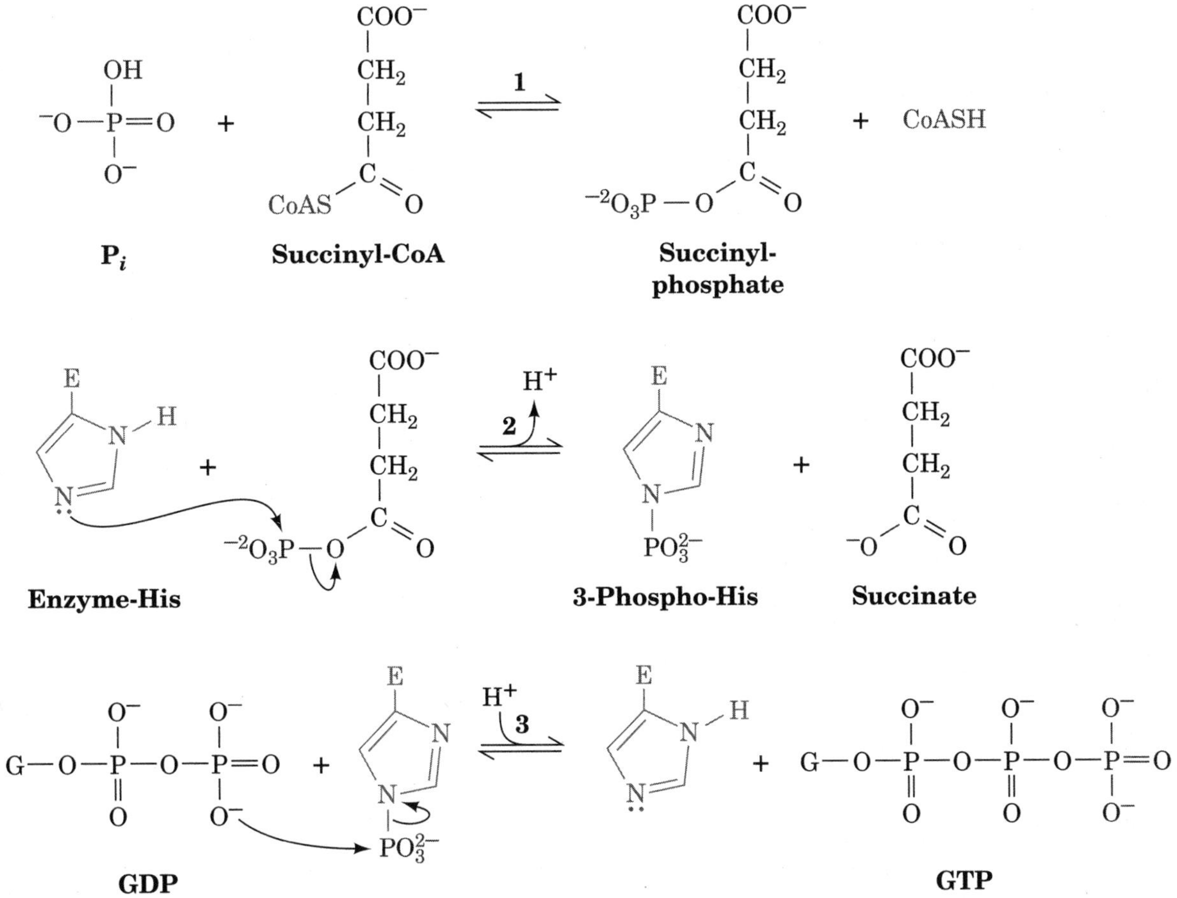

Figure 16-12 The reaction catalyzed by succinyl-CoA synthetase.

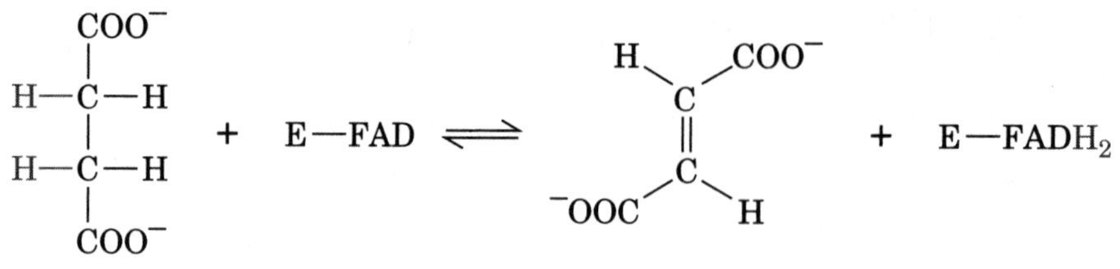

Succinate + E—FAD ⇌ Fumarate + E—FADH₂

$$\text{Succinate} + \text{E—FAD} \rightleftharpoons \text{Fumarate} + \text{E—FADH}_2$$

Succinate

Fumarate

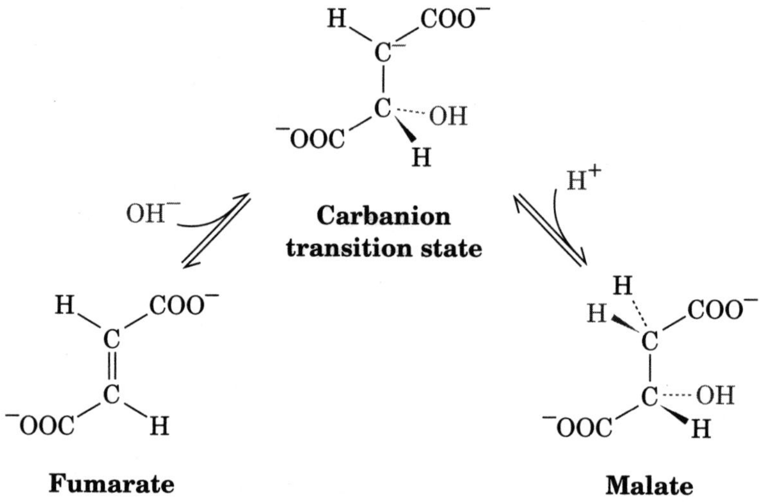

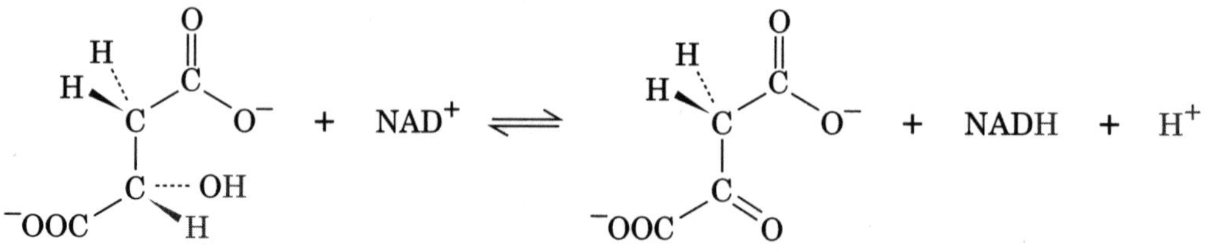

Carbanion transition state

OH⁻

H⁺

Fumarate

Malate

$$\text{Malate} + \text{NAD}^+ \rightleftharpoons \text{Oxaloacetate} + \text{NADH} + \text{H}^+$$

Malate

Oxaloacetate

206

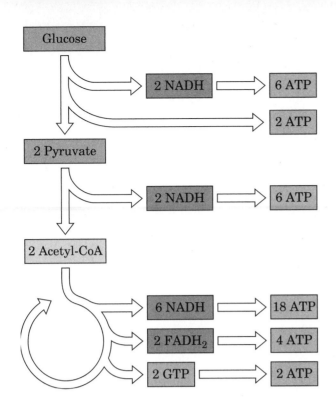

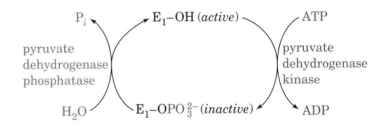

Figure 16-14 Covalent modification of eukaryotic pyruvate dehydrogenase.

Table 16-2 Standard Free Energy Changes ($\Delta G^{\circ\prime}$) and Physiological Free Energy Changes (ΔG) of Citric Acid Cycle Reactions

Reaction	Enzyme	$\Delta G^{\circ\prime}$ (kJ·mol^{-1})	ΔG (kJ·mol^{-1})
1	Citrate synthase	−31.5	Negative
2	Aconitase	~5	~0
3	Isocitrate dehydrogenase	−21	Negative
4	α-Ketoglutarate dehydrogenase multienzyme complex	−33	Negative
5	Succinyl-CoA synthetase	−2.1	~0
6	Succinate dehydrogenase	+6	~0
7	Fumarase	−3.4	~0
8	Malate dehydrogenase	+29.7	~0

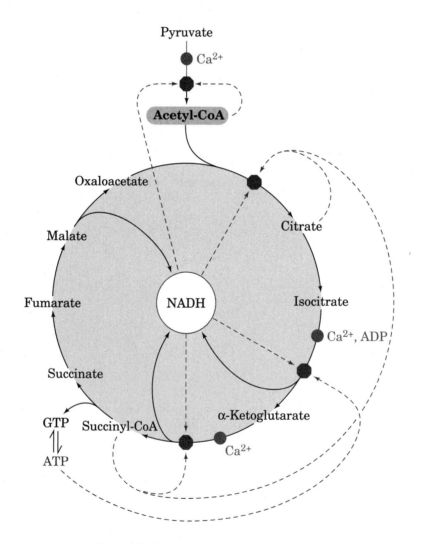

Figure 16-15 Regulation of the citric acid cycle.

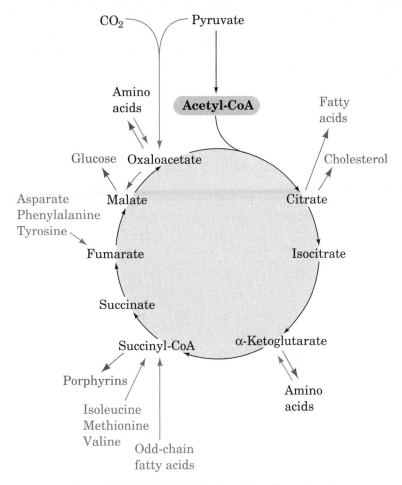

Figure 16-16 Amphibolic functions of the citric acid cycle.

$$
\begin{array}{l}
\text{COO}^- \\
| \\
\text{CH}_2 \\
| \\
\text{CH}_2 \quad + \text{ NADH } + \text{ H}^+ + \text{ NH}_4^+ \rightleftharpoons \\
| \\
\text{C}=\text{O} \\
| \\
\text{COO}^-
\end{array}
\qquad
\begin{array}{l}
\text{COO}^- \\
| \\
\text{CH}_2 \\
| \\
\text{CH}_2 \quad + \text{ NAD}^+ + \text{ H}_2\text{O} \\
| \\
\text{H}-\text{C}-\text{NH}_3^+ \\
| \\
\text{COO}^-
\end{array}
$$

α-Ketoglutarate   **Glutamate**

$$
\begin{array}{l}
\text{COO}^- \\
| \\
\text{C}=\text{O} \\
| \\
\text{CH}_2 \\
| \\
\text{COO}^-
\end{array}
\; + \;
\begin{array}{l}
\text{COO}^- \\
| \\
\text{H}_3\overset{+}{\text{N}}-\text{C}-\text{H} \\
| \\
\text{CH}_3
\end{array}
\; \rightleftharpoons \;
\begin{array}{l}
\text{COO}^- \\
| \\
\text{H}_3\overset{+}{\text{N}}-\text{C}-\text{H} \\
| \\
\text{CH}_2 \\
| \\
\text{COO}^-
\end{array}
\; + \;
\begin{array}{l}
\text{COO}^- \\
| \\
\text{C}=\text{O} \\
| \\
\text{CH}_3
\end{array}
$$

Oxaloacetate   **Alanine**   **Aspartate**   **Pyruvate**

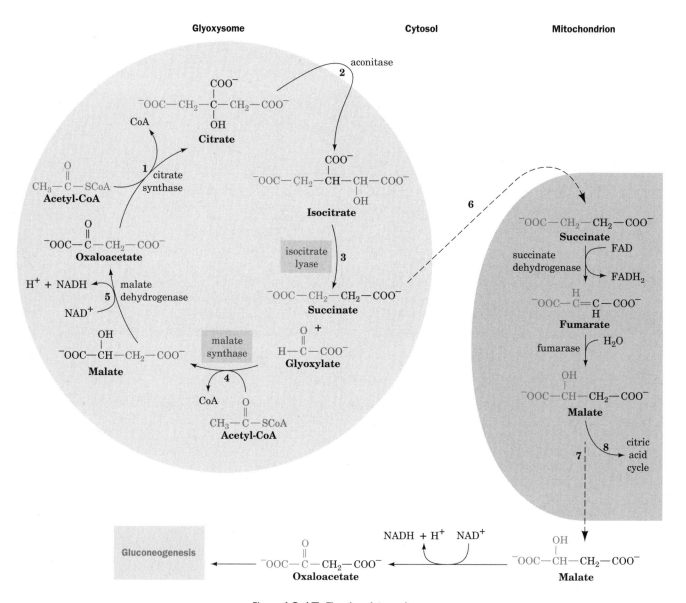

Figure 16-17 The glyoxylate cycle.

CHAPTER 17

Electron Transport and Oxidative Phosphorylation

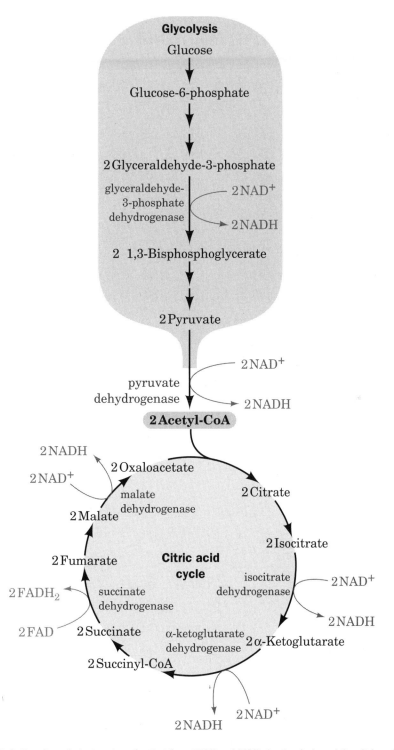

Figure 17-1 The sites of electron transfer that form NADH and FADH₂ in glycolysis and the citric acid cycle.

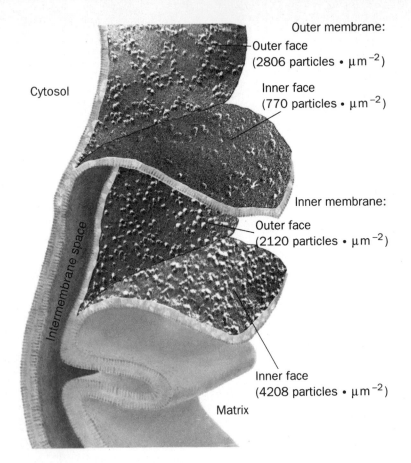

Figure 17-4 Electron micrographs of the inner and outer mitochondrial membranes that have been split to expose the inner surfaces of their bilayer leaflets.

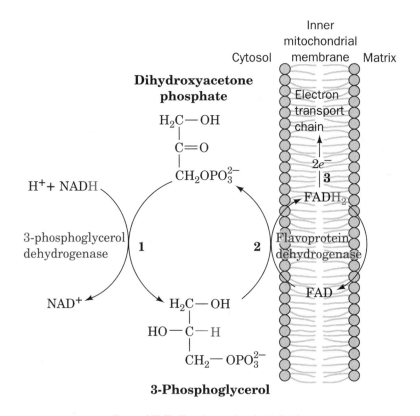

Figure 17-5 The glycerophosphate shuttle.

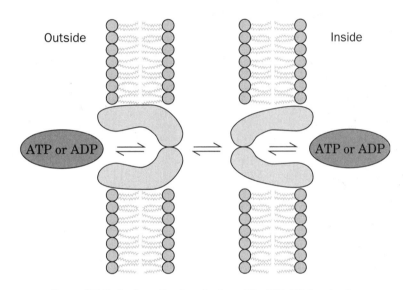

Figure 17-6 Conformational mechanism of the ADP–ATP translocator.

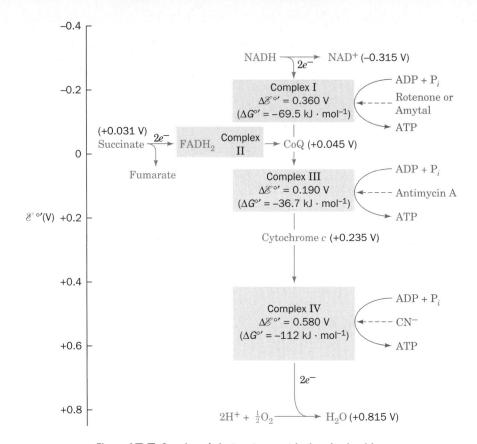

Figure 17-7 Overview of electron transport in the mitochondrion.

$$\tfrac{1}{2}\,O_2 + NADH + H^+ \rightleftharpoons H_2O + NAD^+$$

$$\Delta\mathscr{E}^{\circ\prime} = 0.815\ V - (-0.315\ V) = 1.130\ V$$

$$\Delta G^{\circ\prime} = -n\mathscr{F}\Delta\mathscr{E}^{\circ\prime}$$

$$NADH + CoQ\ (\textit{oxidized}) \rightarrow NAD^+ + CoQ\ (\textit{reduced})$$
$$\Delta\mathscr{E}^{\circ\prime} = 0.360\ V \qquad \Delta G^{\circ\prime} = -69.5\ kJ \cdot mol^{-1}$$

$$CoQ\ (\textit{reduced}) + 2\ cytochrome\ c\ (\textit{oxidized}) \rightarrow$$
$$CoQ\ (\textit{oxidized}) + 2\ cytochrome\ c\ (\textit{reduced})$$
$$\Delta\mathscr{E}^{\circ\prime} = 0.190\ V \qquad \Delta G^{\circ\prime} = -36.7\ kJ \cdot mol^{-1}$$

$$2\ Cytochrome\ c\ (\textit{reduced}) + \tfrac{1}{2}\,O_2 \rightarrow 2\ cytochrome\ c\ (\textit{oxidized}) + H_2O$$
$$\Delta\mathscr{E}^{\circ\prime} = 0.580\ V \qquad \Delta G^{\circ\prime} = -112\ kJ \cdot mol^{-1}$$

Table 17-1 Reduction Potentials of Electron-Transport Chain Components in Resting Mitochondria

Component	$e^{\circ\prime}$ (V)
NADH	−0.315
Complex I (NADH–CoQ oxidoreductase; ~900 kD, 43 subunits):	
FMN	?
(Fe–S)N-1a	−0.380
(Fe–S)N-1b	−0.250
(Fe–S)N-2	−0.030
(Fe–S)N-3,4	−0.245
(Fe–S)N-5,6	−0.270
Succinate	0.031
Complex II (succinate–CoQ oxidoreductase; ~120 kD, 4 subunits):	
FAD	−0.040
[2Fe–2S]	−0.030
[4Fe–4S]	−0.245
[3Fe–4S]	0.060
Heme b_{560}	−0.080
Coenzyme Q	0.045
Complex III (CoQ–cytochrome c oxidoreductase; ~240 kD, 9–11 subunits):	
Heme b_H (b_{562})	0.030
Heme b_L (b_{566})	−0.030
[2Fe–2S]	0.280
Heme c_1	0.215
Cytochrome c	0.235
Complex IV (cytochrome c oxidase; ~205 kD, 8–13 subunits):	
Heme a	0.210
Cu_A	0.245
Cu_B	0.340
Heme a_3	0.385
O_2	0.815

Source: Mainly Wilson, D.F., Erecinska, M., and Dutton, P.L., *Annu. Rev. Biophys. Bioeng.* **3,** 205 and 208 (1974); *and* Wilson, D.F., *in* Bittar, E.E. (Ed.), *Membrane Structure and Function,* Vol. 1, *p.* 160, Wiley (1980).

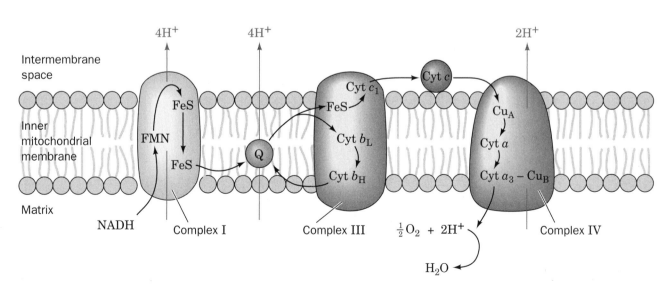

Figure 17-8 *Key to Function.* The mitochondrial electron-transport chain.

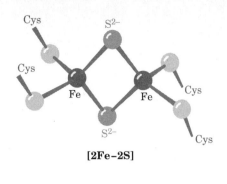

[2Fe–2S]

[4Fe–4S]

(a)

Flavin mononucleotide (FMN)
(oxidized or quinone form)

[H•]

FMNH• (radical or semiquinone form)

[H•]

FMNH₂ (reduced or hydroquinone form)

(b)

Coenzyme Q (CoQ) or ubiquinone
(oxidized or quinone form)

Isoprenoid units

[H•]

Coenzyme QH• or ubisemiquinone
(radical or semiquinone form)

[H•]

Coenzyme QH₂ or ubiquinol
(reduced or hydroquinone form)

Figure 17-11 The oxidation states of FMN and coenzyme Q.

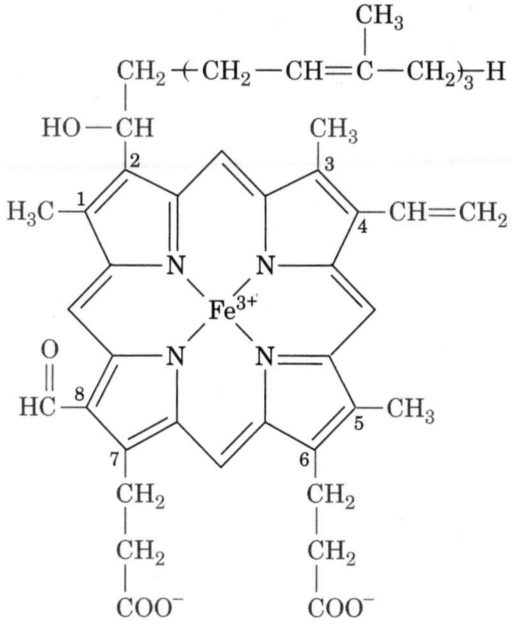

Heme *a*

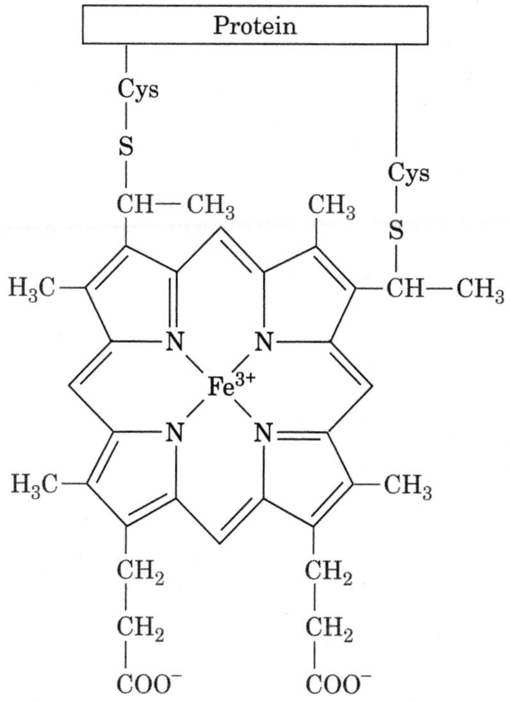

Heme *c*

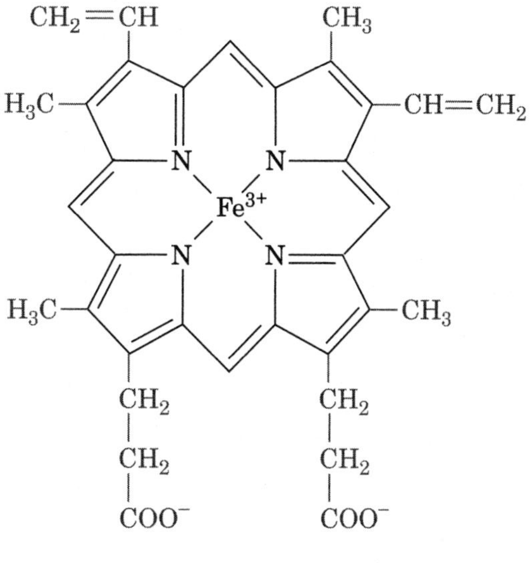

Protein

Heme *b*
(iron–protoporphyrin IX)

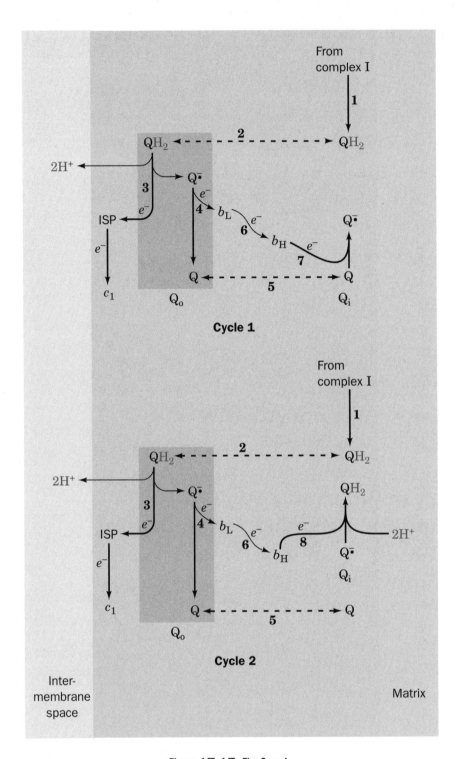

Figure 17-15 The Q cycle.

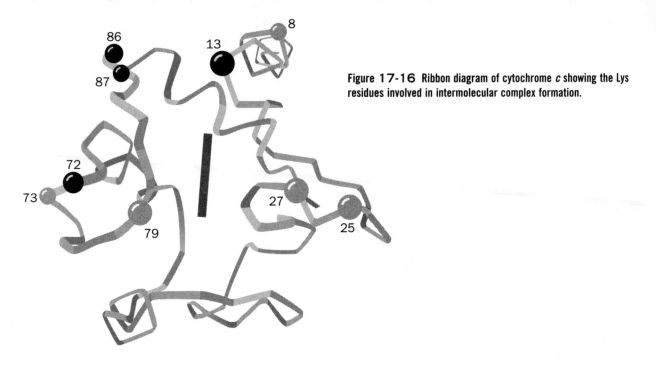

Figure 17-16 Ribbon diagram of cytochrome *c* showing the Lys residues involved in intermolecular complex formation.

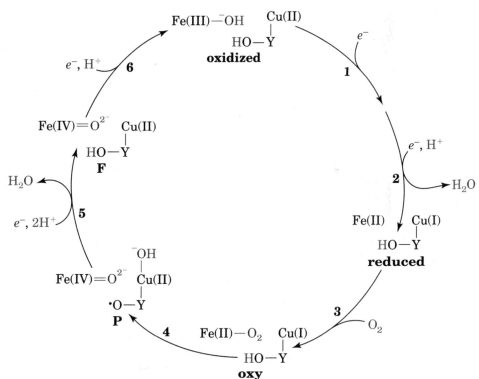

Figure 17-19 Proposed reaction sequence for cytochrome *c* oxidase.

$$8 \text{ H}^+_{matrix} + \text{O}_2 + 4 \text{ cytochrome } c \text{ (Fe}^{2+}) \rightarrow$$
$$4 \text{ cytochrome } c \text{ (Fe}^{3+}) + 2 \text{ H}_2\text{O} + 4 \text{ H}^+_{intermembrane}$$

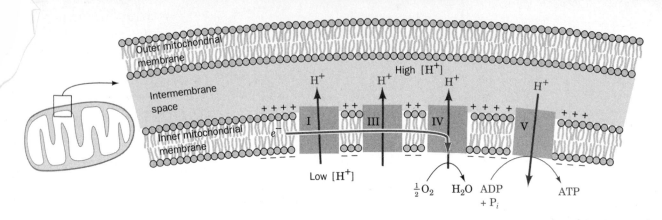

Figure 17-20 The coupling of electron transport and ATP synthesis.

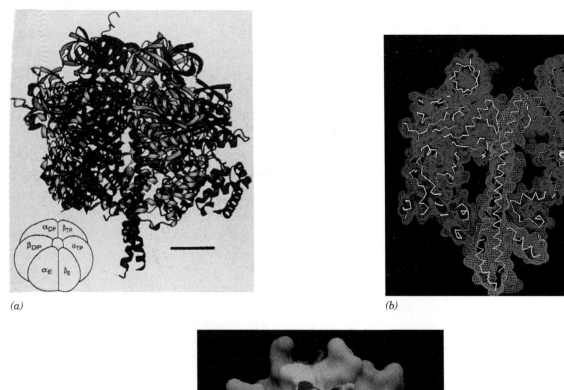

(a)

(b)

(c)

Figure 17-22 X-Ray structure of F$_1$-ATPase from bovine heart mitochondria.

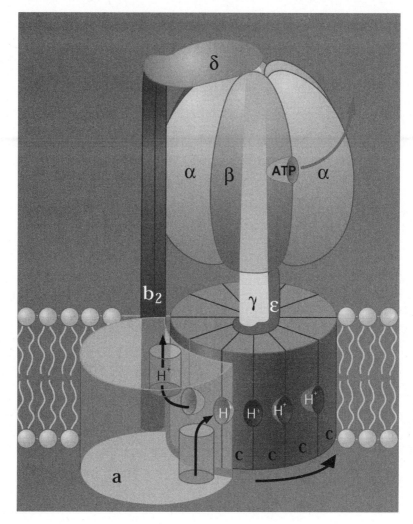

Figure 17-25 Model of the *E. coli* F_1F_0-ATPase.

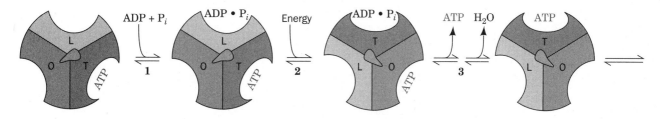

Figure 17-26 *Key to Function.* The binding change mechanism for ATP synthase.

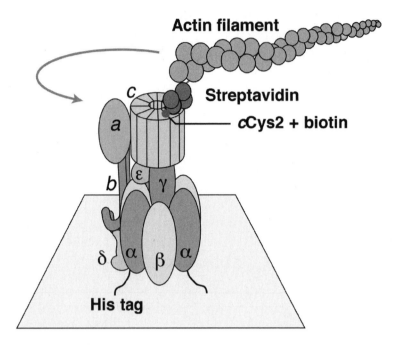

Figure 17-27 The rotation of the *c*-ring in *E. coli* F_1F_0-ATPase.

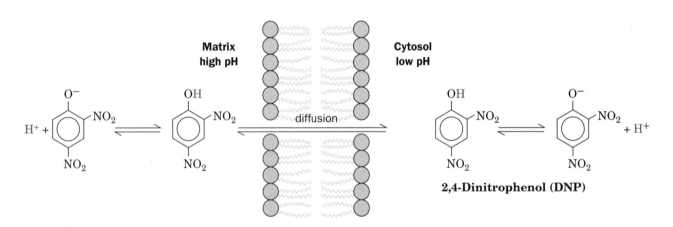

2,4-Dinitrophenol (DNP)

Figure 17-28 Action of 2,4-dinitrophenol.

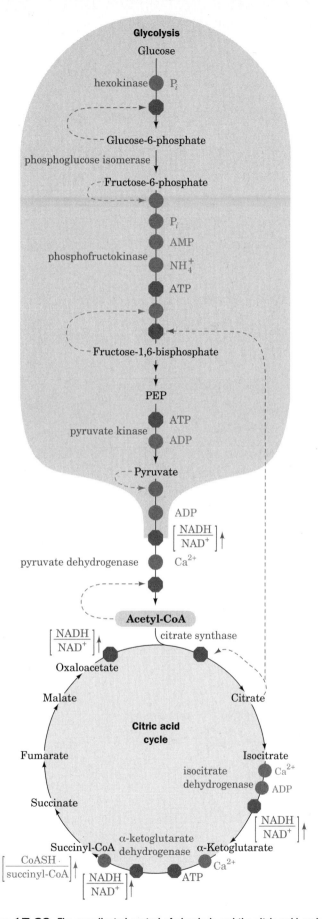

Figure 17-29 The coordinated control of glycolysis and the citric acid cycle.

CHAPTER 18

Photosysthesis

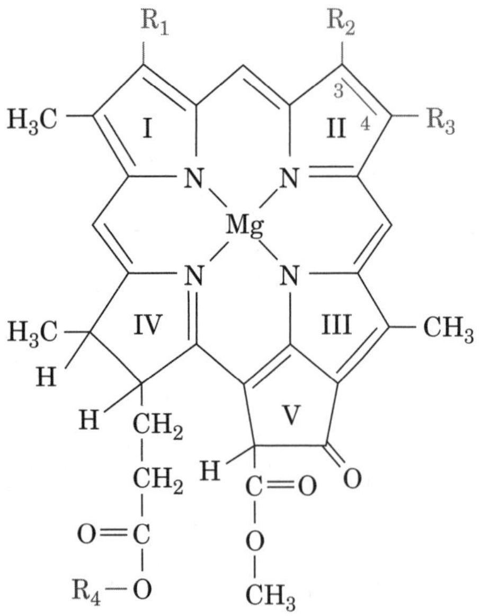

Chlorophyll 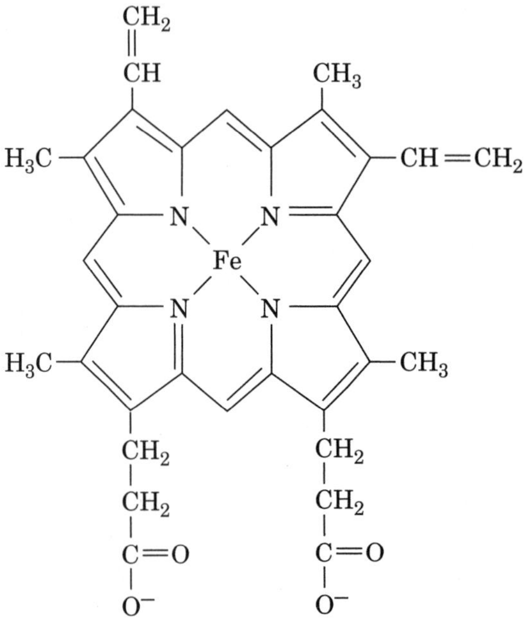   **Iron–protoporphyrin IX**

	R_1	R_2	R_3	R_4
Chlorophyll a	$-CH=CH_2$	$-CH_3$	$-CH_2-CH_3$	P
Chlorophyll b	$-CH=CH_2$	$\overset{O}{\overset{\|}{-C}}-H$	$-CH_2-CH_3$	P
Bacteriochlorophyll a	$\overset{O}{\overset{\|}{-C}}-CH_3$	$-CH_3$ [a]	$-CH_2-CH_3$ [a]	P or G
Bacteriochlorophyll b	$\overset{O}{\overset{\|}{-C}}-CH_3$	$-CH_3$ [a]	$=CH-CH_3$ [a]	P

[a] No double bond between positions C3 and C4.

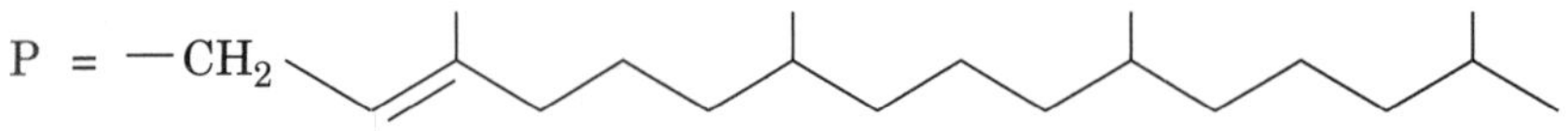

$P = -CH_2$

Phytyl side chain

$G = -CH_2$

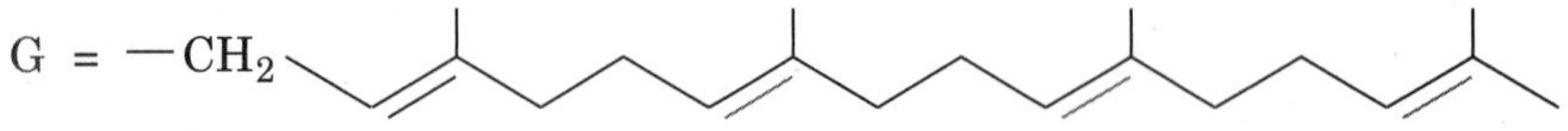

Geranylgeranyl side chain

Figure 18-2 Chlorophyll structures.

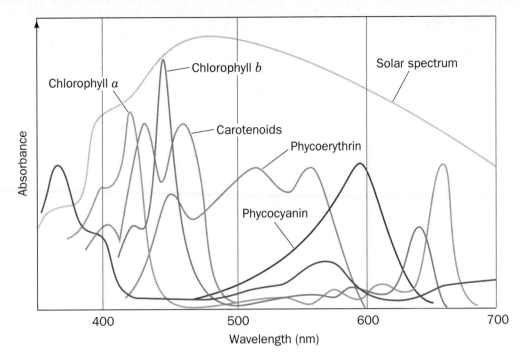

Figure 18-3 Absorption spectra of various photosynthetic pigments.

β-Carotene

Phycoerythrobilin and Phycocyanobilin

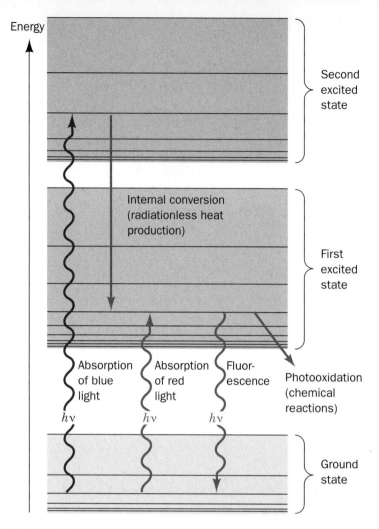

Figure 18-6 An energy diagram indicating the electronic states of chlorophyll and their most important modes of interconversion.

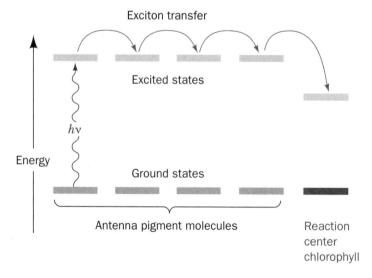

Figure 18-7 Excitation energy trapping by the photosynthetic reaction center.

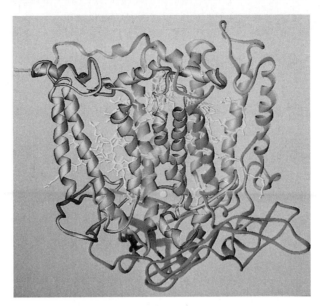

Figure 18-8 X-Ray structure of the photosynthetic reaction center from *Rb. sphaeroides.*

Menaquinone

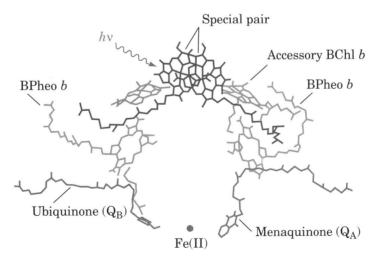

Figure 18-9 Disposition of prosthetic groups in the photosynthetic reaction center of *Rps. viridis.*

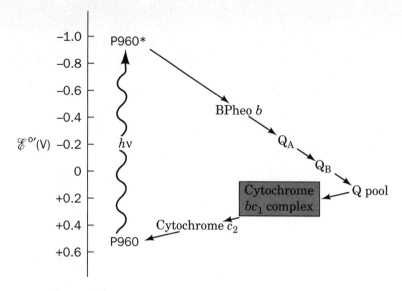

Figure 18-10 The photosynthetic electron-transport system of purple photosynthetic bacteria.

**3-(3,4-Dichlorophenyl)-1,1-dimethylurea
(DCMU)**

Plastoquinone

2 [H•]

Plastoquinol

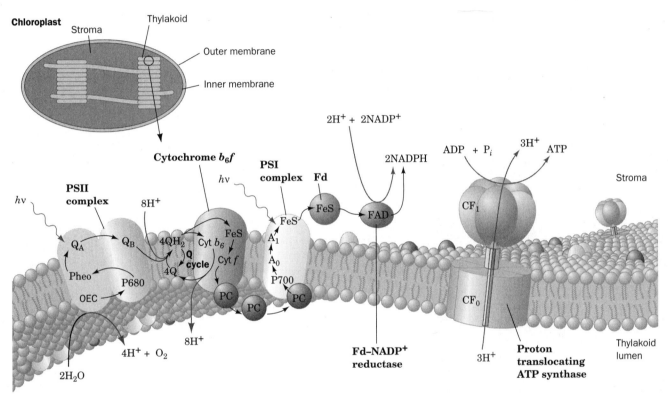

Figure 18-11 *Key to Function.* A model of the thylakoid membrane.

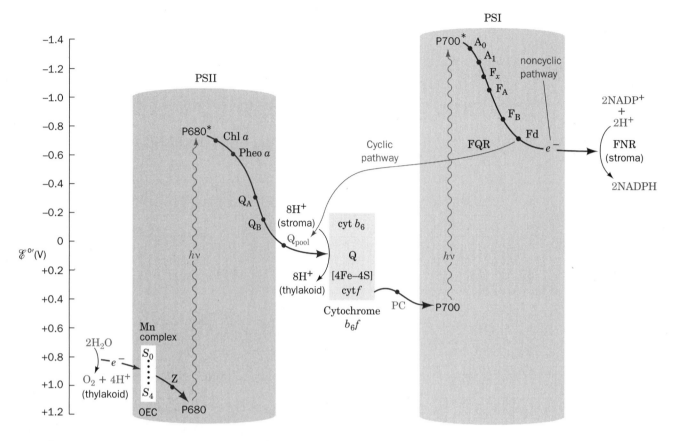

Figure 18-12 The Z-scheme of photosynthesis.

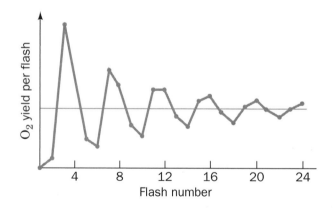

Figure 18-13 The O_2 yield per flash in dark-adapted spinach chloroplasts.

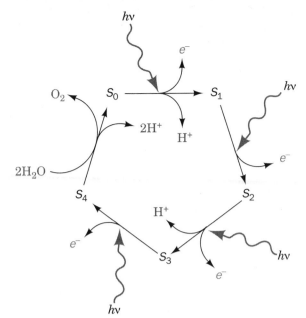

Figure 18-14 The schematic mechanism of O_2 generation in chloroplasts.

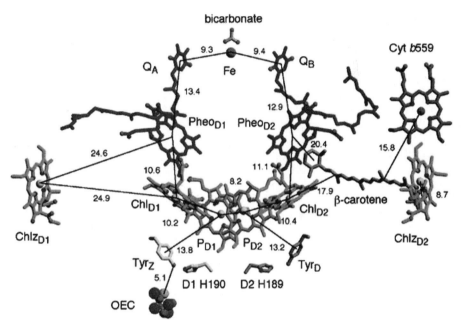

Figure 18-16 The arrangement of electron transfer cofactors in PSII from *S. elongatus.*

Phylloquinone

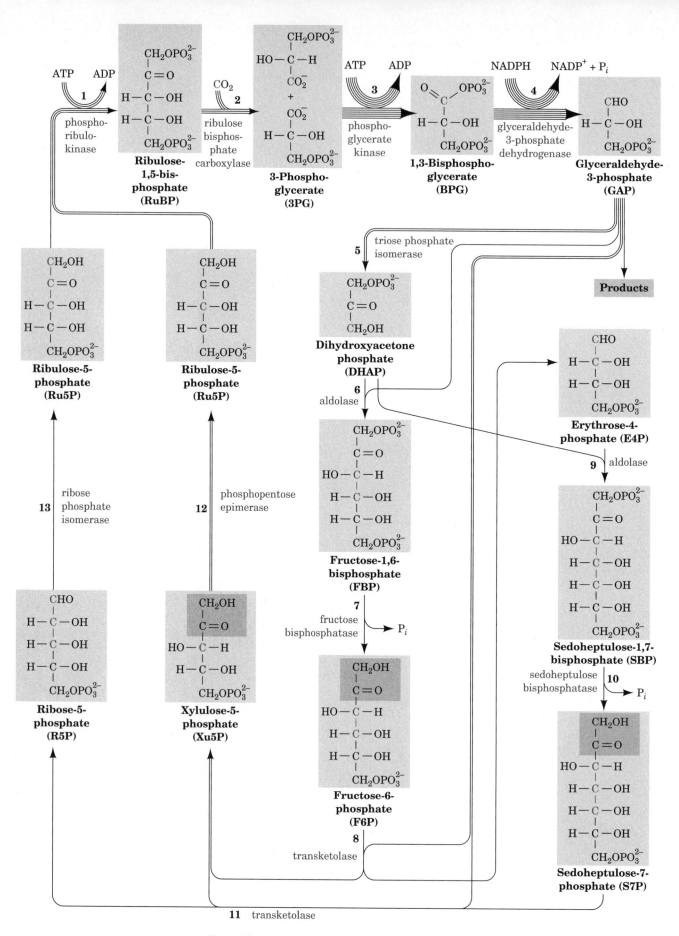

Figure 18-24 *Key to Metabolism.* The Calvin cycle.

6. $C_3 + C_3 \rightarrow C_6$

8. $C_3 + C_6 \rightarrow C_5 + C_4$

9. $C_3 + C_4 \rightarrow C_7$

11. $C_3 + C_7 \rightarrow C_5 + C_5$

$$5\,C_3 \rightarrow 3\,C_5$$

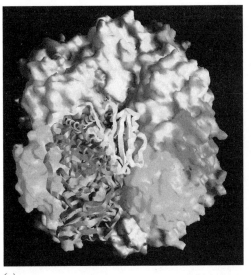

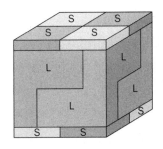

(a)

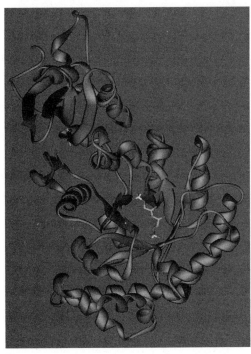

(b)

Figure 18-25 X-Ray structure of RuBP carboxylase.

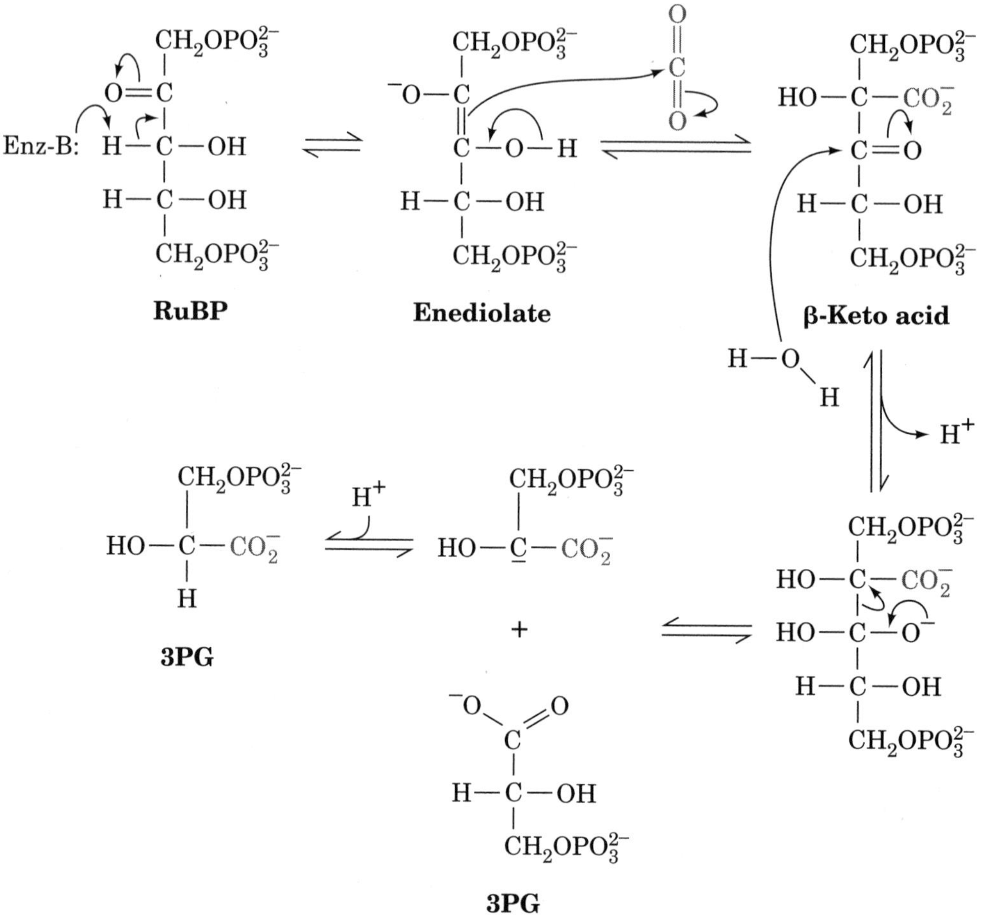

Figure 18-26 Mechanism of the RuBP carboxylase reaction.

$$3\ CO_2 + 9\ ATP + 6\ NADH \rightarrow GAP + 9\ ADP + 8\ P_i + 6\ NADP^+$$

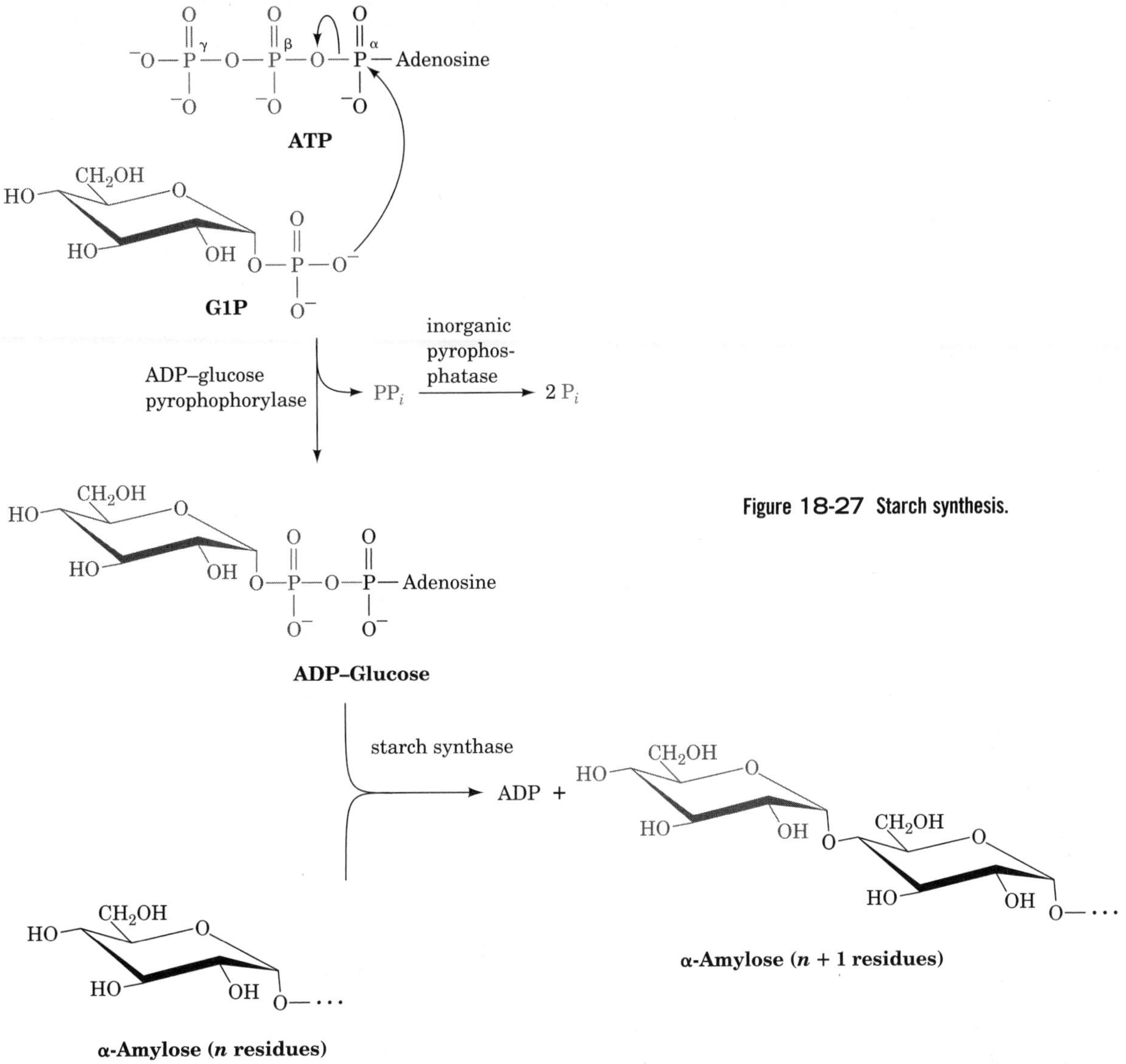

Adenosine
ATP
G1P
ADP–glucose pyrophophorylase
inorganic pyrophos-phatase
PP$_i$
2 P$_i$
Adenosine
ADP–Glucose
starch synthase
ADP +
α-Amylose (n + 1 residues)
α-Amylose (n residues)

Figure 18-27 Starch synthesis.

Table 18-1 Standard and Physiological Free Energy Changes for the Reactions of the Calvin Cycle

Step[a]	Enzyme	$\Delta G^{\circ\prime}$ ($kJ \cdot mol^{-1}$)	ΔG ($kJ \cdot mol^{-1}$)
1	Phosphoribulokinase	−21.8	−15.9
2	Ribulose bisphosphate carboxylase	−35.1	−41.0
3 + 4	Phosphoglycerate kinase + glyceraldehyde-3-phosphate dehydrogenase	+18.0	−6.7
5	Triose phosphate isomerase	−7.5	−0.8
6	Aldolase	−21.8	−1.7
7	Fructose bisphosphatase	−14.2	−27.2
8	Transketolase	+6.3	−3.8
9	Aldolase	−23.4	−0.8
10	Sedoheptulose bisphosphatase	−14.2	−29.7
11	Transketolase	+0.4	−5.9
12	Phosphopentose epimerase	+0.8	−0.4
13	Ribose phosphate isomerase	+2.1	−0.4

[a]Refer to Fig. 18-24.

Source: Bassham, J.A. and Buchanan, B.B., *in* Govindjee (Ed.), *Photosynthesis,* Vol. II, p. 155, Academic Press (1982).

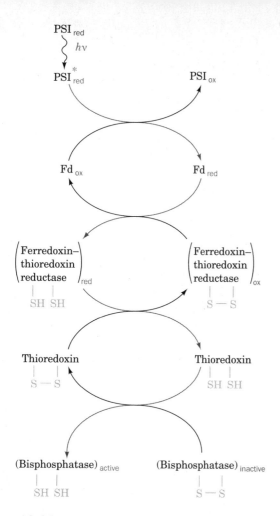

Figure 18-28 The light-activation mechanism of FBPase and SBPase.

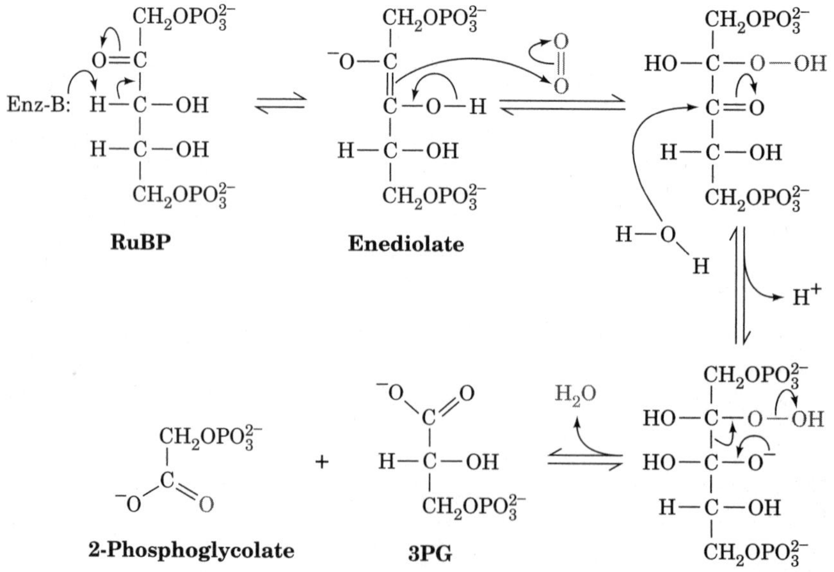

CH$_2$OPO$_3^{2-}$ ... (structural mechanism diagram)

RuBP　　　　**Enediolate**

2-Phosphoglycolate　　　**3PG**

Figure 18-29 Probable mechanism of the oxygenase reaction catalyzed by RuBP carboxylase–oxygenase.

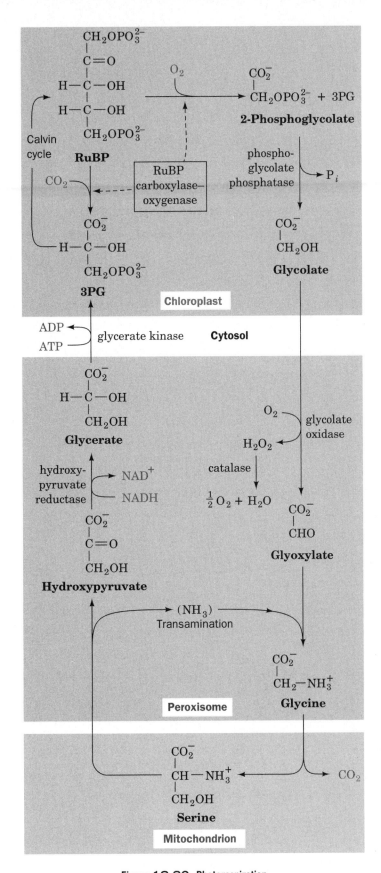

Figure 18-30 Photorespiration.

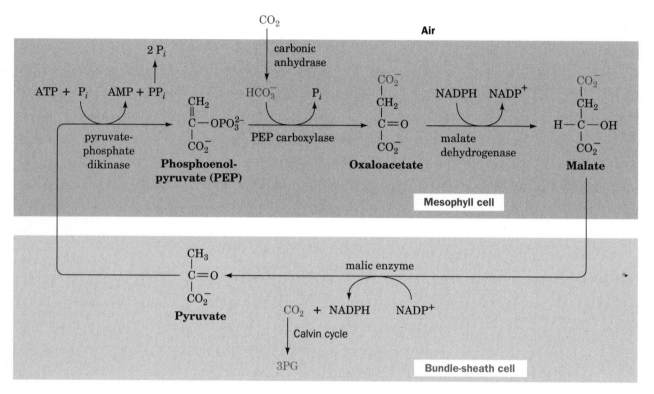

Figure 18-31 The C$_4$ pathway.

CHAPTER 19

Lipid Metabolism

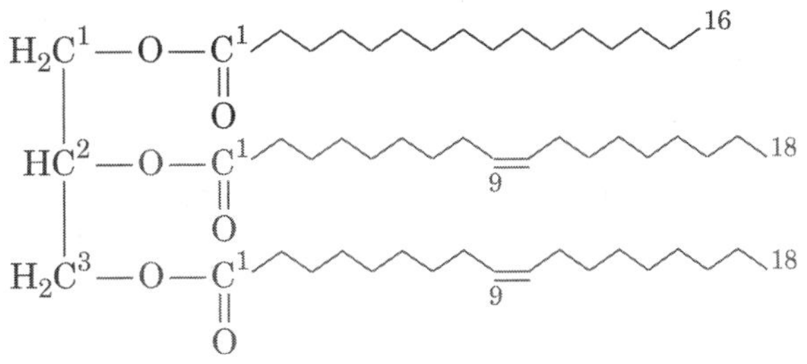

1-Palmitoyl-2,3-dioleoyl-glycerol

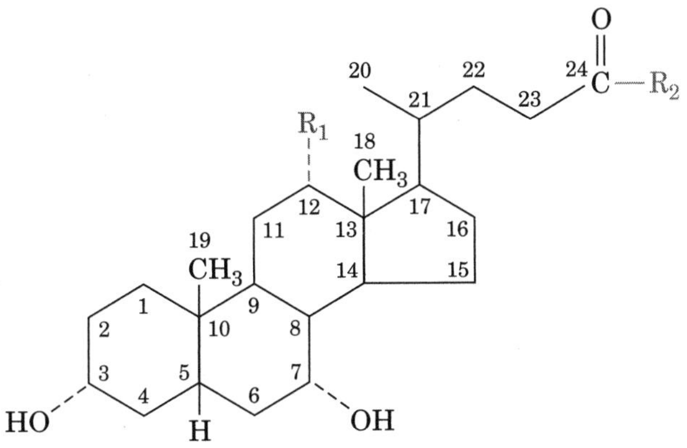

	$R_1 = OH$	$R_1 = H$
$R_2 = OH$	**Cholic acid**	**Chenodeoxycholic acid**
$R_2 = NH-CH_2-COOH$	**Glycocholic acid**	**Glycochenodeoxycholic acid**
$R_2 = NH-CH_2-CH_2-SO_3H$	**Taurocholic acid**	**Taurochenodeoxycholic acid**

Figure 19-1 Structures of the major bile acids and their glycine and taurine conjugates.

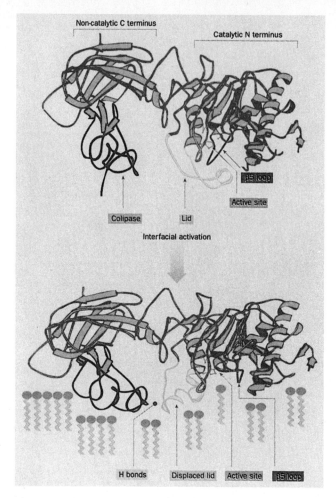

Figure 19-2 The mechanism of interfacial activation of triacylglycerol lipase.

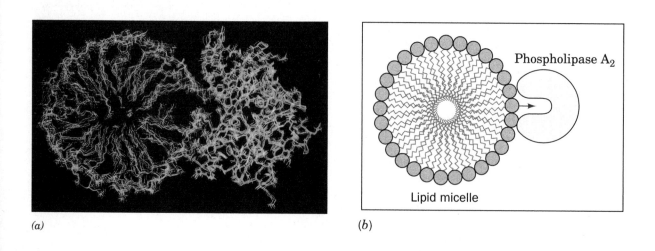

(a)   *(b)*

Figure 19-3 Substrate binding to phospholipase A$_2$.

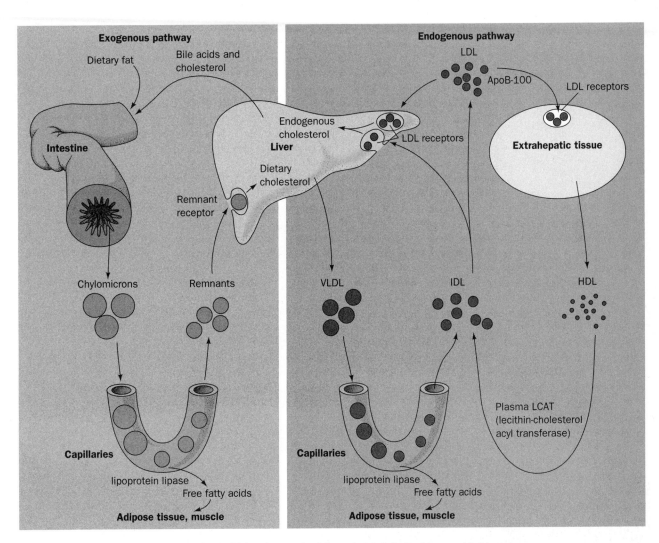

Figure 19-7 Model for plasma triacylglycerol and cholesterol transport in humans.

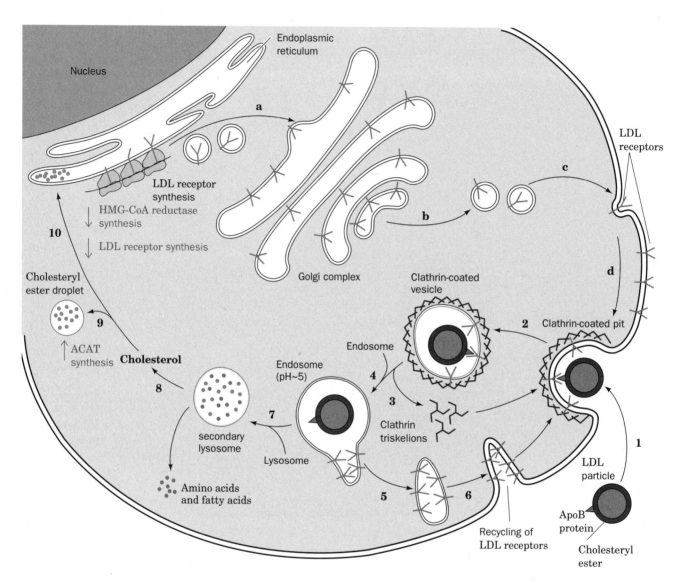

Figure 19-8 *Key to Function.* Receptor-mediated endocytosis of LDL.

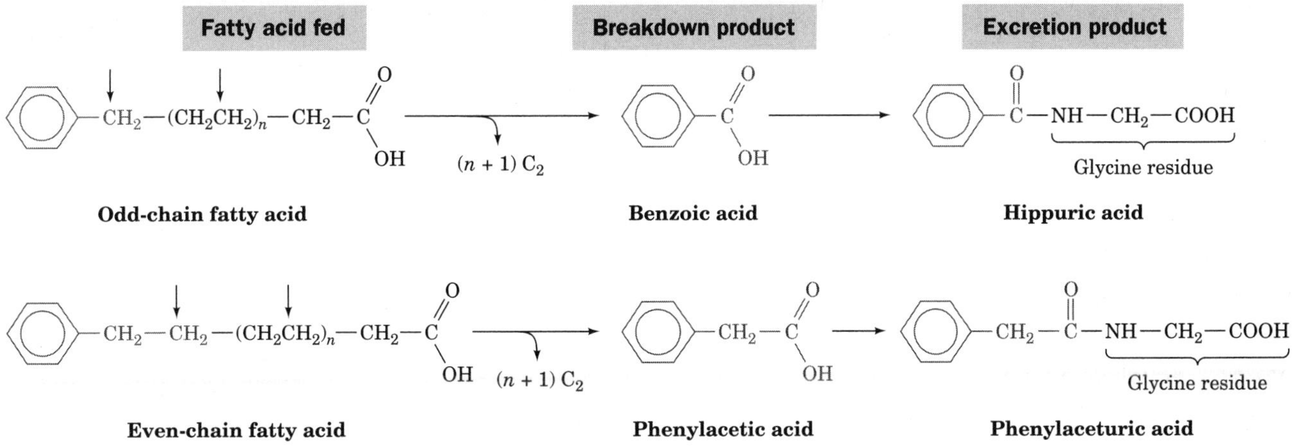

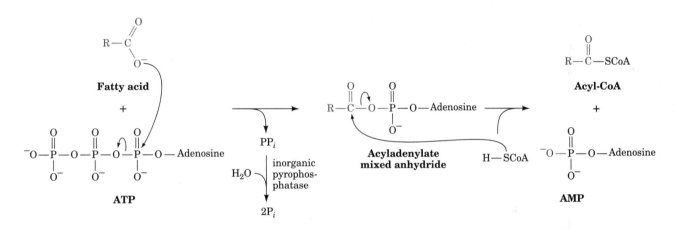

Fatty acid fed | **Breakdown product** | **Excretion product**

Odd-chain fatty acid → Benzoic acid → Hippuric acid

Glycine residue

Even-chain fatty acid → Phenylacetic acid → Phenylaceturic acid

Glycine residue

Figure 19-9 Franz Knoop's classic experiment indicating that fatty acids are metabolically oxidized at their β-carbon atom.

Fatty acid + ATP → PP$_i$ (inorganic pyrophosphatase, H$_2$O → 2P$_i$) → Acyladenylate mixed anhydride + H—SCoA → Acyl-CoA + AMP

Figure 19-10 The mechanism of fatty acid activation catalyzed by acyl-CoA synthetase.

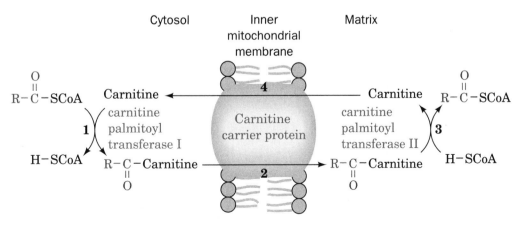

Figure 19-11 The transport of fatty acids into the mitochondrion.

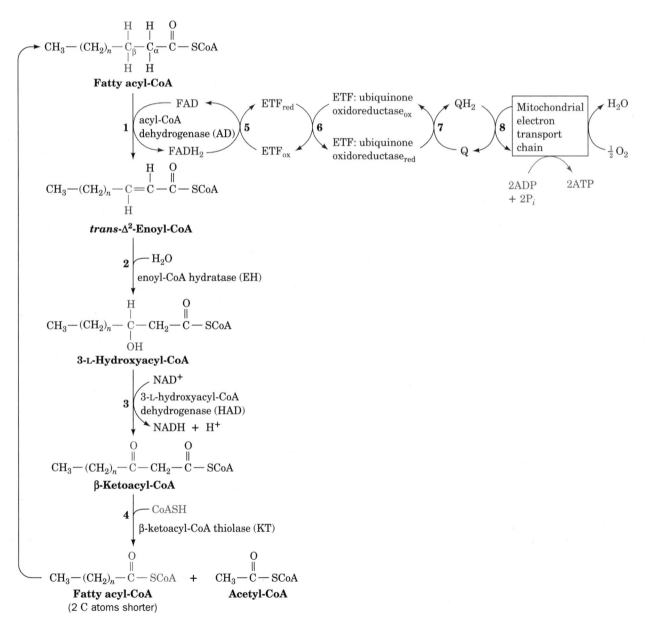

The top portion of the figure shows:

$(CH_3)_3\overset{+}{N}-CH_2-\overset{\overset{\displaystyle OH}{|}}{CH}-CH_2-COO^-$ + $R-\overset{\overset{\displaystyle O}{||}}{C}-SCoA$

Carnitine (4-trimethylamino-3-hydroxybutyrate)

carnitine palmitoyl transferase

$(CH_3)_3\overset{+}{N}-CH_2-\overset{\overset{\displaystyle R-\overset{\overset{\displaystyle O}{||}}{C}-O}{|}}{CH}-CH_2-COO^-$ + $H-SCoA$

Acyl-carnitine

Fatty acyl-CoA

trans-Δ^2-Enoyl-CoA

3-L-Hydroxyacyl-CoA

β-Ketoacyl-CoA

Fatty acyl-CoA (2 C atoms shorter) + **Acetyl-CoA**

Figure 19-12 *Key to Metabolism.* The β-oxidation pathway of fatty acyl-CoA.

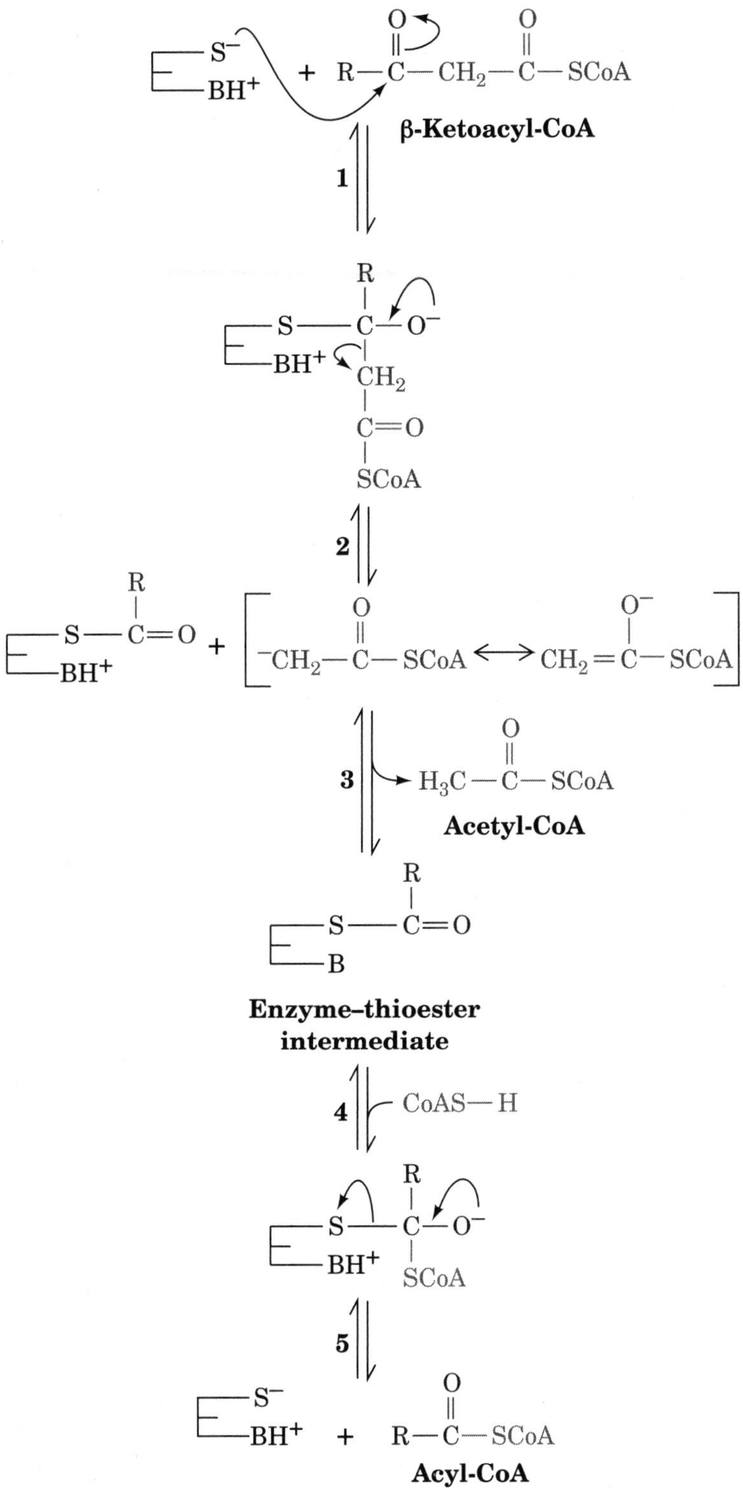

Figure 19-14 Mechanism of action of β-ketoacyl-CoA thiolase.

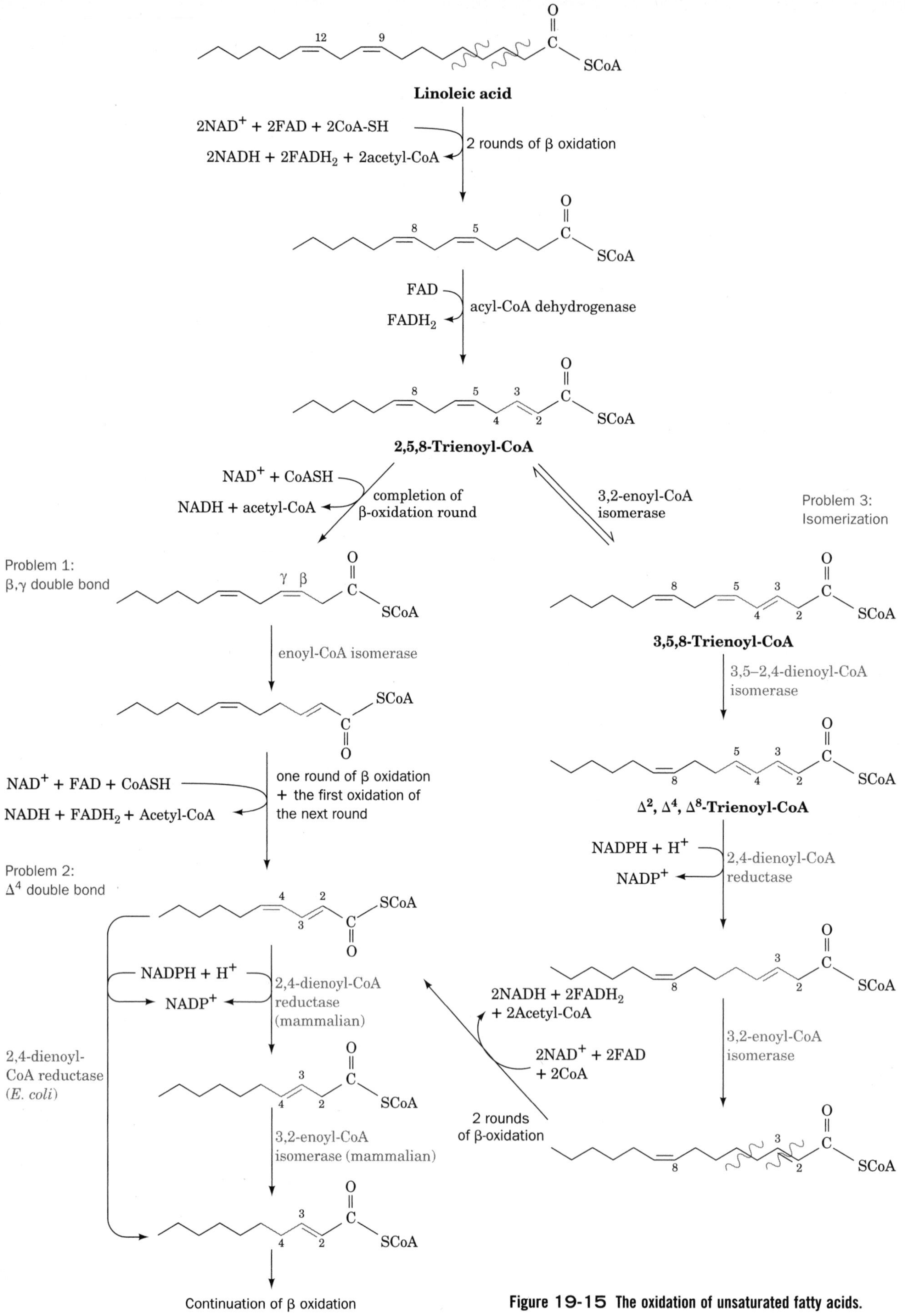

Figure 19-15 The oxidation of unsaturated fatty acids.

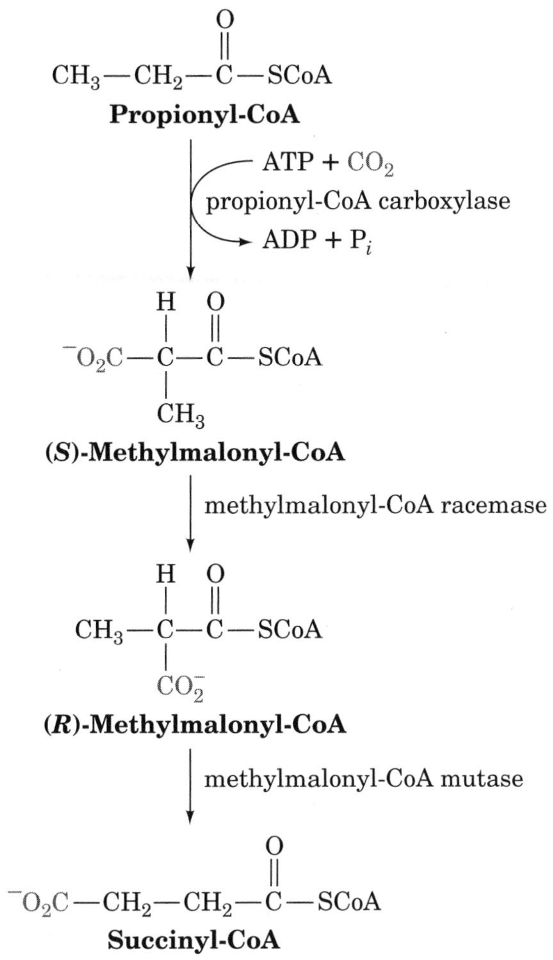

Figure 19-16 Conversion of propionyl-CoA to succinyl-CoA.

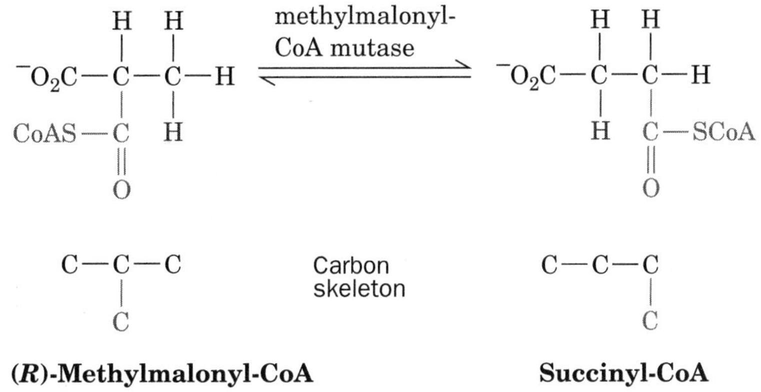

(R)-Methylmalonyl-CoA   Succinyl-CoA

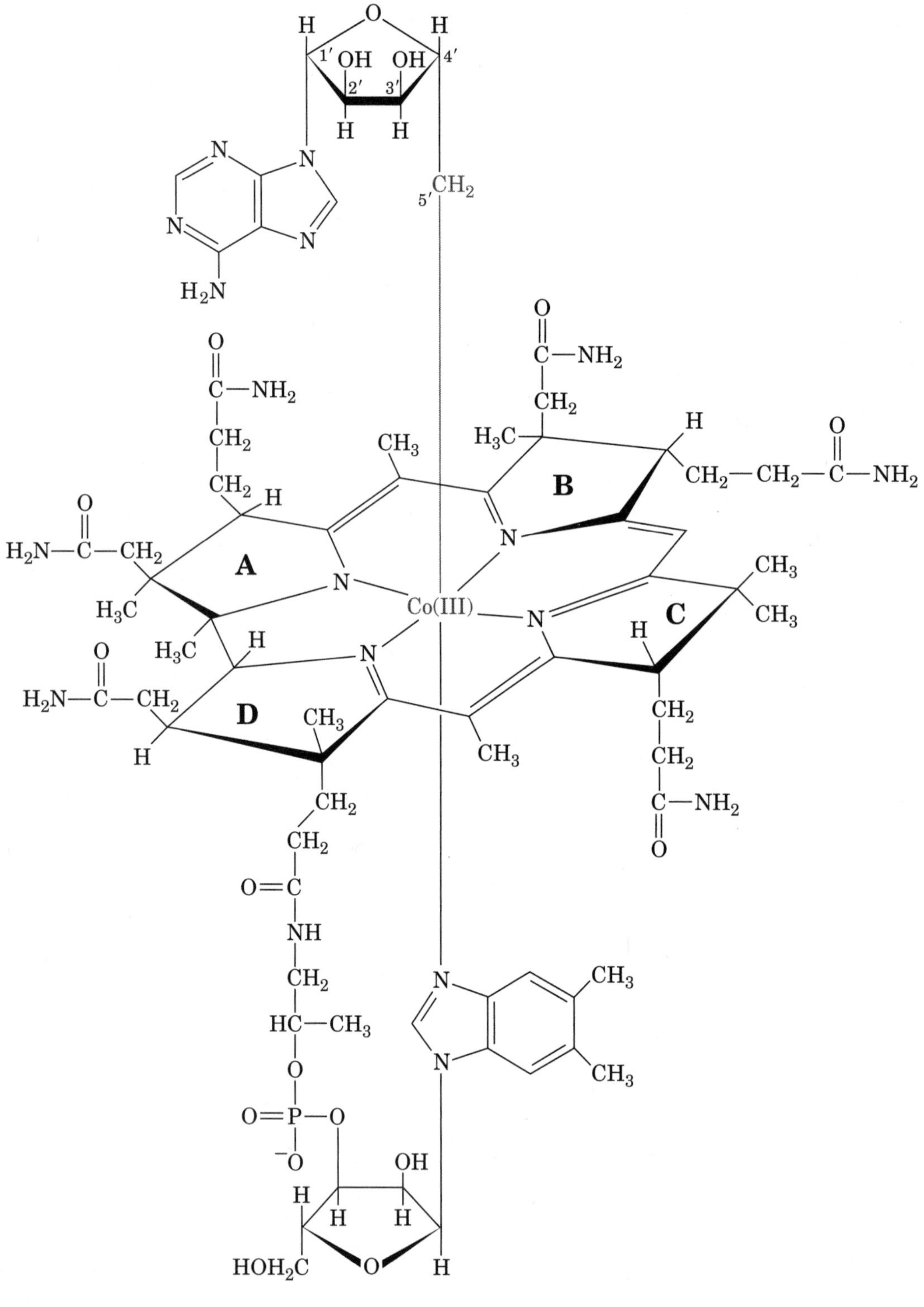

5′-Deoxyadenosylcobalamin (coenzyme B$_{12}$)

Figure 19-17 Structure of 5′-deoxyadenosylcobalamin.

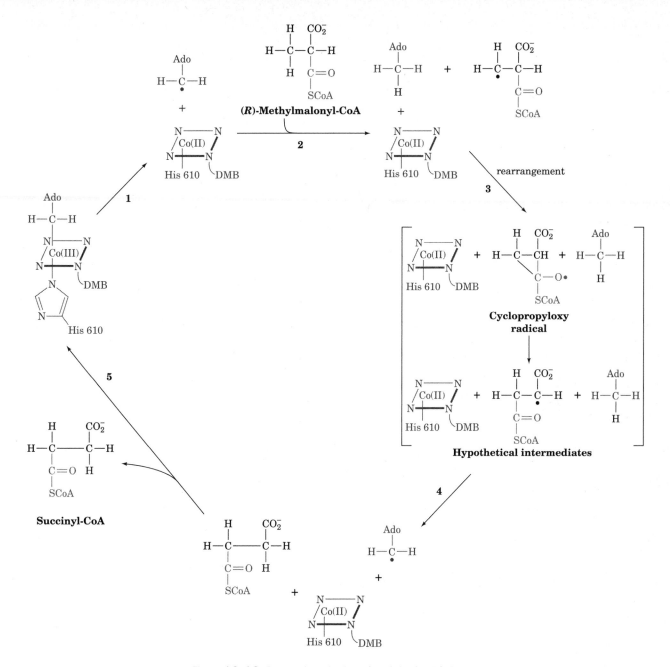

Figure 19-19 Proposed mechanism of methylmalonyl-CoA mutase.

KETONE BODIES

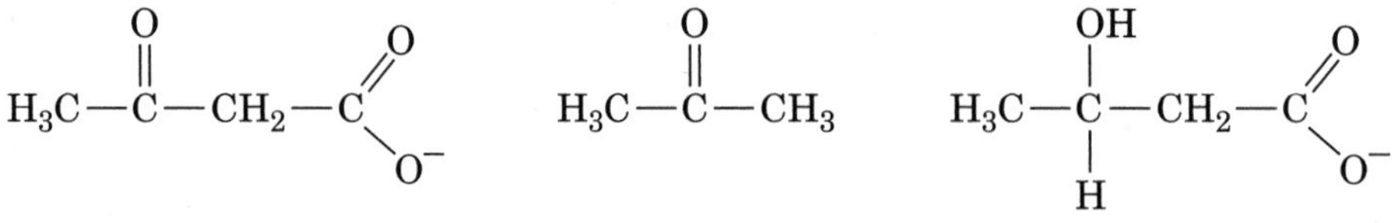

Acetoacetate · Acetone · D-β-Hydroxybutyrate

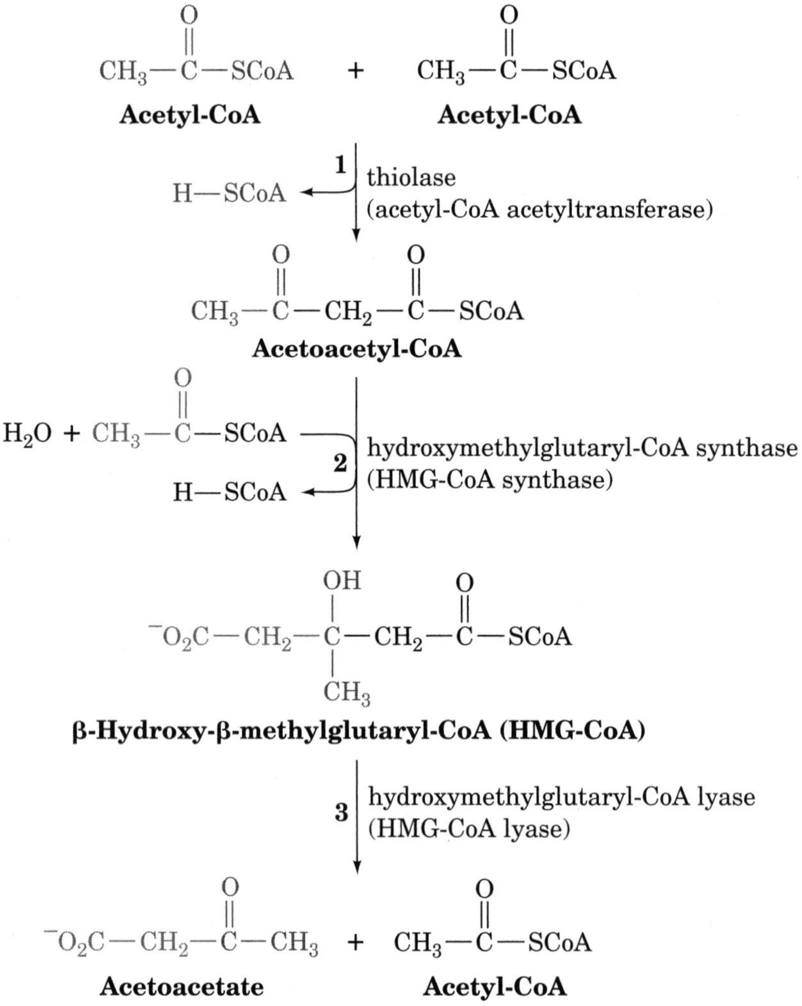

$$CH_3-\overset{\overset{\displaystyle O}{\|}}{C}-SCoA \quad + \quad CH_3-\overset{\overset{\displaystyle O}{\|}}{C}-SCoA$$

Acetyl-CoA   **Acetyl-CoA**

1 | thiolase
(acetyl-CoA acetyltransferase)

H—SCoA

$$CH_3-\overset{\overset{\displaystyle O}{\|}}{C}-CH_2-\overset{\overset{\displaystyle O}{\|}}{C}-SCoA$$

Acetoacetyl-CoA

$$H_2O + CH_3-\overset{\overset{\displaystyle O}{\|}}{C}-SCoA$$

H—SCoA

2 | hydroxymethylglutaryl-CoA synthase
(HMG-CoA synthase)

$$^-O_2C-CH_2-\overset{\overset{\displaystyle OH}{|}}{\underset{\underset{\displaystyle CH_3}{|}}{C}}-CH_2-\overset{\overset{\displaystyle O}{\|}}{C}-SCoA$$

β-Hydroxy-β-methylglutaryl-CoA (HMG-CoA)

3 | hydroxymethylglutaryl-CoA lyase
(HMG-CoA lyase)

$$^-O_2C-CH_2-\overset{\overset{\displaystyle O}{\|}}{C}-CH_3 \quad + \quad CH_3-\overset{\overset{\displaystyle O}{\|}}{C}-SCoA$$

Acetoacetate   **Acetyl-CoA**

Figure 19-21 Ketogenesis.

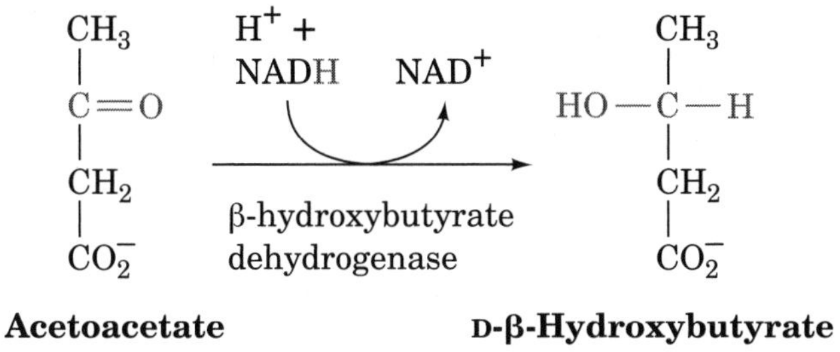

H⁺ +
NADH
NAD⁺
β-hydroxybutyrate
dehydrogenase
Acetoacetate
D-β-Hydroxybutyrate

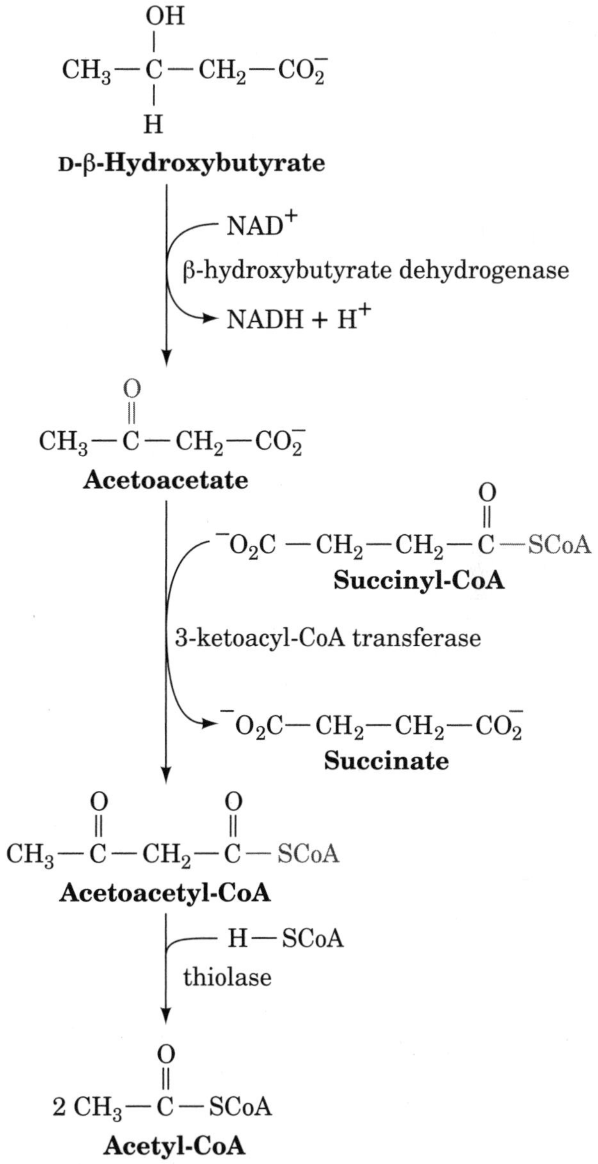

D-β-Hydroxybutyrate
NAD⁺
β-hydroxybutyrate dehydrogenase
NADH + H⁺
Acetoacetate
Succinyl-CoA
3-ketoacyl-CoA transferase
Succinate
Acetoacetyl-CoA
H—SCoA
thiolase
Acetyl-CoA

Figure 19-22 Metabolic conversion of ketone bodies to acetyl-CoA.

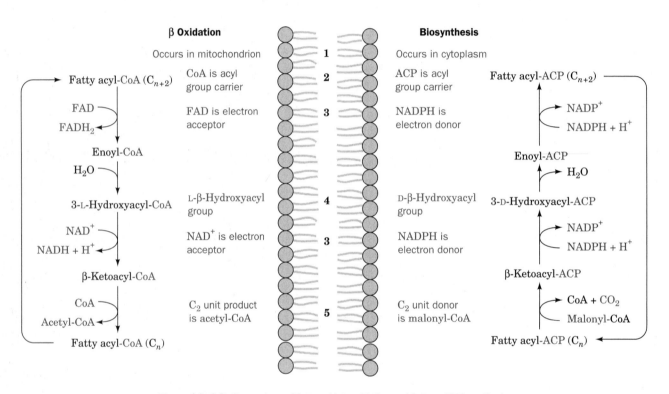

Figure 19-23 Comparison of fatty acid β oxidation and fatty acid biosynthesis.

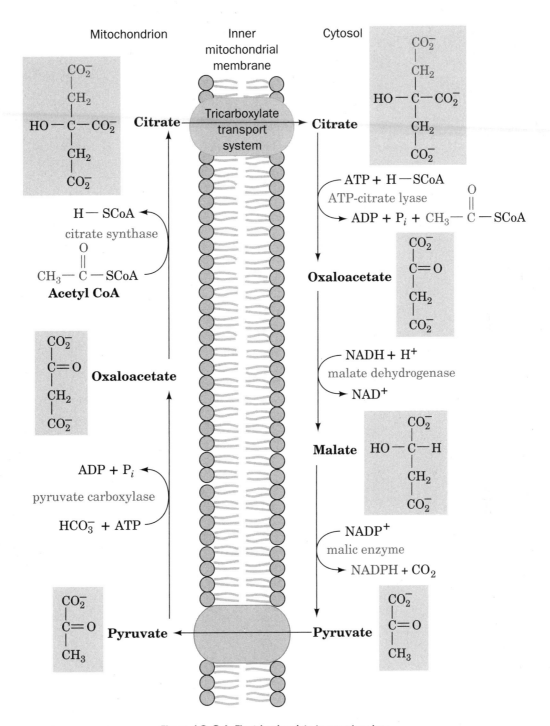

Figure 19-24 The tricarboxylate transport system.

ACETYL-CoA CARBOXYLASE

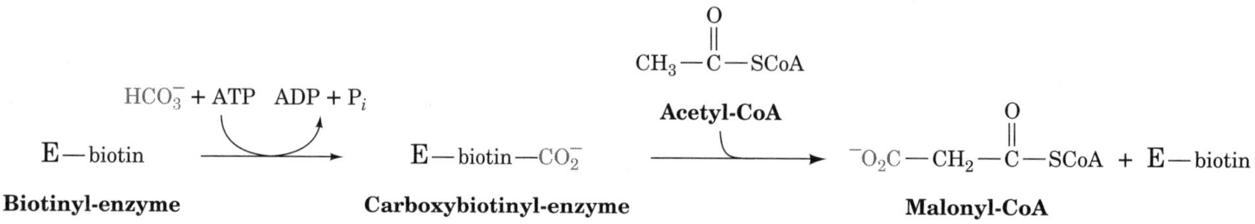

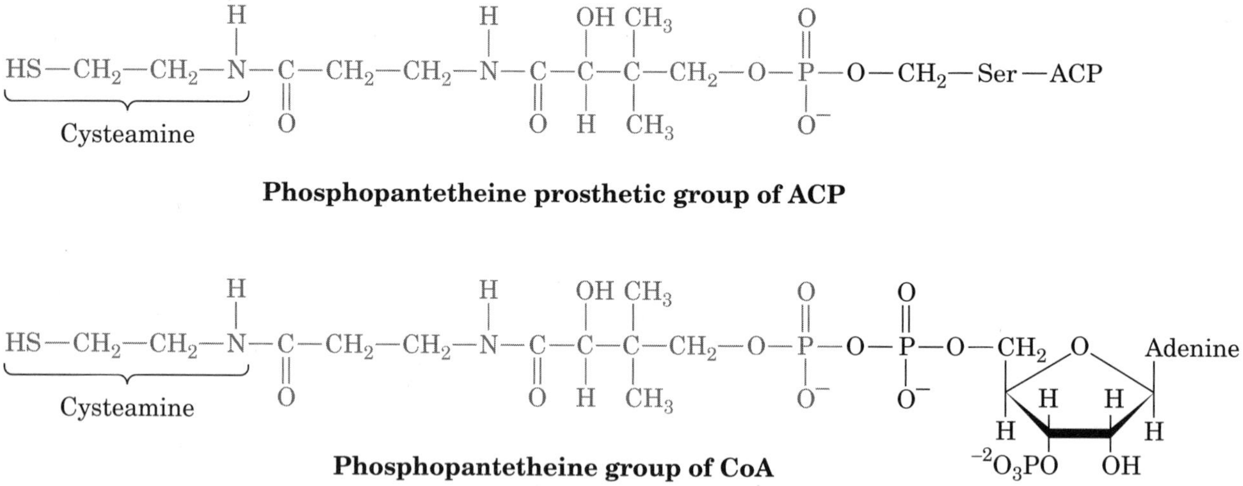

$$CH_3-\overset{\overset{\displaystyle O}{\|}}{C}-SCoA$$

Acetyl-CoA

$$\text{HCO}_3^- + \text{ATP} \quad \text{ADP} + \text{P}_i$$

$$\text{E}-\text{biotin} \xrightarrow{\hspace{2cm}} \text{E}-\text{biotin}-\text{CO}_2^- \xrightarrow{\hspace{2cm}} {}^-\text{O}_2\text{C}-\text{CH}_2-\overset{\overset{\displaystyle O}{\|}}{C}-SCoA \; + \; \text{E}-\text{biotin}$$

Biotinyl-enzyme   **Carboxybiotinyl-enzyme**   **Malonyl-CoA**

$$\text{HS}-\text{CH}_2-\text{CH}_2-\overset{\overset{\displaystyle H}{|}}{N}-\overset{\overset{\displaystyle}{\underset{\underset{\displaystyle O}{\|}}{C}}}-\text{CH}_2-\text{CH}_2-\overset{\overset{\displaystyle H}{|}}{N}-\overset{\overset{\displaystyle OH}{|}}{\underset{\underset{\displaystyle H}{|}}{C}}-\overset{\overset{\displaystyle CH_3}{|}}{\underset{\underset{\displaystyle CH_3}{|}}{C}}-\text{CH}_2-\text{O}-\overset{\overset{\displaystyle O}{\|}}{\underset{\underset{\displaystyle O^-}{|}}{P}}-\text{O}-\text{CH}_2-\text{Ser}-\text{ACP}$$

Cysteamine

Phosphopantetheine prosthetic group of ACP

$$\text{HS}-\text{CH}_2-\text{CH}_2-\overset{\overset{\displaystyle H}{|}}{N}-\overset{\overset{\displaystyle}{\underset{\underset{\displaystyle O}{\|}}{C}}}-\text{CH}_2-\text{CH}_2-\overset{\overset{\displaystyle H}{|}}{N}-\overset{\overset{\displaystyle OH}{|}}{\underset{\underset{\displaystyle H}{|}}{C}}-\overset{\overset{\displaystyle CH_3}{|}}{\underset{\underset{\displaystyle CH_3}{|}}{C}}-\text{CH}_2-\text{O}-\overset{\overset{\displaystyle O}{\|}}{\underset{\underset{\displaystyle O^-}{|}}{P}}-\text{O}-\overset{\overset{\displaystyle O}{\|}}{\underset{\underset{\displaystyle O^-}{|}}{P}}-\text{O}-\text{CH}_2$$

Adenine

Cysteamine

$${}^{-2}\text{O}_3\text{PO} \qquad \text{OH}$$

Phosphopantetheine group of CoA

Figure 19-25 The phosphopantetheine group in acyl-carrier protein (ACP) and in CoA.

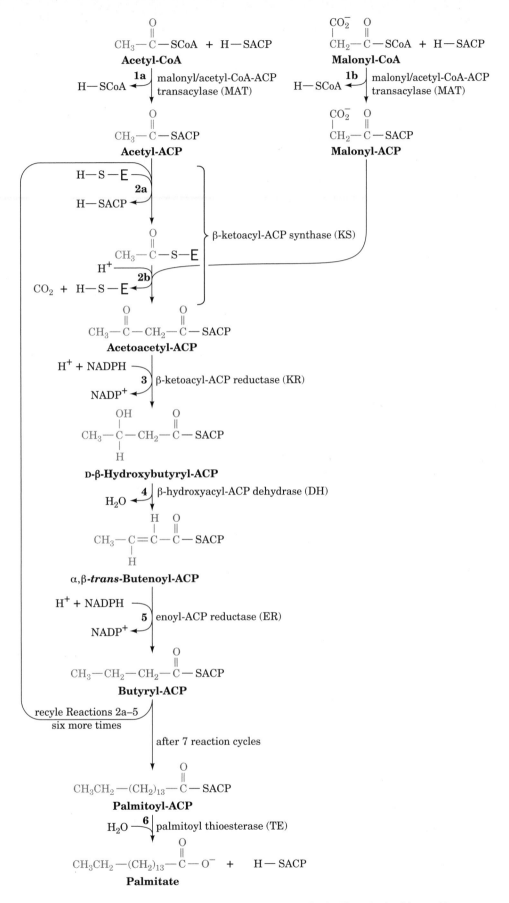

Figure 19-26 *Key to Metabolism.* Reaction sequence for the biosynthesis of fatty acids.

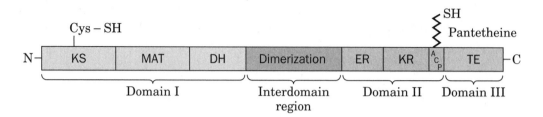

Figure 19-27 Order of enzymatic activity along the polypeptide chain of the fatty acid synthase monomer.

$$8 \text{ Acetyl-CoA} + 14 \text{ NADPH} + 7 \text{ ATP} \rightarrow$$
$$\text{palmitate} + 14 \text{ NADP}^+ + 8 \text{ CoA} + 6 \text{ H}_2\text{O} + 7 \text{ ADP} + 7 \text{ P}_i$$

FATTY ACYL-CoA DESATURASES

$$\text{CH}_3-(\text{CH}_2)_x-\overset{\overset{\displaystyle H}{|}}{\underset{\underset{\displaystyle H}{|}}{C}}-\overset{\overset{\displaystyle H}{|}}{\underset{\underset{\displaystyle H}{|}}{C}}-(\text{CH}_2)_y-\overset{\overset{\displaystyle O}{\|}}{C}-\text{SCoA} + \text{NADH} + \text{H}^+ + \text{O}_2$$

$$\downarrow$$

$$\text{CH}_3-(\text{CH}_2)_x-\overset{\overset{\displaystyle H}{|}}{C}=\overset{\overset{\displaystyle H}{|}}{C}-(\text{CH}_2)_y-\overset{\overset{\displaystyle O}{\|}}{C}-\text{SCoA} + 2\text{H}_2\text{O} + \text{NAD}^+$$

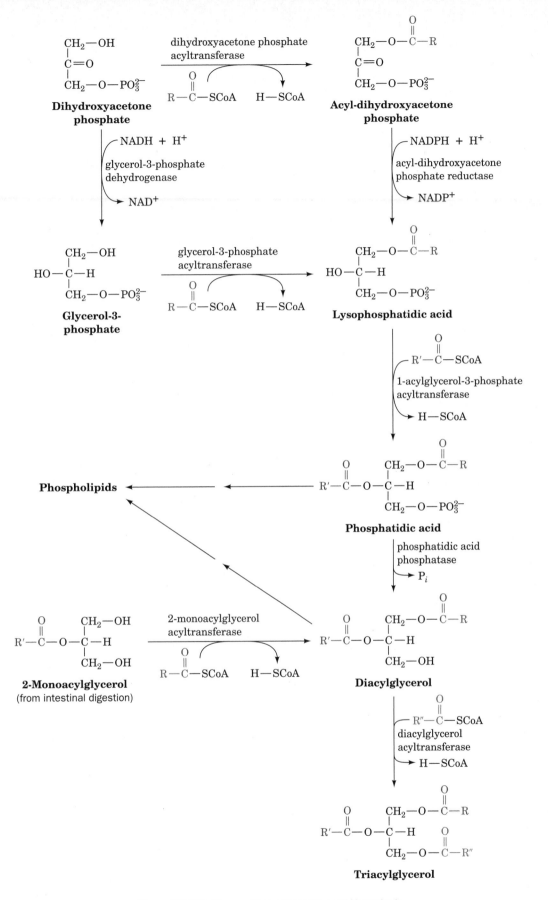

Figure 19-30 The reactions of triacylglycerol biosynthesis.

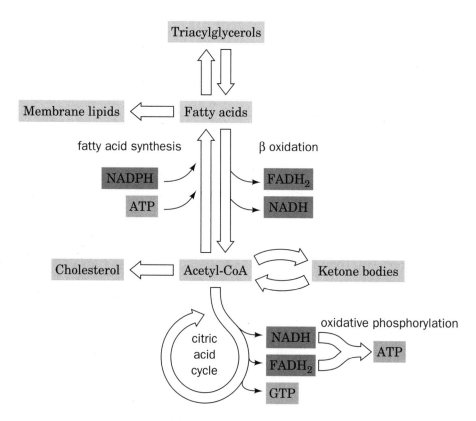

Figure 19-31 A summary of lipid metabolism.

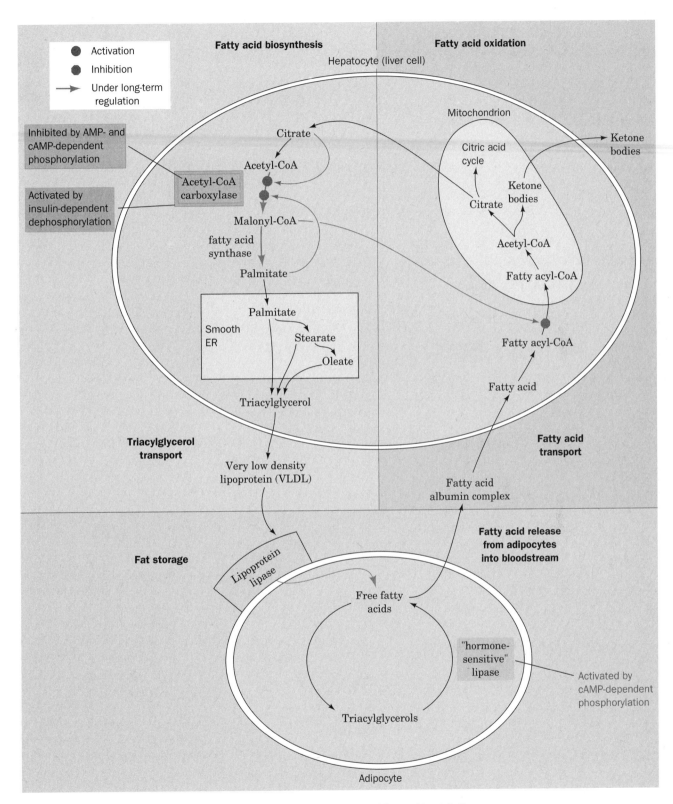

Figure 19-32 Sites of regulation of fatty acid metabolism.

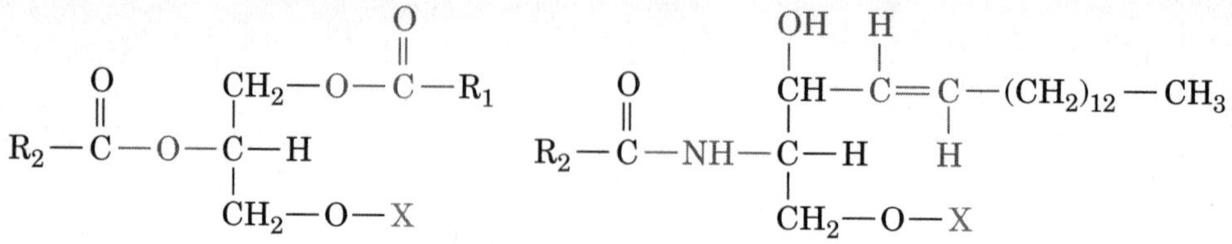

Figure 19-33

Glycerolipid

R₂ structure (left) and **Sphingolipid** (right)

	Glycerolipid	Sphingolipid
X = H	**1,2-Diacylglycerol**	**N-Acylsphingosine (ceramide)**
X = Carbohydrate	**Glyceroglycolipid**	**Sphingoglycolipid (glycosphingolipid)**
X = Phosphate ester	**Glycerophospholipid**	**Sphingophospholipid**

Figure 19-33 Glycerolipids and sphingolipids.

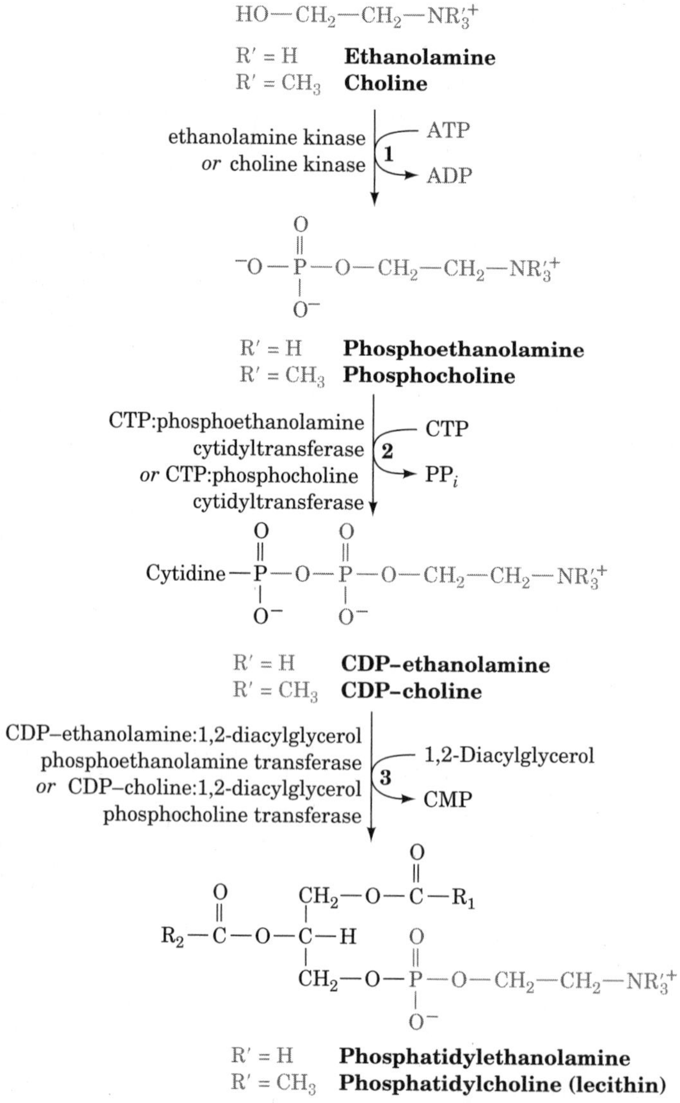

$HO-CH_2-CH_2-NR_3'^+$

R′ = H   **Ethanolamine**
R′ = CH_3 **Choline**

ethanolamine kinase *or* choline kinase   **1**   ATP → ADP

R′ = H   **Phosphoethanolamine**
R′ = CH_3 **Phosphocholine**

CTP:phosphoethanolamine cytidyltransferase *or* CTP:phosphocholine cytidyltransferase   **2**   CTP → PP_i

R′ = H   **CDP–ethanolamine**
R′ = CH_3 **CDP–choline**

CDP–ethanolamine:1,2-diacylglycerol phosphoethanolamine transferase *or* CDP–choline:1,2-diacylglycerol phosphocholine transferase   **3**   1,2-Diacylglycerol → CMP

R′ = H   **Phosphatidylethanolamine**
R′ = CH_3 **Phosphatidylcholine (lecithin)**

Figure 19-34 Biosynthesis of phosphatidylethanolamine and phosphatidylcholine.

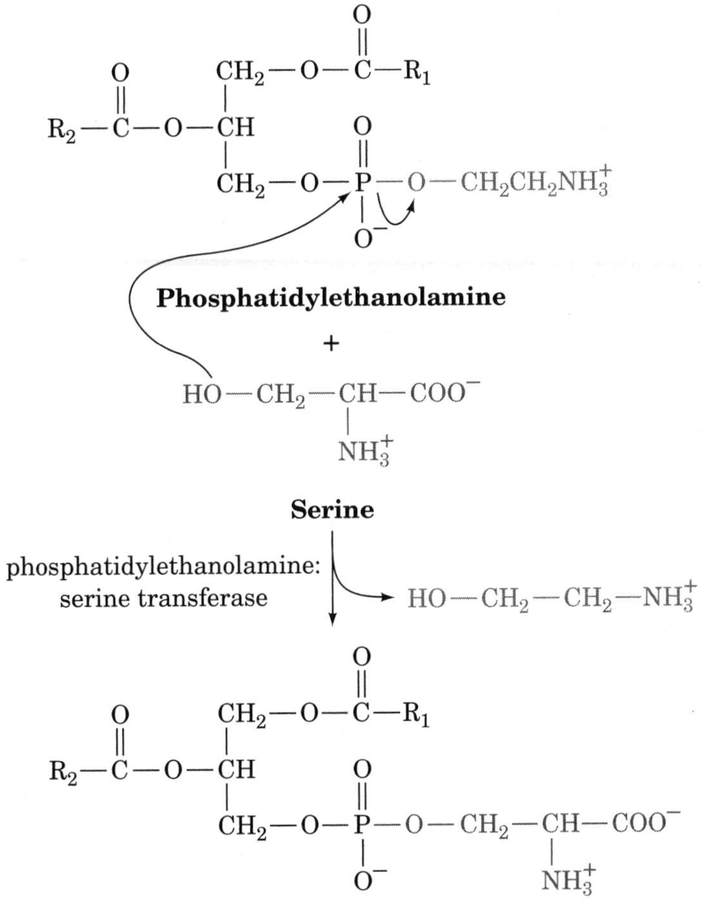

Phosphatidylethanolamine

+

Serine

phosphatidylethanolamine:
serine transferase

$$HO-CH_2-CH_2-NH_3^+$$

Phosphatidylserine

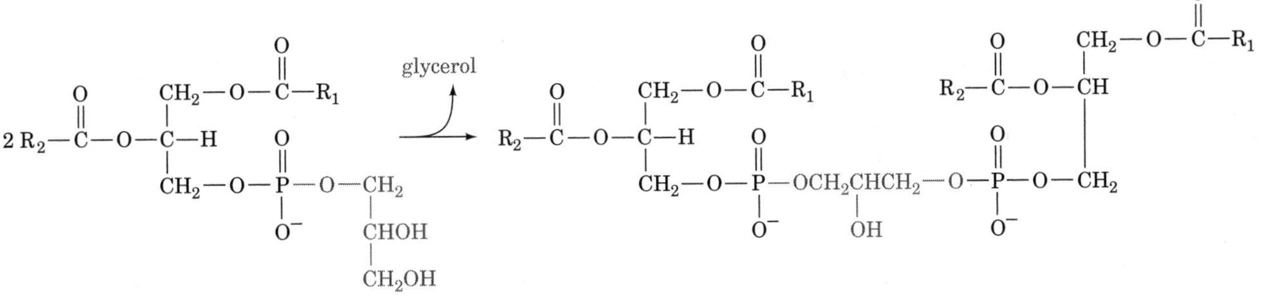

Phosphatidylglycerol

glycerol

Cardiolipin

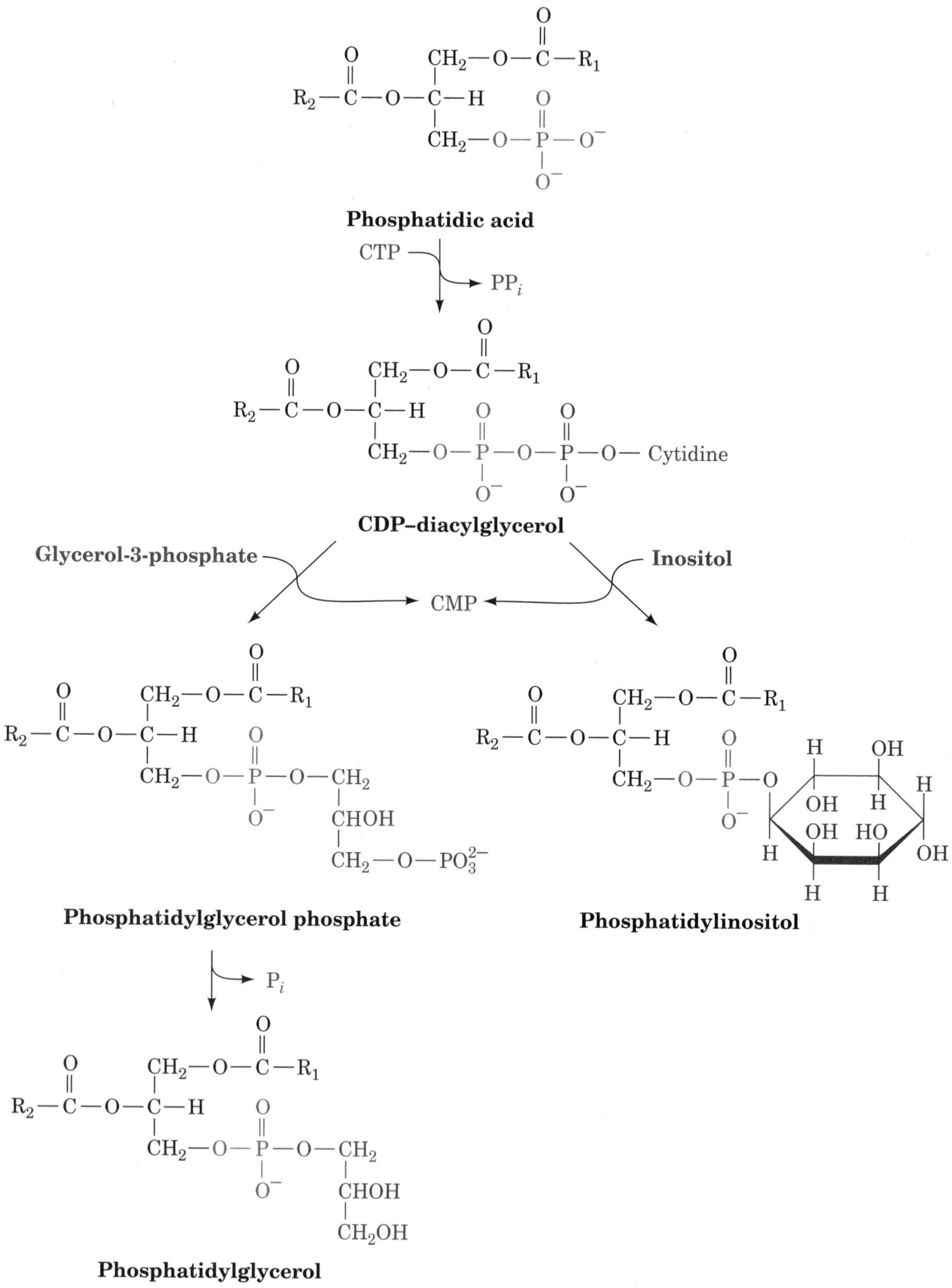

Phosphatidic acid

CTP $\longrightarrow$ PP$_i$

CDP–diacylglycerol

Glycerol-3-phosphate

Inositol

CMP

Phosphatidylglycerol phosphate

Phosphatidylinositol

P$_i$

Phosphatidylglycerol

Figure 19-35 Biosynthesis of phosphatidylinositol and phosphatidylglycerol.

$$\text{A plasmalogen}$$

$$\text{An alkylacyl-} \atop \text{glycerophospholipid}$$

$$\text{Sphingomyelin}$$

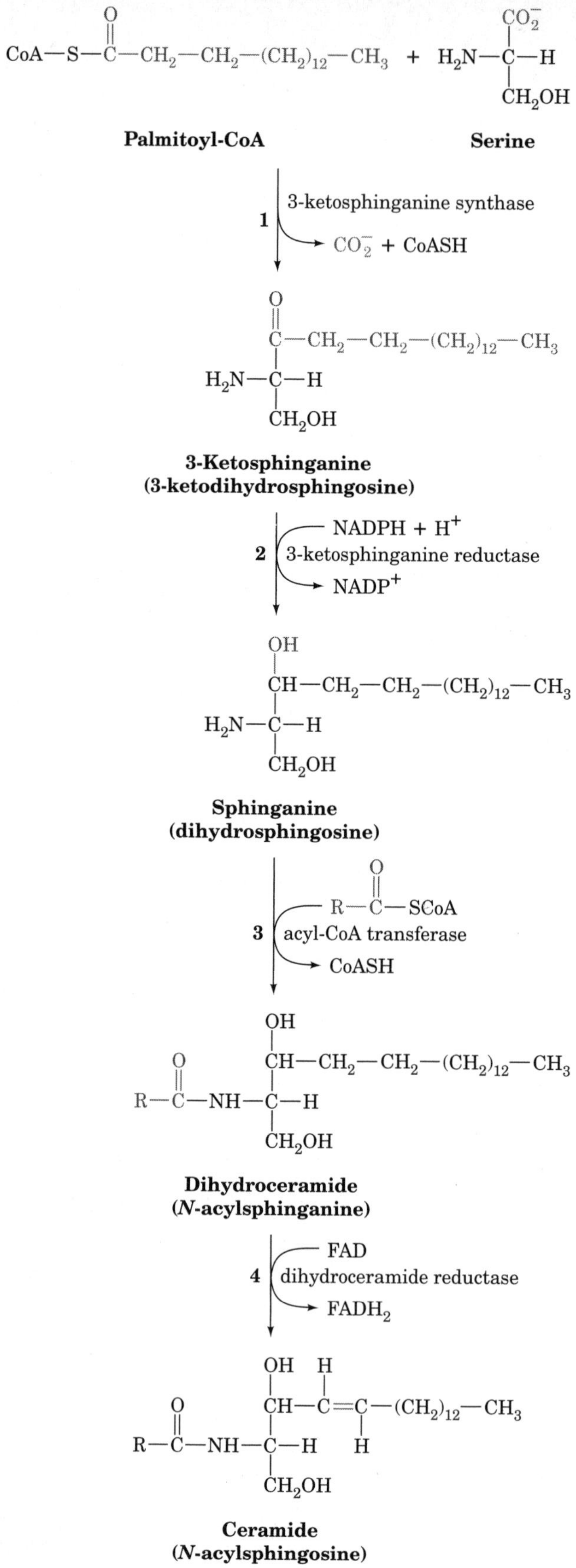

$$\text{CoA}-\text{S}-\overset{\overset{\displaystyle O}{\|}}{\text{C}}-\text{CH}_2-\text{CH}_2-(\text{CH}_2)_{12}-\text{CH}_3 \quad + \quad \text{H}_2\text{N}-\overset{\overset{\displaystyle \text{CO}_2^-}{|}}{\underset{\underset{\displaystyle \text{CH}_2\text{OH}}{|}}{\text{C}}}-\text{H}$$

Palmitoyl-CoA   **Serine**

1   3-ketosphinganine synthase
$\longrightarrow \text{CO}_2^- + \text{CoASH}$

$$\overset{\overset{\displaystyle O}{\|}}{\underset{\underset{\displaystyle \text{H}_2\text{N}-\text{C}-\text{H}}{}}{\text{C}}}-\text{CH}_2-\text{CH}_2-(\text{CH}_2)_{12}-\text{CH}_3$$

$$\text{CH}_2\text{OH}$$

3-Ketosphinganine
(3-ketodihydrosphingosine)

2   NADPH + H$^+$
3-ketosphinganine reductase
$\longrightarrow \text{NADP}^+$

$$\overset{\overset{\displaystyle \text{OH}}{|}}{\text{CH}}-\text{CH}_2-\text{CH}_2-(\text{CH}_2)_{12}-\text{CH}_3$$
$$\text{H}_2\text{N}-\text{C}-\text{H}$$
$$\text{CH}_2\text{OH}$$

Sphinganine
(dihydrosphingosine)

3 $\text{R}-\overset{\overset{\displaystyle O}{\|}}{\text{C}}-\text{SCoA}$
acyl-CoA transferase
$\longrightarrow \text{CoASH}$

$$\overset{\overset{\displaystyle \text{OH}}{|}}{\text{CH}}-\text{CH}_2-\text{CH}_2-(\text{CH}_2)_{12}-\text{CH}_3$$
$$\text{R}-\overset{\overset{\displaystyle O}{\|}}{\text{C}}-\text{NH}-\text{C}-\text{H}$$
$$\text{CH}_2\text{OH}$$

Dihydroceramide
(*N*-acylsphinganine)

4   FAD
dihydroceramide reductase
$\longrightarrow \text{FADH}_2$

$$\overset{\overset{\displaystyle \text{OH} \quad \text{H}}{| \quad |}}{\text{CH}-\text{C}}=\text{C}-(\text{CH}_2)_{12}-\text{CH}_3$$
$$\text{R}-\overset{\overset{\displaystyle O}{\|}}{\text{C}}-\text{NH}-\text{C}-\text{H} \quad \text{H}$$
$$\text{CH}_2\text{OH}$$

Ceramide
(*N*-acylsphingosine)

Figure 19-36 Biosynthesis of ceramide (*N*-acylsphingosine).

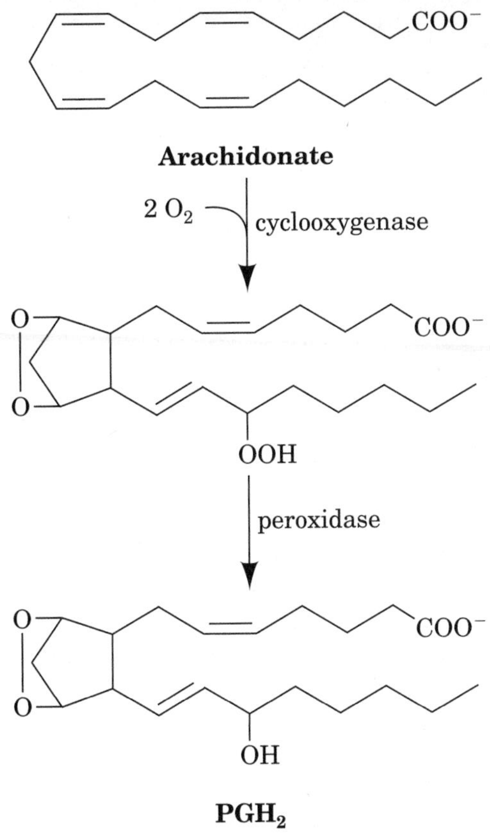

Arachidonate

$2 O_2$ — cyclooxygenase

PGH$_2$

Figure 19-37 The prostaglandin H$_2$ synthase reaction.

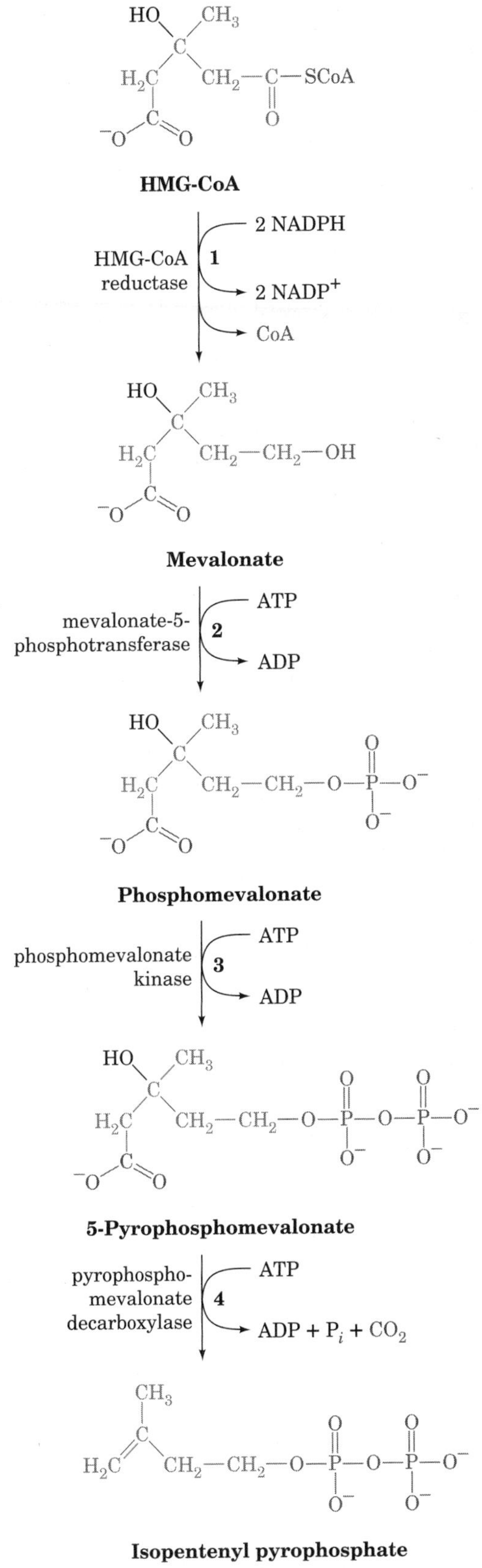

HMG-CoA

2 NADPH

HMG-CoA reductase **1**

2 NADP$^+$

CoA

Mevalonate

mevalonate-5-phosphotransferase **2**

ATP

ADP

Phosphomevalonate

phosphomevalonate kinase **3**

ATP

ADP

5-Pyrophosphomevalonate

pyrophospho-mevalonate decarboxylase **4**

ATP

ADP + P$_i$ + CO$_2$

Isopentenyl pyrophosphate

Figure 19-38 Formation of isopentenyl pyrophosphate from HMG-CoA.

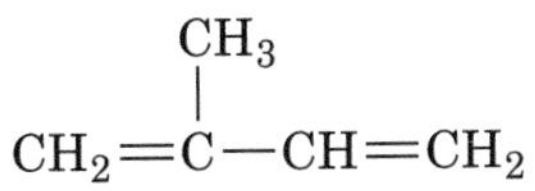

CH_2=C—CH=CH$_2$ with CH$_3$

Isoprene
(2-methyl-1,3-butadiene)

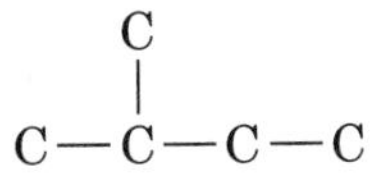

C—C—C—C with C

An isoprene unit

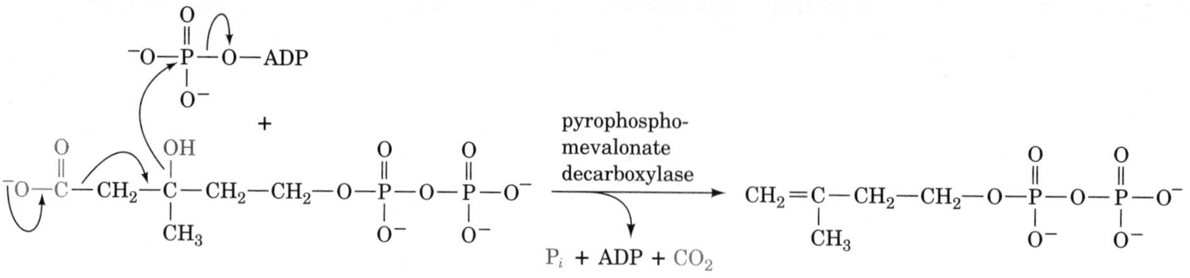

5-Pyrophosphomevalonate **Isopentenyl pyrophosphate**

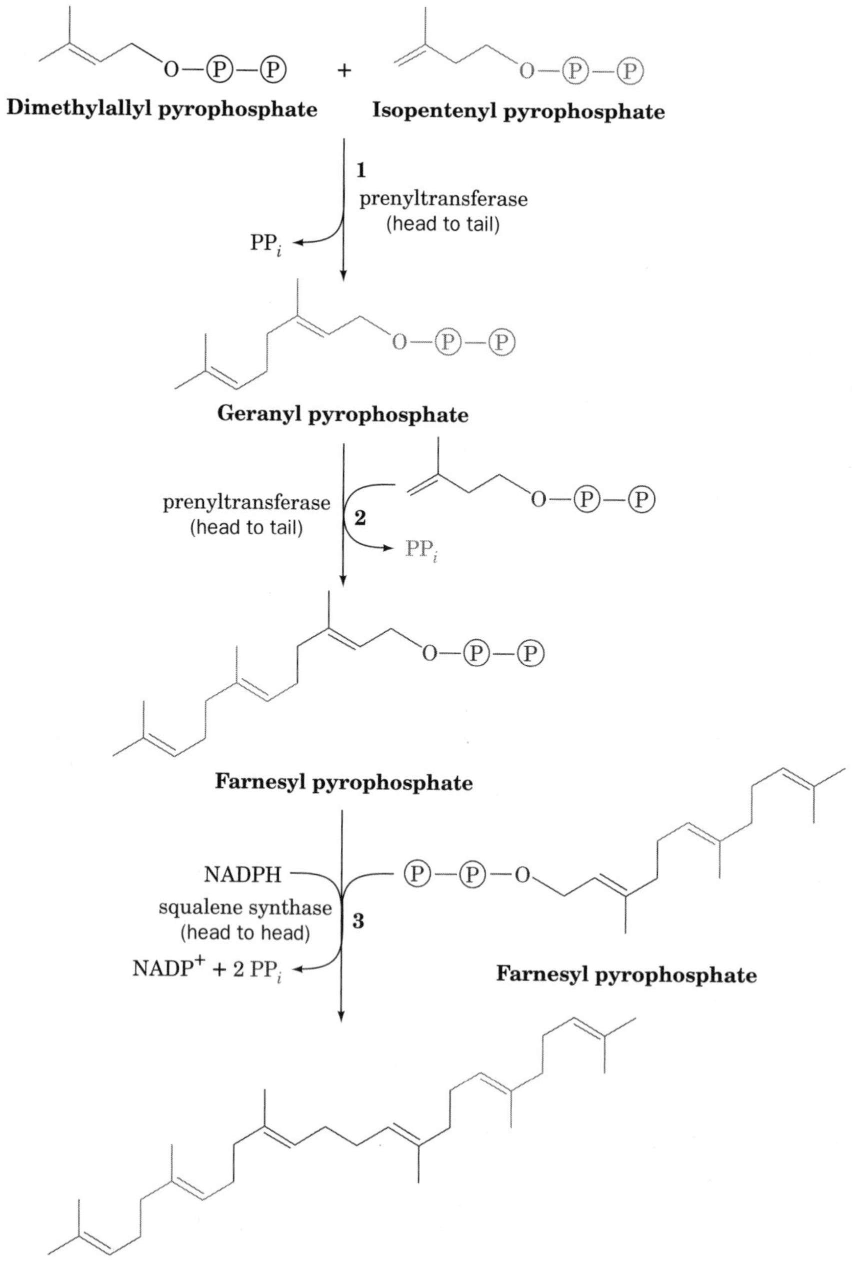

Squalene

Figure 19-39 **Formation of squalene from isopentenyl pyrophosphate and dimethylallyl pyrophosphate.**

PRENYL TRANSFERASE REACTION MECHANISM

Ionization–condensation–elimination

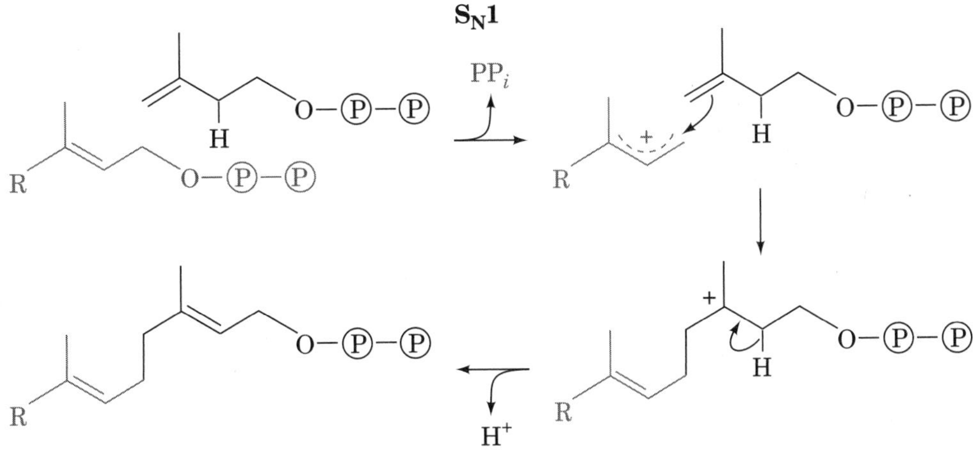

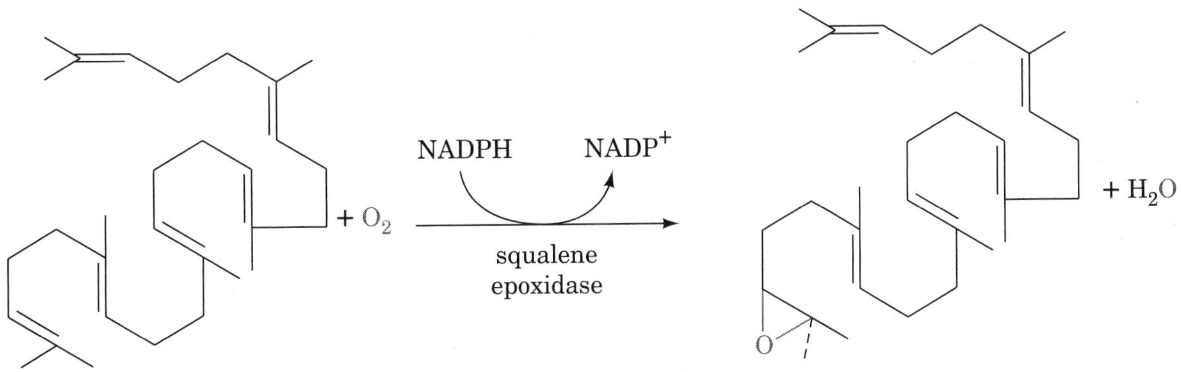

S_N1

Figure 19-40 The squalene epoxidase reaction.

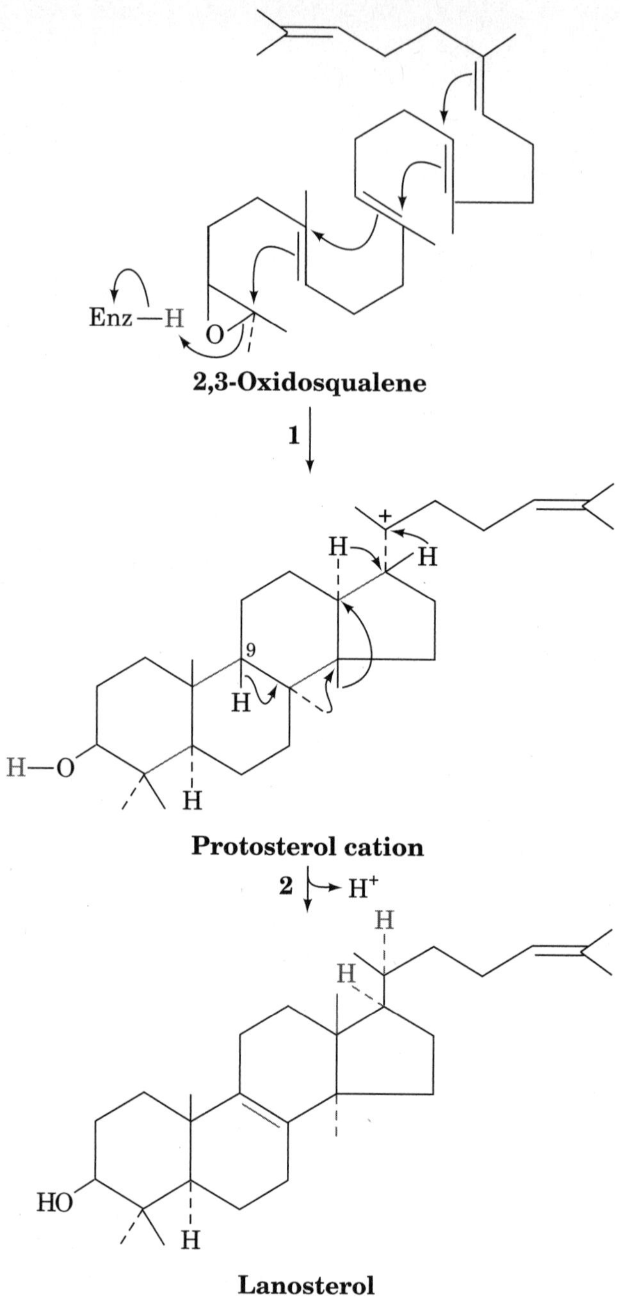
Enz—H
O
2,3-Oxidosqualene
1
H
H
9
H
H—O
H
Protosterol cation
2
H⁺
H
H
HO
H
Lanosterol

Figure 19-41 The oxidosqualene cyclase reaction.

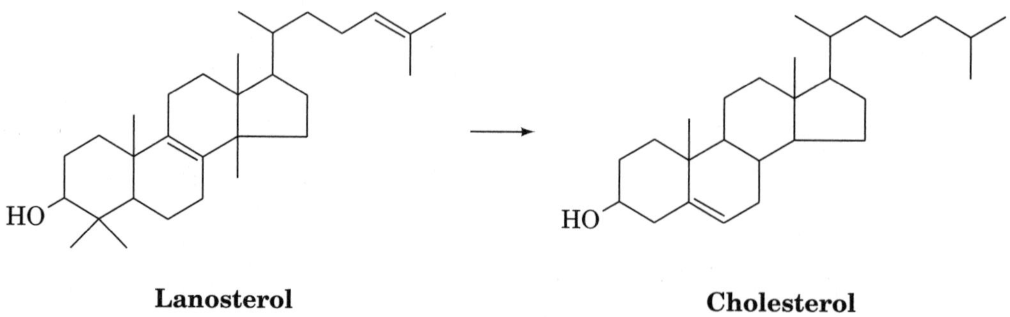
HO
Lanosterol
HO
Cholesterol

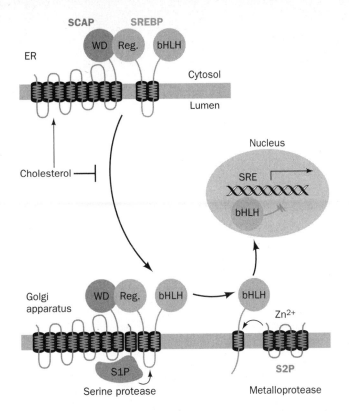

Figure 19-42 The cholesterol-mediated proteolytic activation of SREBP.

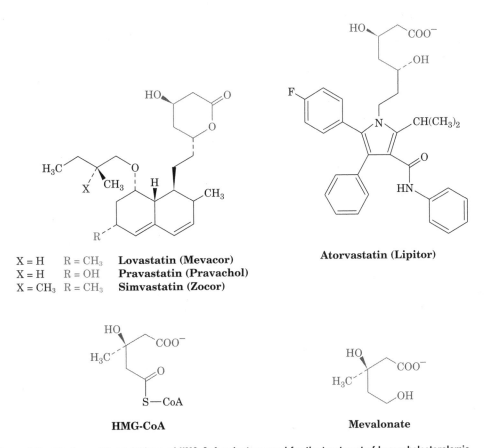

X = H   R = CH₃   **Lovastatin (Mevacor)**
X = H   R = OH   **Pravastatin (Pravachol)**
X = CH₃ R = CH₃ **Simvastatin (Zocor)**

Atorvastatin (Lipitor)

HMG-CoA

Mevalonate

Figure 19-43 Competitive inhibitors of HMG-CoA reductase used for the treatment of hypercholesterolemia.

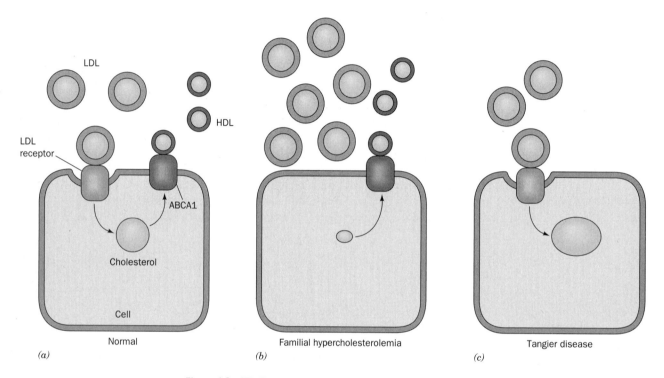

LDL

LDL
receptor

HDL

ABCA1

Cholesterol

Cell

Normal

Familial hypercholesterolemia

Tangier disease

(a)

(b)

(c)

Figure 19-45 The role of LDL and HDL in cholesterol metabolism.

CHAPTER 20

Amino Acid Metabolism

Table 20-1 Half-Lives of Some Rat Liver Enzymes

Enzyme	Half-Life (h)
Short-Lived Enzymes	
Ornithine decarboxylase	0.2
RNA polymerase I	1.3
Tyrosine aminotransferase	2.0
Serine dehydratase	4.0
PEP carboxylase	5.0
Long-Lived Enzymes	
Aldolase	118
GAPDH	130
Cytochrome b	130
LDH	130
Cytochrome c	150

Source: Dice, J.F. and Goldberg, A.L., *Arch. Biochem. Biophys.* **170**, 214 (1975).

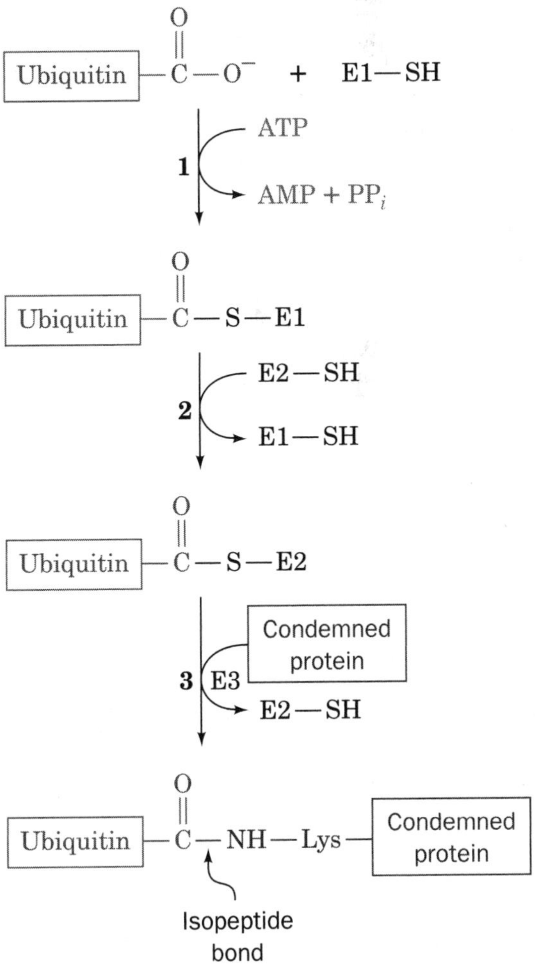

Figure 20-2 Reactions involved in protein ubiquitination.

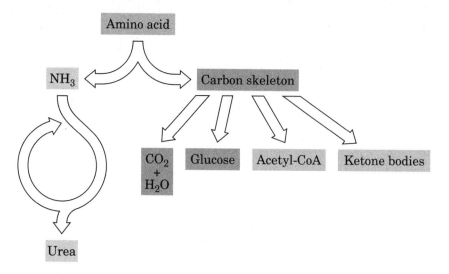

Figure 20-6 Overview of amino acid catabolism.

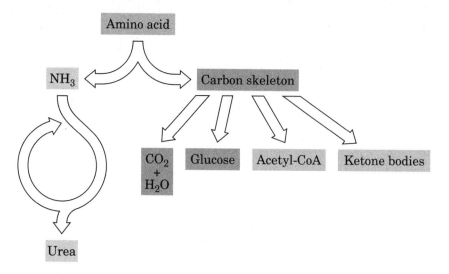

Amino acid + **α-Ketoglutarate**

⇅

α-Keto acid + **Glutamate**

TRANSAMINATION

Glutamate + **Oxaloacetate**

⇅

α-Ketoglutarate + **Aspartate**

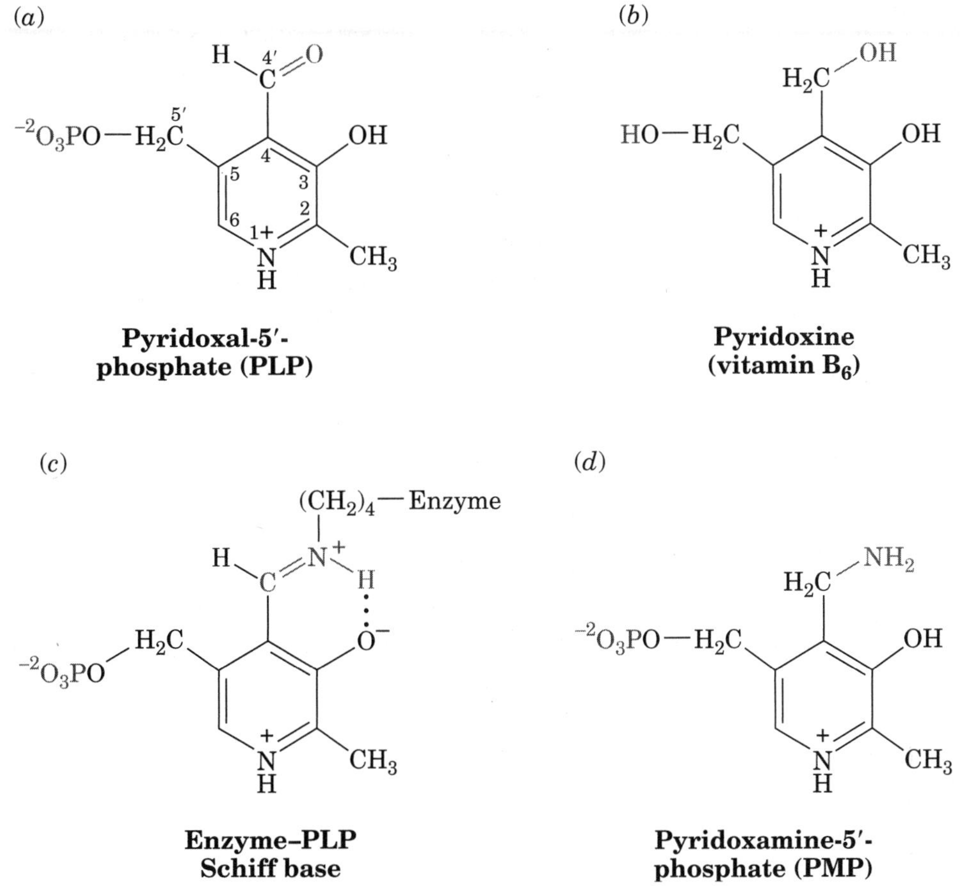

Figure 20-7 Forms of pyridoxal-5′-phosphate.

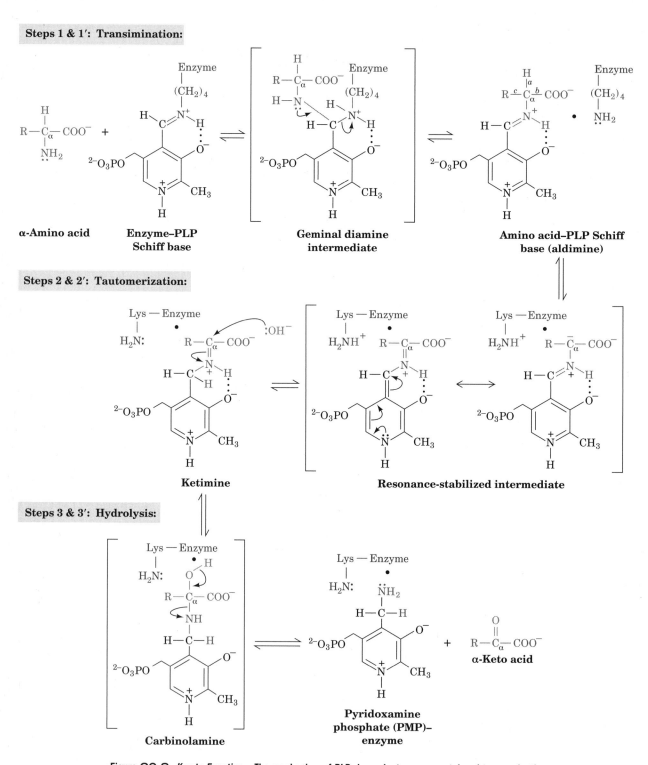

Steps 1 & 1′: Transimination:

α-Amino acid · Enzyme–PLP Schiff base · Geminal diamine intermediate · Amino acid–PLP Schiff base (aldimine)

Steps 2 & 2′: Tautomerization:

Ketimine · Resonance-stabilized intermediate

Steps 3 & 3′: Hydrolysis:

Carbinolamine · Pyridoxamine phosphate (PMP)–enzyme · α-Keto acid

Figure 20-8 *Key to Function.* The mechanism of PLP-dependent enzyme-catalyzed transamination.

GLUTAMATE DEHYDROGENASE

$$^-OOC-CH_2-CH_2-\overset{\overset{NH_3^+}{|}}{\underset{\underset{H}{|}}{C}}-COO^-$$

Glutamate

NAD(P)$^+$ NAD(P)H + H$^+$

$$\left[^-OOC-CH_2-CH_2-\overset{\overset{NH_2^+}{\|}}{C}-COO^- \right]$$

α-Iminoglutarate

H$_2$O   NH$_4^+$

$$^-OOC-CH_2-CH_2-\overset{\overset{O}{\|}}{C}-COO^-$$

α-Ketoglutarate

NH_3

Ammonia

$$H_2N-\overset{\overset{O}{\|}}{C}-NH_2$$

Urea

Uric acid

THE UREA CYCLE

$$NH_3 + HCO_3^- + \ ^-OOC-CH_2-\overset{\overset{NH_3^+}{|}}{CH}-COO^-$$

Aspartate

3 ATP

2 ADP + 2 P$_i$ + AMP + PP$_i$

$$H_2N-\overset{\overset{O}{\|}}{C}-NH_2 + \ ^-OOC-CH=CH-COO^-$$

Urea   **Fumarate**

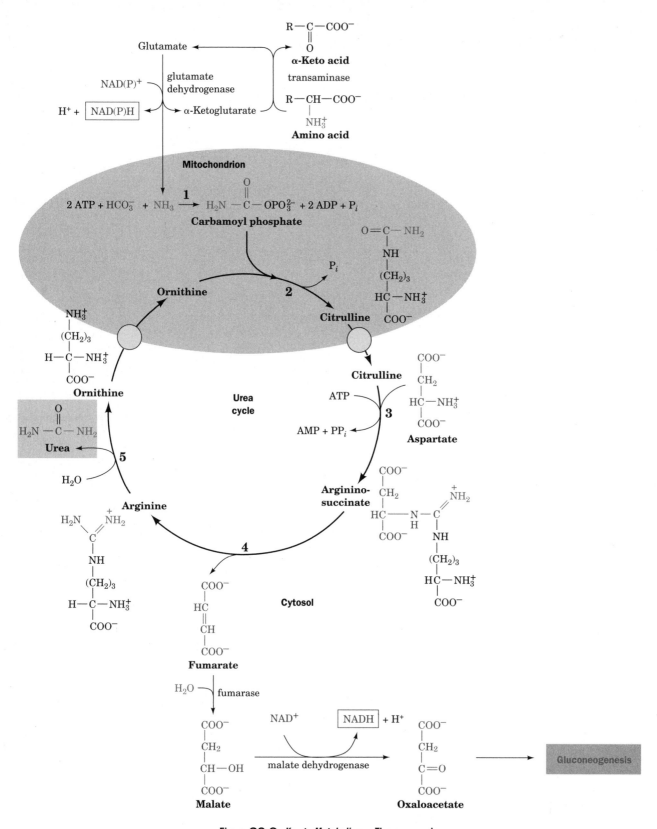

Figure 20-9 *Key to Metabolism.* The urea cycle.

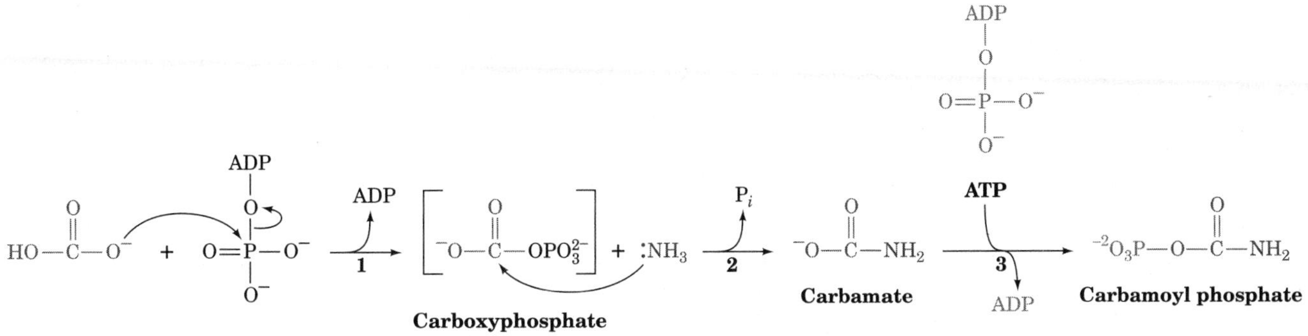

Figure 20-10 The mechanism of action of CPS I.

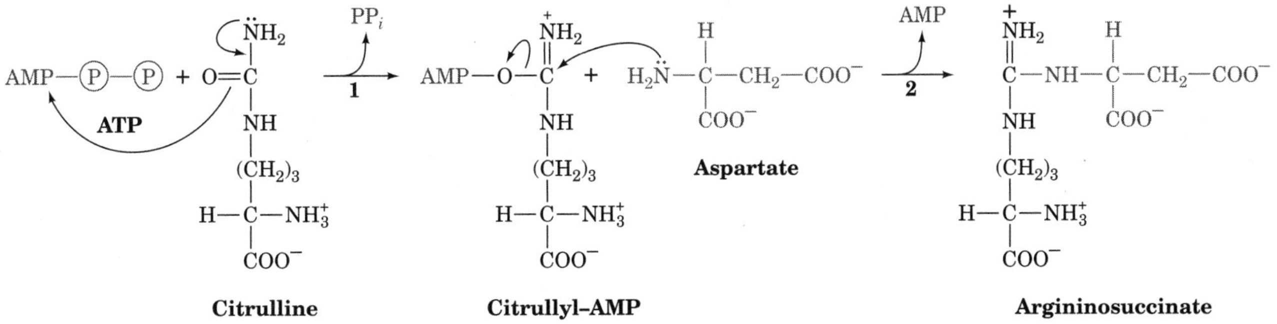

Figure 20-12 The mechanism of action of argininosuccinate synthetase.

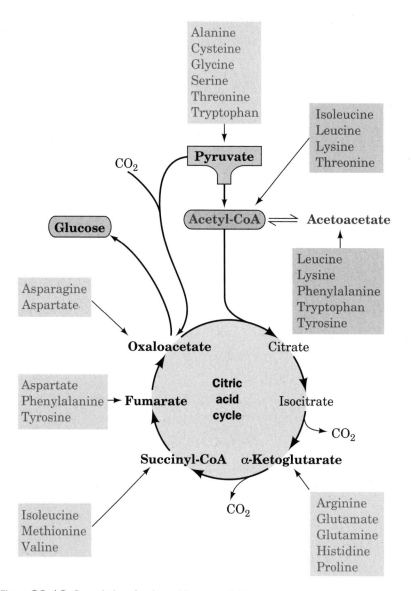

Figure 20-13 Degradation of amino acids to one of seven common metabolic intermediates.

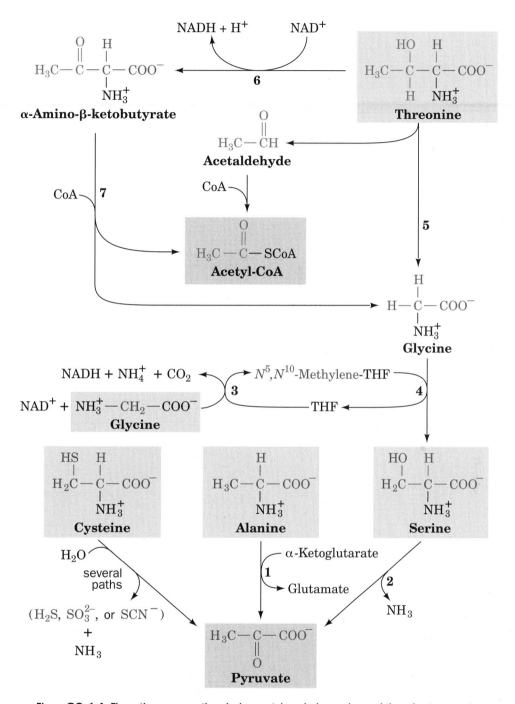

Figure 20-14 The pathways converting alanine, cysteine, glycine, serine, and threonine to pyruvate.

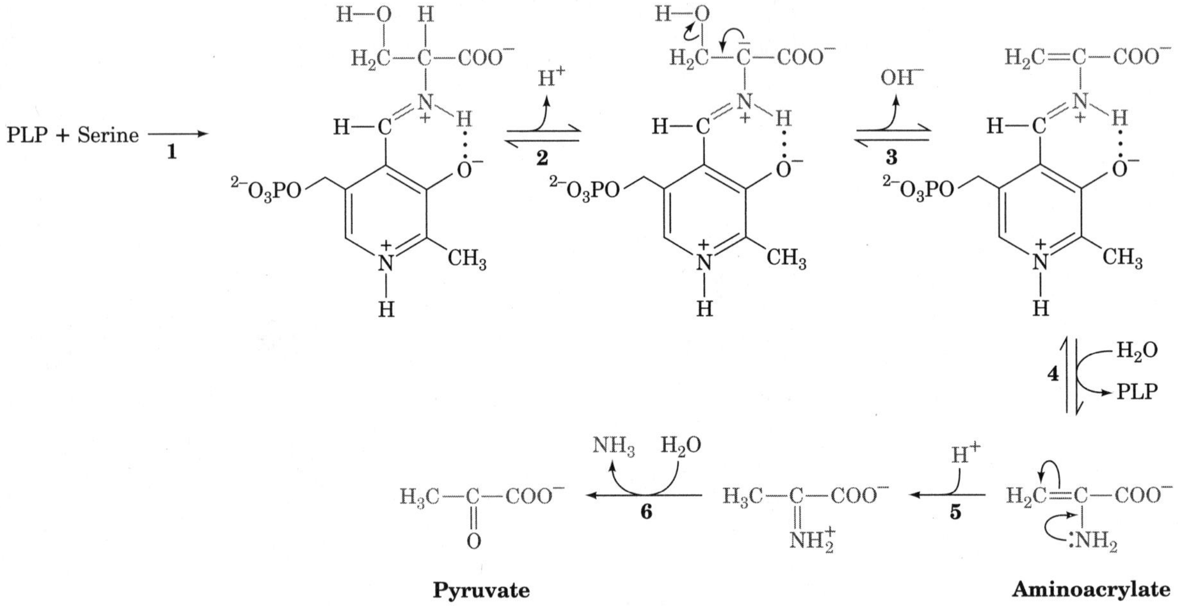

Figure 20-15 The serine dehydratase reaction.

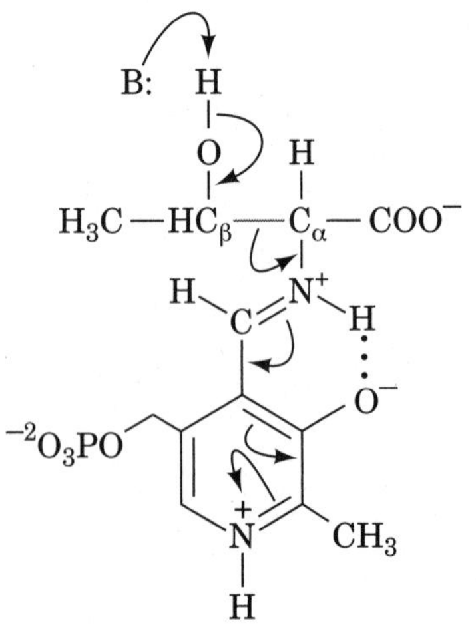

Serine Hydroxymethyltransferase Cleavage of threonine

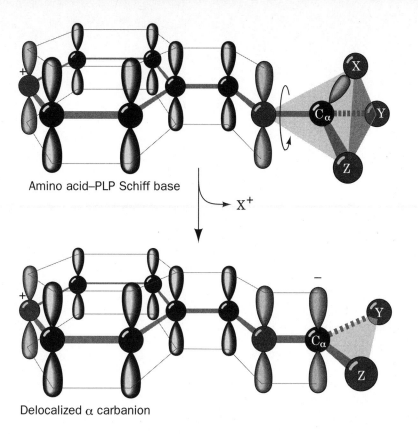

Amino acid–PLP Schiff base

$\longrightarrow X^+$

Delocalized α carbanion

Figure 20-16 The π-orbital framework of a PLP–amino acid Schiff base.

Aspartate

α-Ketoglutarate ⟶ | aminotransferase

Glutamate ⟵

Oxaloacetate

Asparagine

H_2O ⟶ | L-asparaginase

NH_4^+ ⟵

Aspartate

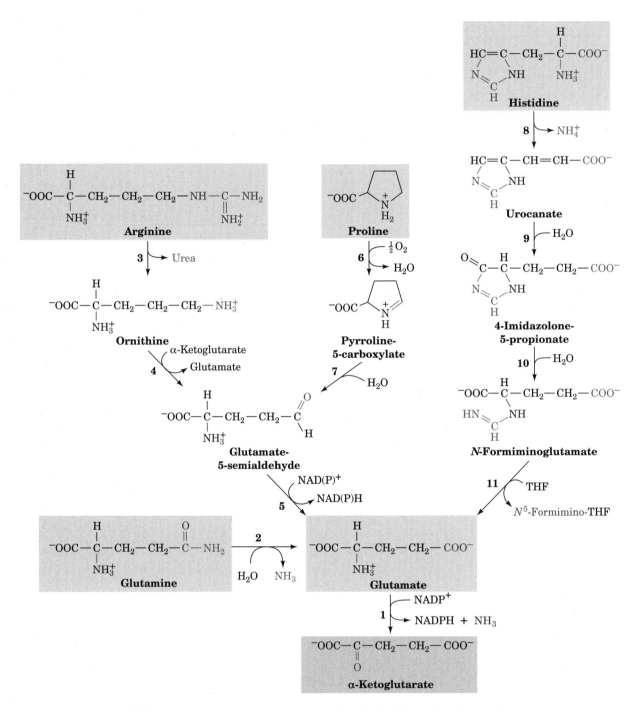

Figure 20-17 The degradation of arginine, glutamate, glutamine, histidine, and proline to α-ketoglutarate.

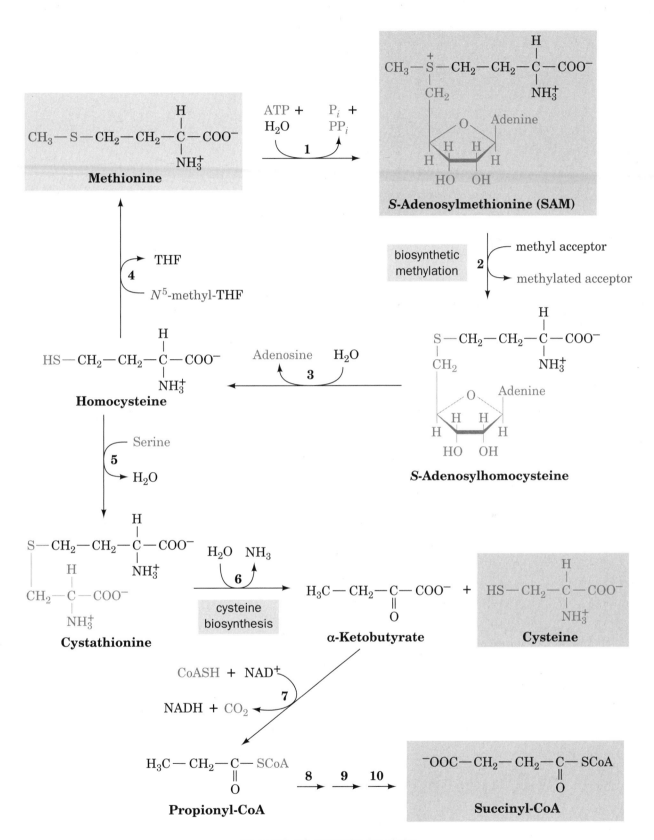

Figure 20-18 Methionine degradation.

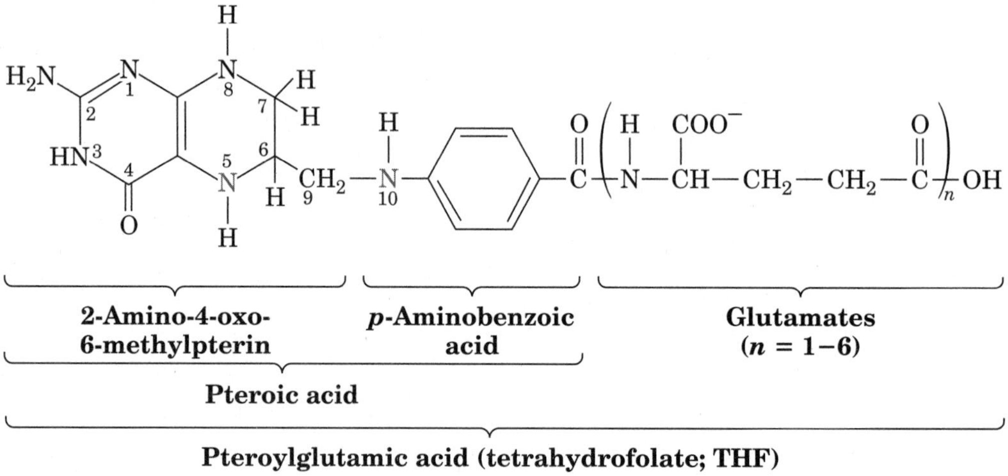
2-Amino-4-oxo-6-methylpterin
p-Aminobenzoic acid
Glutamates (n = 1–6)
Pteroic acid
Pteroylglutamic acid (tetrahydrofolate; THF)

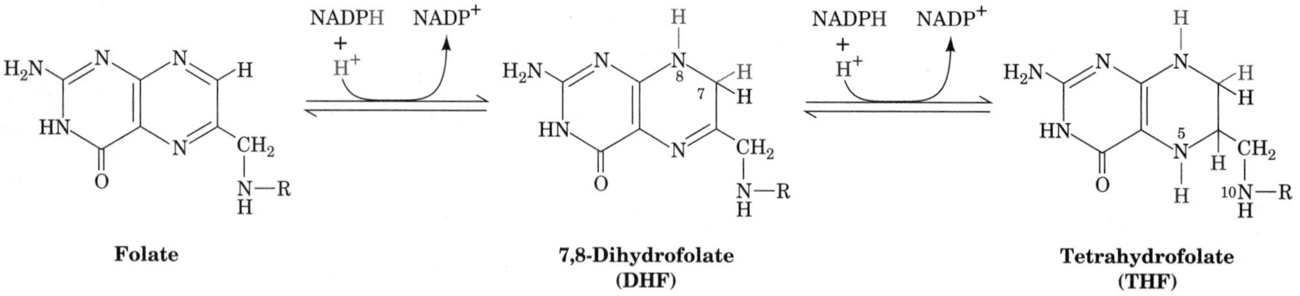
NADPH + H+
NADP+
Folate
7,8-Dihydrofolate (DHF)
Tetrahydrofolate (THF)

Figure 20-19 The two-stage reduction of folate to THF.

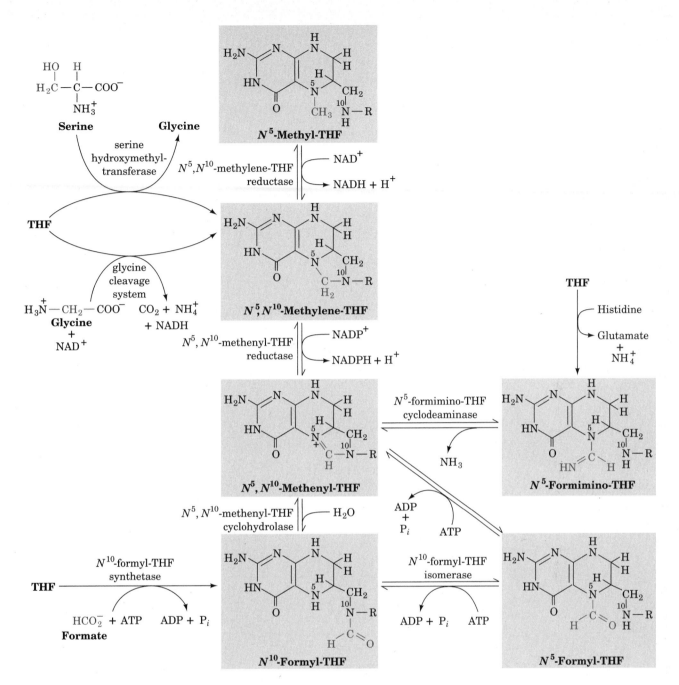

Figure 20-20 Interconversion of the C$_1$ units carried by THF.

Table 20-2 Oxidation Levels of C$_1$ Groups Carried by THF

Oxidation Level	Group Carried	THF Derivative(s)
Methanol	Methyl (—CH$_3$)	N^5-Methyl-THF
Formaldehyde	Methylene (—CH$_2$—)	N^5,N^{10}-Methylene-THF
Formate	Formyl (—CH=O)	N^5-Formyl-THF, N^{10}-formyl-THF
	Formimino (—CH=NH)	N^5-Formimino-THF
	Methenyl (—CH=)	N^5,N^{10}-Methenyl-THF

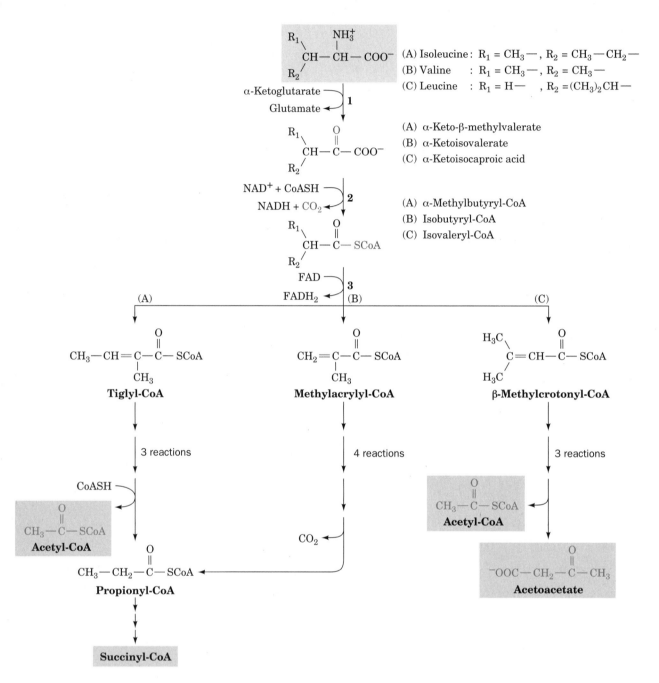

Figure 20-21 The degradation of the branched-chain amino acids.

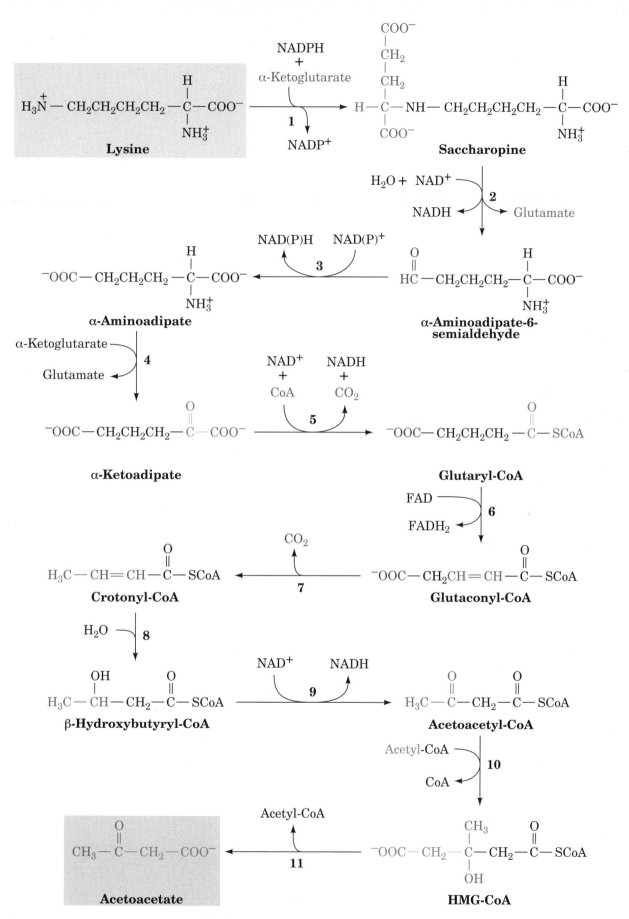

Figure 20-22 The pathway of lysine degradation in mammalian liver.

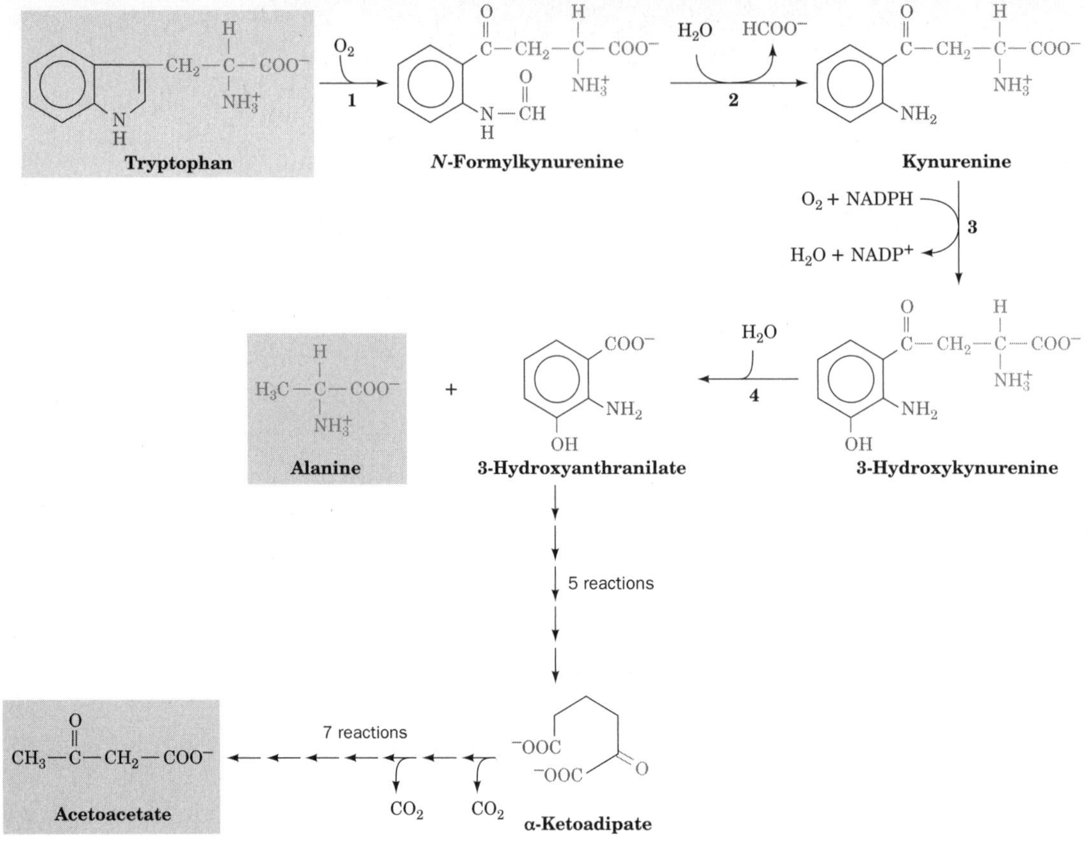

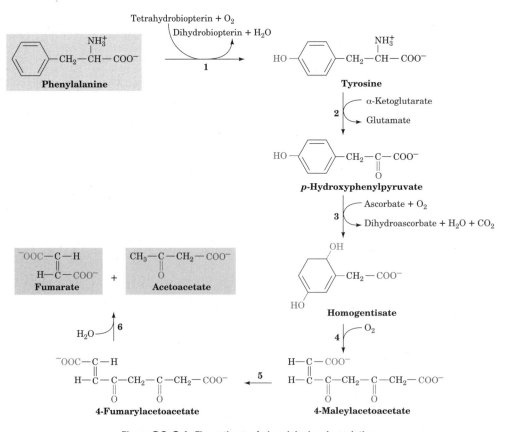

Figure 20-23 The pathway of tryptophan degradation.

Figure 20-24 The pathway of phenylalanine degradation.

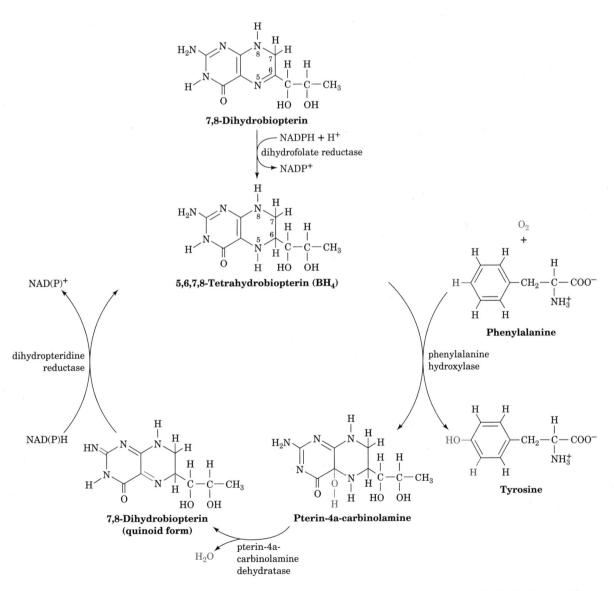

Figure 20-25 The pteridine ring nucleus of biopterin and folate.

Biopterin: R =

Folate: R =

7,8-Dihydrobiopterin

NADPH + H⁺
dihydrofolate reductase
NADP⁺

5,6,7,8-Tetrahydrobiopterin (BH₄)

O₂
+

Phenylalanine

NAD(P)⁺

dihydropteridine
reductase

NAD(P)H

phenylalanine
hydroxylase

Tyrosine

**7,8-Dihydrobiopterin
(quinoid form)**

Pterin-4a-carbinolamine

H₂O

pterin-4a-
carbinolamine
dehydratase

Figure 20-26 The formation, utilization, and regeneration of 5,6,7,8-tetrahydrobiopterin in the phenylalanine hydroxylase reaction.

289

Table 20-3 Essential and Nonessential Amino Acids in Humans

Essential	Nonessential
Arginine[a]	Alanine
Histidine	Asparagine
Isoleucine	Aspartate
Leucine	Cysteine
Lysine	Glutamate
Methionine	Glutamine
Phenylalanine	Glycine
Threonine	Proline
Tryptophan	Serine
Valine	Tyrosine

[a]Although mammals synthesize arginine, they cleave most of it to form urea (Section 20-3A).

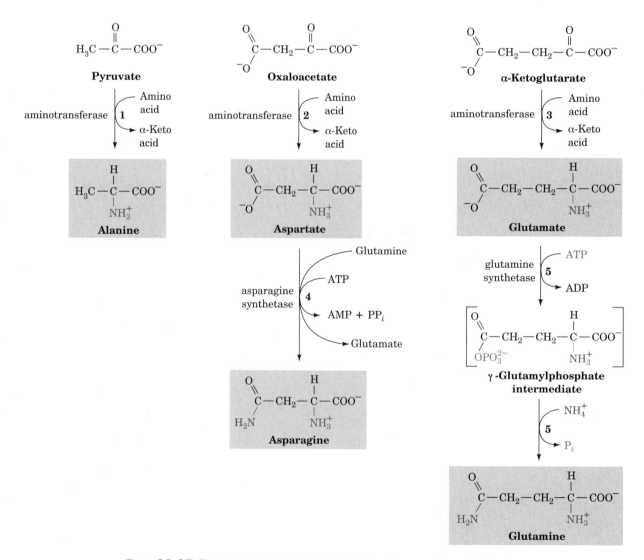

Figure 20-27 The syntheses of alanine, aspartate, glutamate, asparagine, and glutamine.

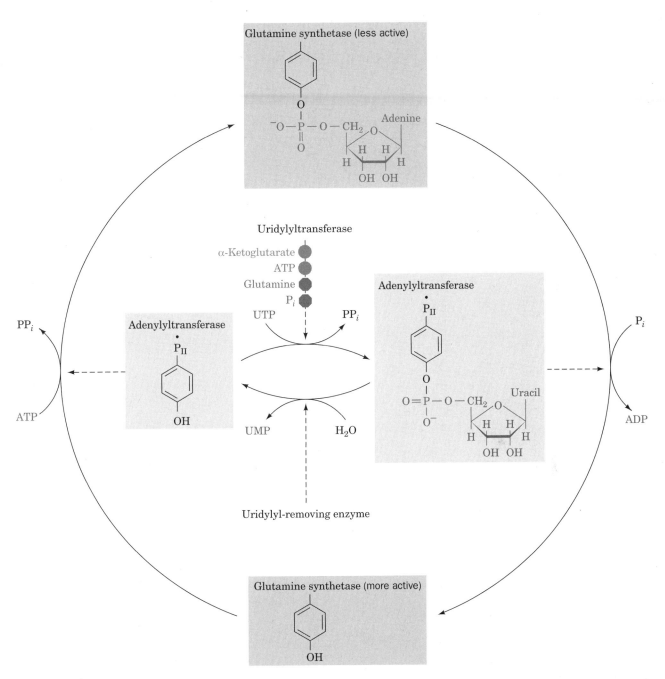

Figure 20-29 The regulation of bacterial glutamine synthetase.

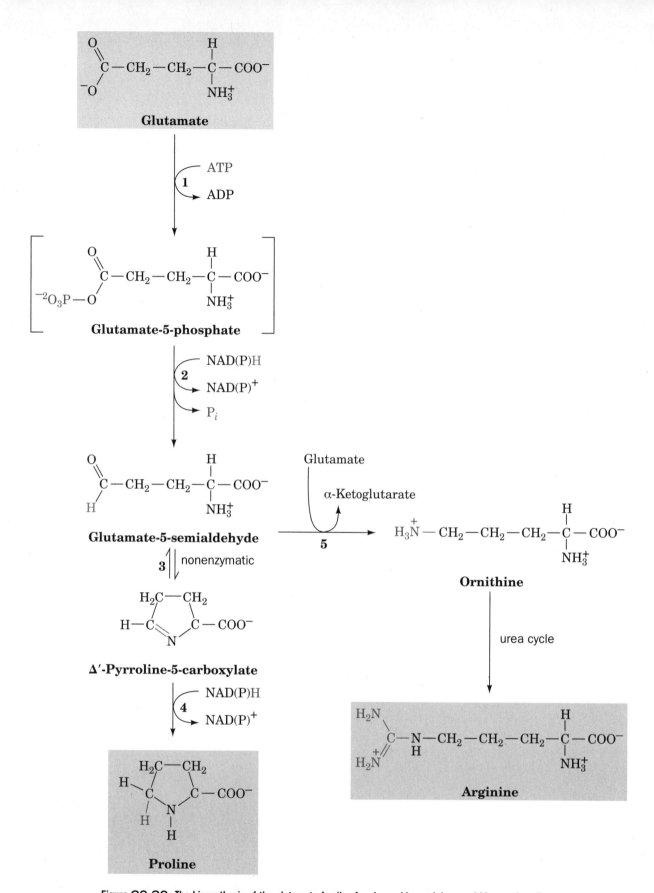

Figure 20-30 The biosynthesis of the glutamate family of amino acids: arginine, ornithine, and proline.

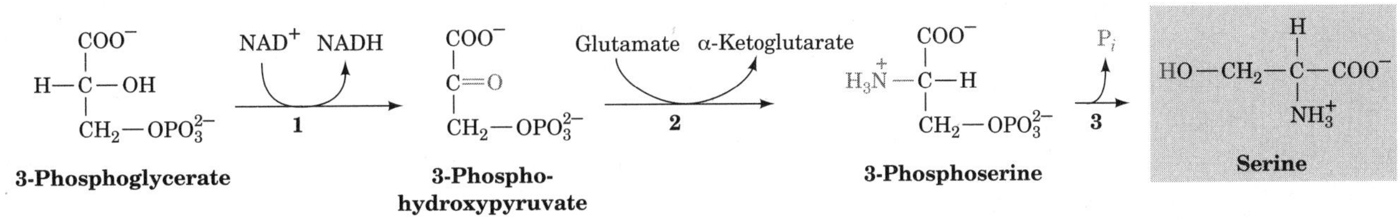

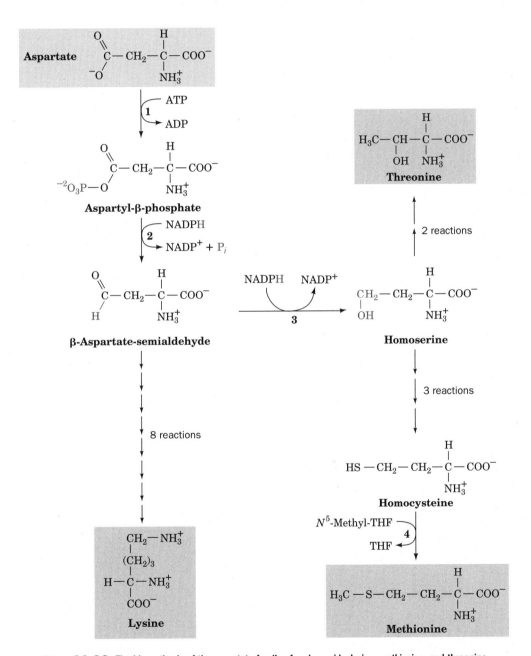

Figure 20-31 The conversion of 3-phosphoglycerate to serine.

Figure 20-32 The biosynthesis of the aspartate family of amino acids: lysine, methionine, and threonine.

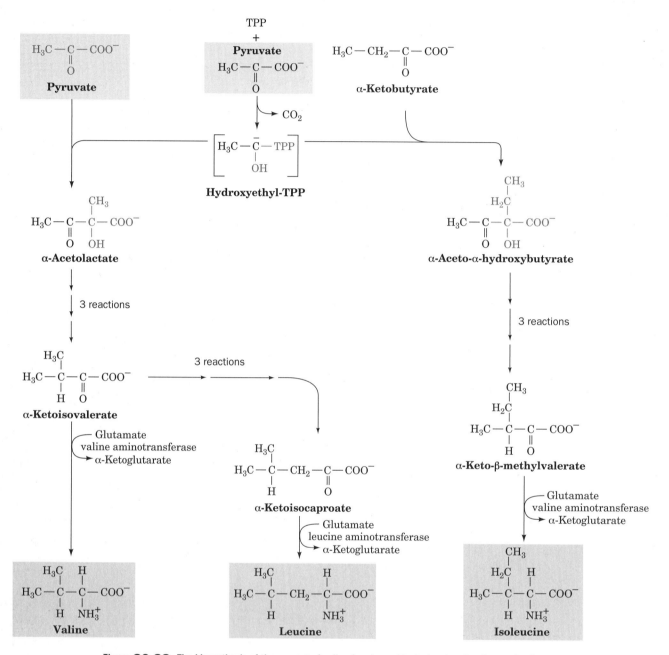

Figure 20-33 The biosynthesis of the pyruvate family of amino acids: isoleucine, leucine, and valine.

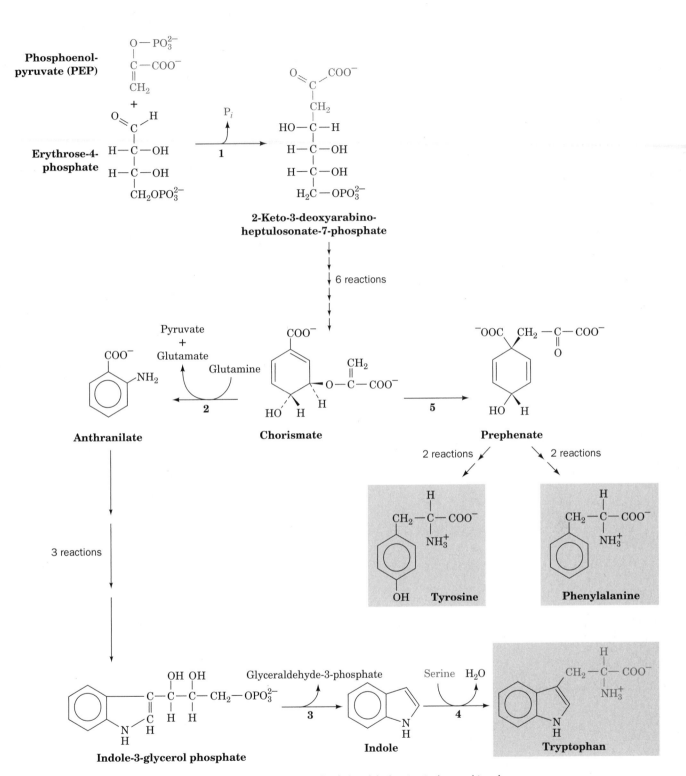

Figure 20-34 The biosynthesis of phenylalanine, tryptophan, and tyrosine.

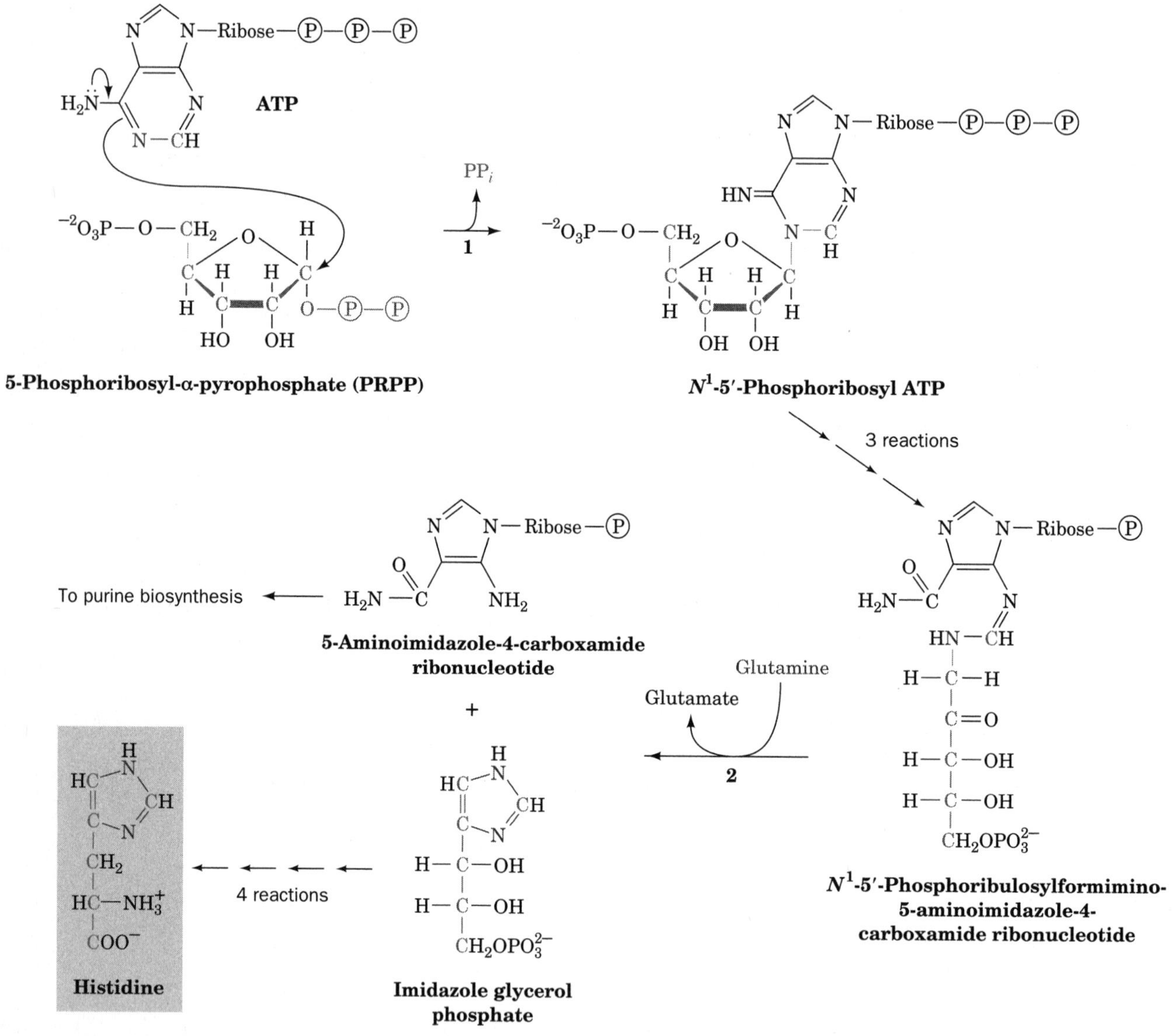

ATP

$^{-2}O_3P$—O—CH_2

HO OH

5-Phosphoribosyl-α-pyrophosphate (PRPP)

PP$_i$

1

$^{-2}O_3P$—O—CH_2

OH OH

N^1-5′-Phosphoribosyl ATP

3 reactions

H_2N—C

HN—CH

H—C—H

C=O

H—C—OH

H—C—OH

$CH_2OPO_3^{2-}$

N^1-5′-Phosphoribulosylformimino-
5-aminoimidazole-4-
carboxamide ribonucleotide

To purine biosynthesis ◄— H_2N—C NH_2

N—Ribose—Ⓟ

5-Aminoimidazole-4-carboxamide
ribonucleotide

+

HC

CH

H—C—OH

H—C—OH

$CH_2OPO_3^{2-}$

Imidazole glycerol
phosphate

Glutamine

Glutamate

2

HC

CH

CH_2

HC—NH_3^+

COO^-

Histidine

4 reactions

Figure 20-36 The biosynthesis of histidine.

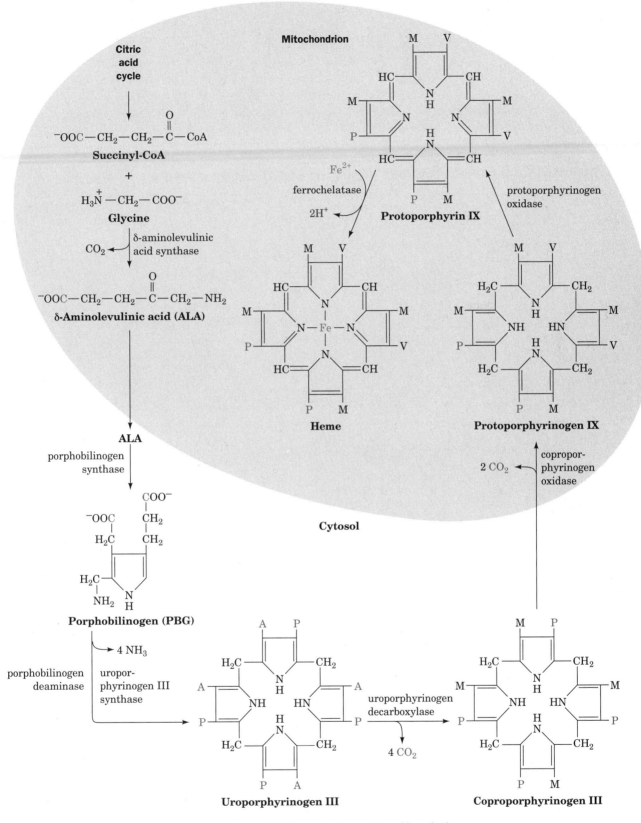

Figure 20-37 The pathway of heme biosynthesis.

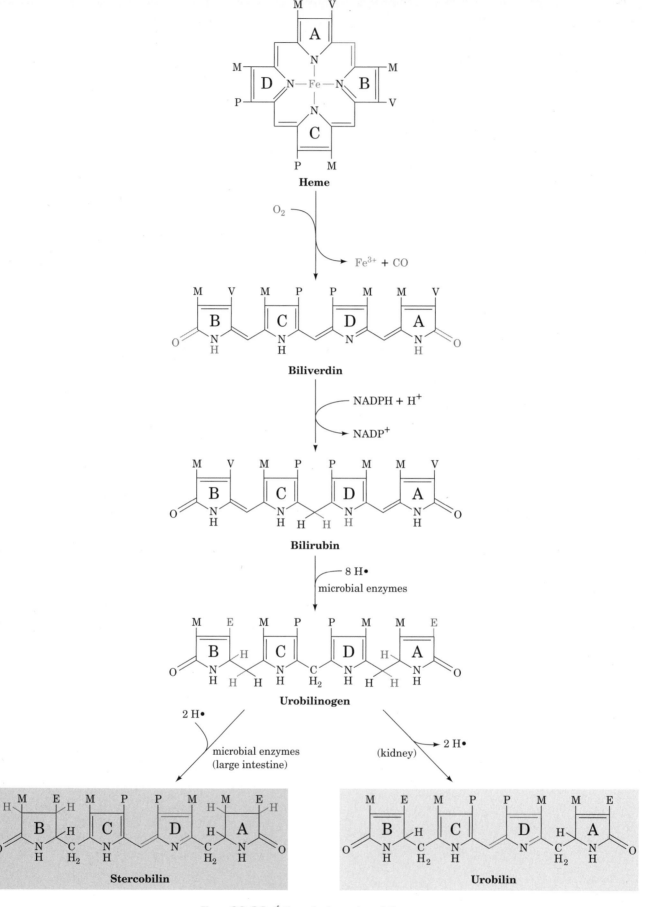

Figure 20-38 Pathway for heme degradation.

HO—⟨benzene ring with HO⟩—C(X)(H)—CH₂—NH—R

X = OH, R = CH₃ **Epinephrine (Adrenalin)**
X = OH, R = H   **Norepinephrine**
X = H,   R = H   **Dopamine**

HO—⟨indole ring structure⟩—CH₂—CH₂—NH₃⁺

**Serotonin
(5-hydroxytryptamine)**

⁻OOC—CH₂—CH₂—CH₂—NH₃⁺

γ-Aminobutyric acid (GABA)

⟨imidazole ring structure⟩—CH₂—CH₂—NH₃⁺ **Histamine**

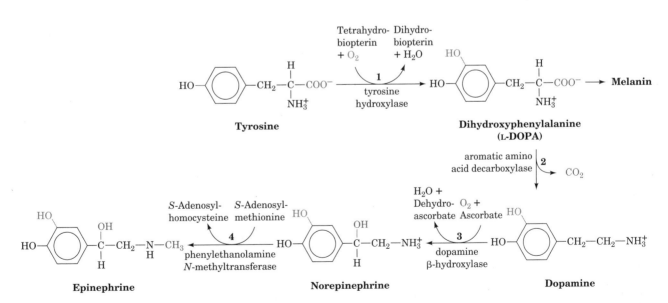

Figure 20-39 The sequential synthesis of L-DOPA, dopamine, norepinephrine, and epinephrine from tyrosine.

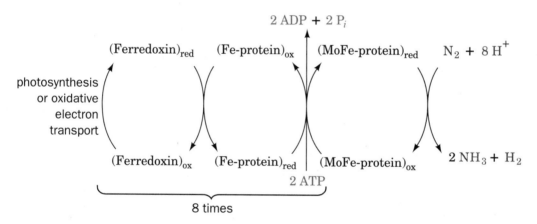

Figure 20-43 The flow of electrons in the nitrogenase-catalyzed reduction of N₂.

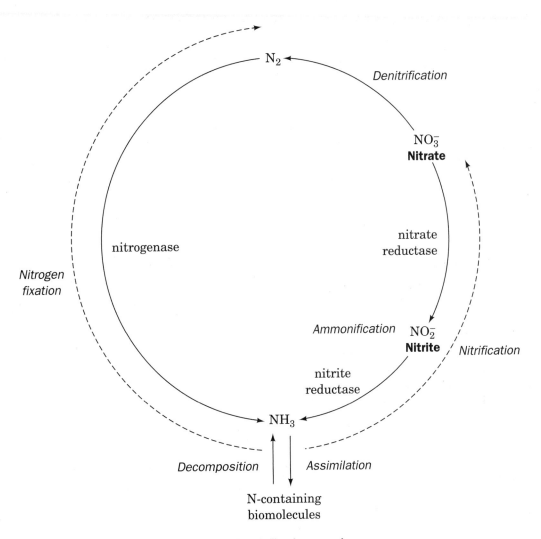

Figure 20-44 The nitrogen cycle.

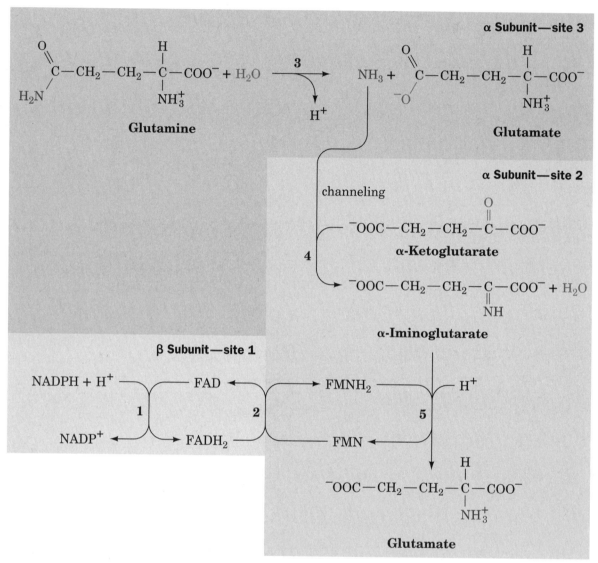

Overall: NADPH + H$^+$ + glutamine + α-ketoglutarate $\longrightarrow$ 2 glutamate + NADP$^+$

Figure 20-45 The glutamate synthase reaction.

CHAPTER 21

Mammalian Fuel Metabolism: Integration and Regulation

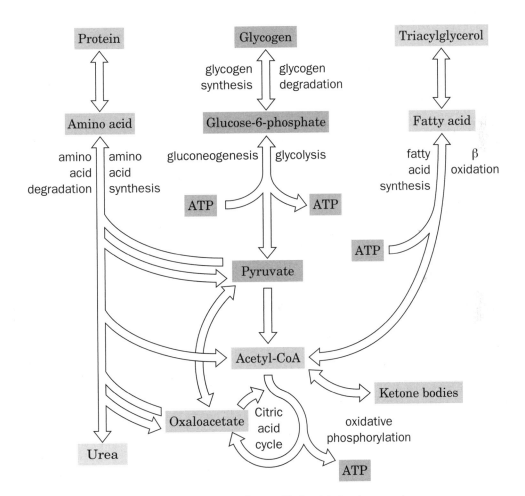

Figure 21-1 The major pathways of fuel metabolism in mammals.

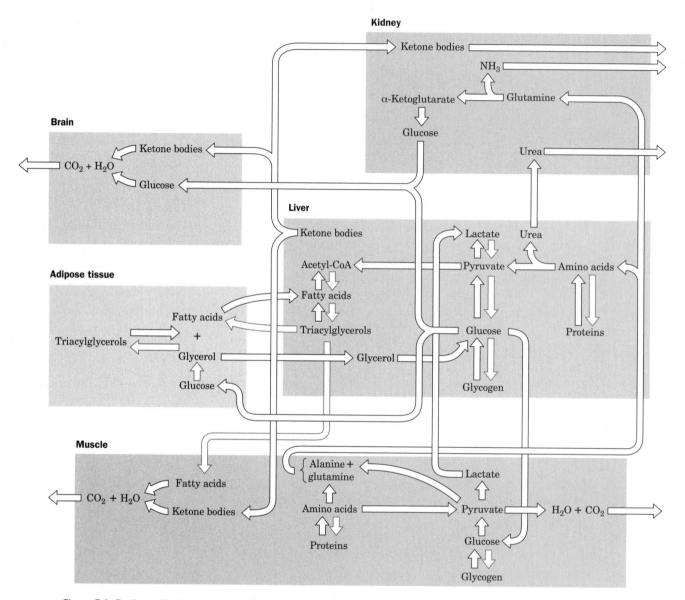

Figure 21-2 *Key to Metabolism.* The metabolic interrelationships among brain, adipose tissue, muscle, liver, and kidney.

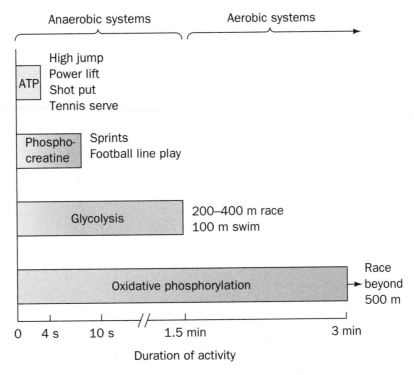

Figure 21-3 Source of ATP during exercise in humans.

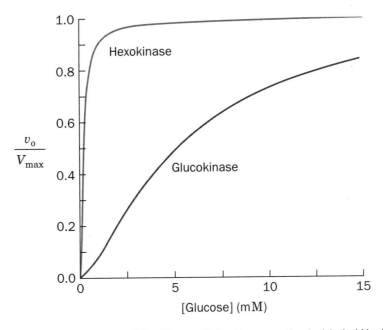

Figure 21-4 Relative enzymatic activities of hexokinase and glucokinase over the physiological blood glucose range.

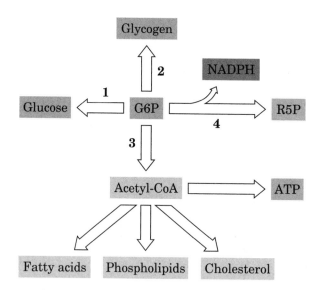

Figure 21-5 Metabolic fate of glucose-6-phosphate in liver.

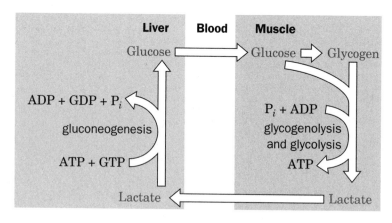

Figure 21-6 The Cori cycle.

$$\underset{\text{Amino acid}}{R-\overset{\overset{+}{N}H_3}{\underset{|}{CH}}-COO^-} + \underset{\text{Pyruvate}}{H_3C-\overset{O}{\overset{\|}{C}}-COO^-} \rightleftharpoons \underset{\alpha\text{-Keto acid}}{R-\overset{O}{\overset{\|}{C}}-COO^-} + \underset{\text{Alanine}}{H_3C-\overset{\overset{+}{N}H_3}{\underset{|}{CH}}-COO^-}$$

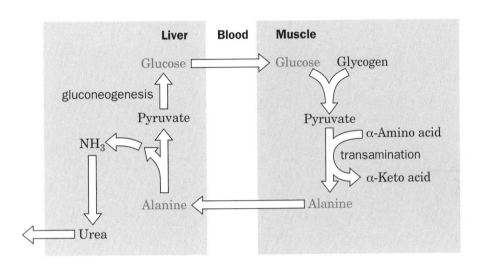

Figure 21-7 The glucose–alanine cycle.

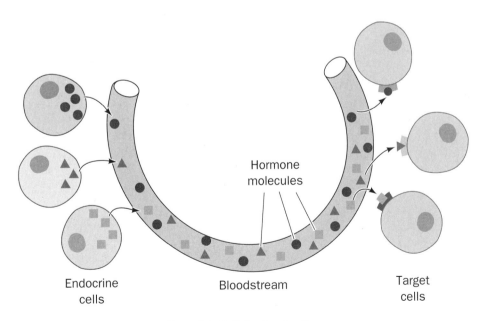

Figure 21-8 Endocrine signaling.

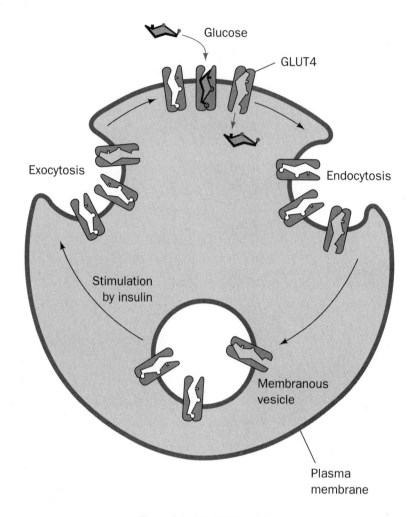

Figure 21-10 GLUT4 activity.

Table 21-1 Hormonal Effects on Fuel Metabolism

Tissue	Insulin	Glucagon	Epinephrine
Muscle	↑ Glucose uptake ↑ Glycogen synthesis	No effect	↑ Glycogenolysis
Adipose tissue	↑ Glucose uptake ↑ Lipogenesis ↓ Lipolysis	↑ Lipolysis	↑ Lipolysis
Liver	↑ Glycogen synthesis ↑ Lipogenesis ↓ Gluconeogenesis	↓ Glycogen synthesis ↑ Glycogenolysis	↓ Glycogen synthesis ↑ Glycogenolysis ↑ Gluconeogenesis

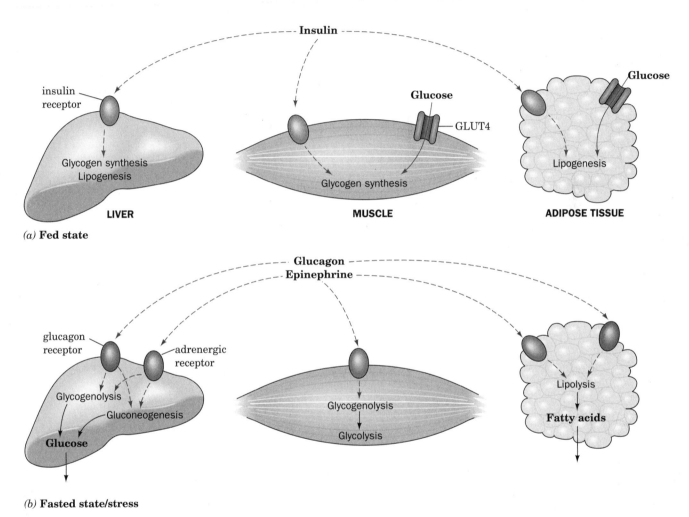

(a) **Fed state**

(b) **Fasted state/stress**

Figure 21-11 Overview of hormonal control of fuel metabolism.

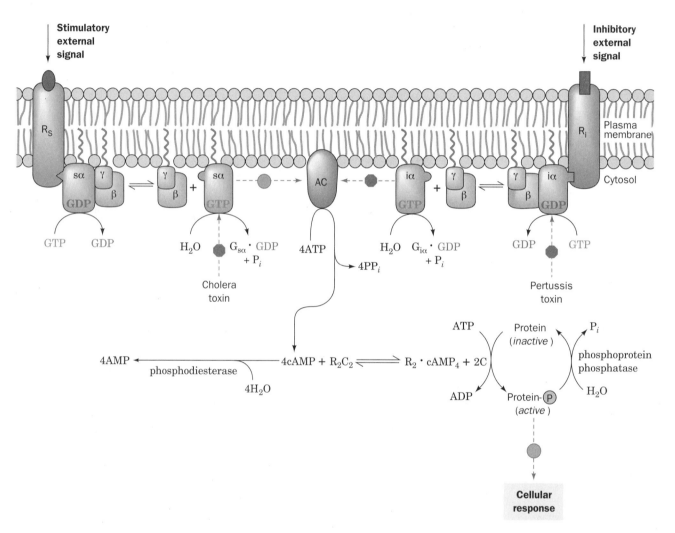

Figure 21-15 *Key to Function.* The adenylate cyclase signaling system.

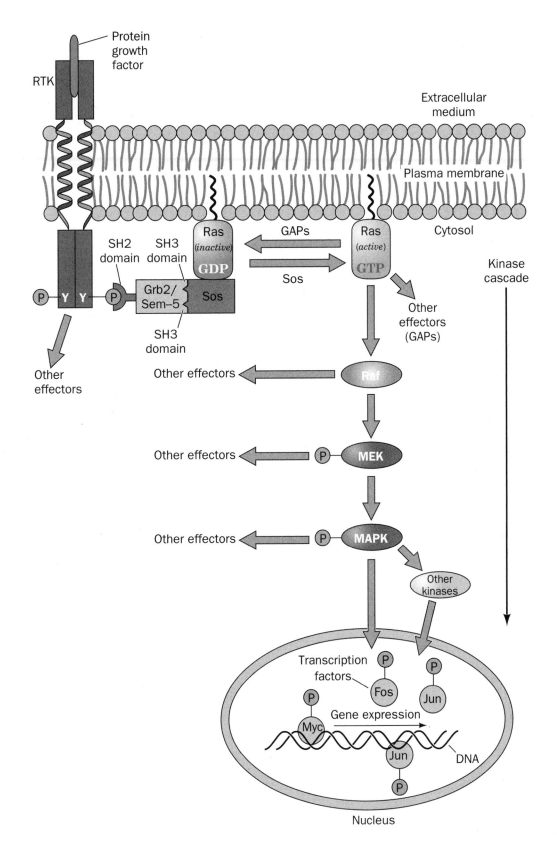

Figure 21-19 The Ras signaling cascade.

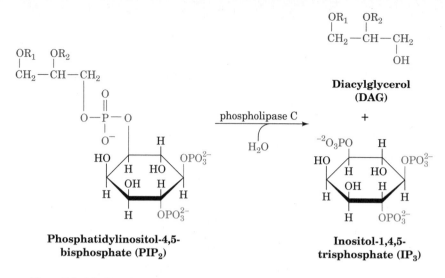

Figure 21-23 Phosphatidylinositol-4,5-bisphosphate (PIP$_2$) and its hydrolysis products.

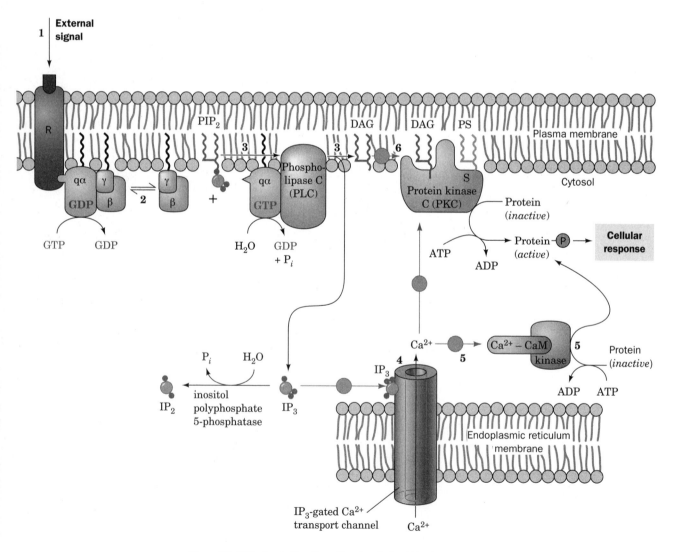

Figure 21-22 *Key to Function.* The phosphoinositide signaling system.

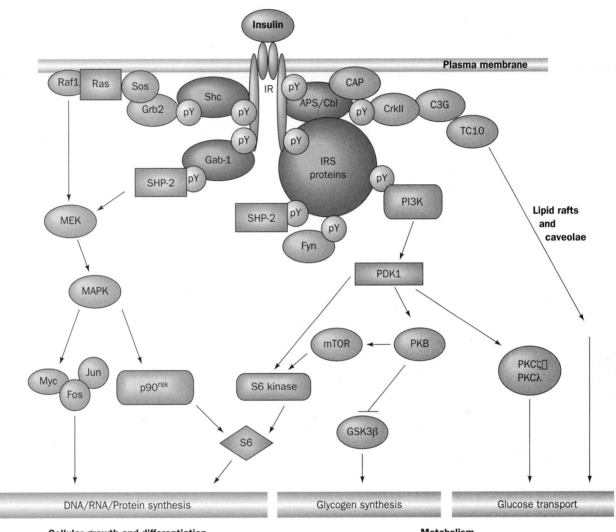

Figure 21-25 Insulin signal transduction.

Table 21-2 Fuel Reserves for a Normal 70-kg Man

Fuel	Mass (kg)	Calories[a]
Tissues		
Fat (adipose triacyglycerols)	15	141,000
Protein (mainly muscle)	6	24,000
Glycogen (muscle)	0.150	600
Glycogen (liver)	0.075	300
Circulating fuels		
Glucose (extracellular fluid)	0.020	80
Free fatty acids (plasma)	0.0003	3
Triacylglycerols (plasma)	0.003	30
Total		***166,000***

[a]1 (dieter's) Calorie = 1 kcal = 4.184 kJ.

Source: Cahill, G.E., Jr., *New Engl. J. Med.* **282,** 669 (1970).

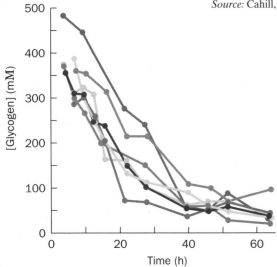

Figure 21-26 Liver glycogen depletion during fasting.

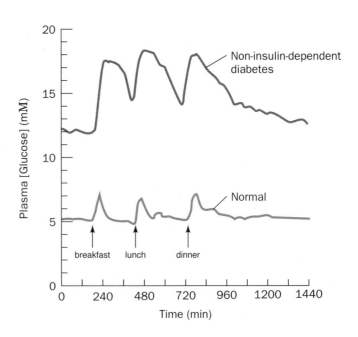

Figure 21-28 Twenty-four-hour plasma glucose profiles in normal and non-insulin-dependent diabetic subjects.

CHAPTER 22

Nucleotide Metabolism

Uric acid

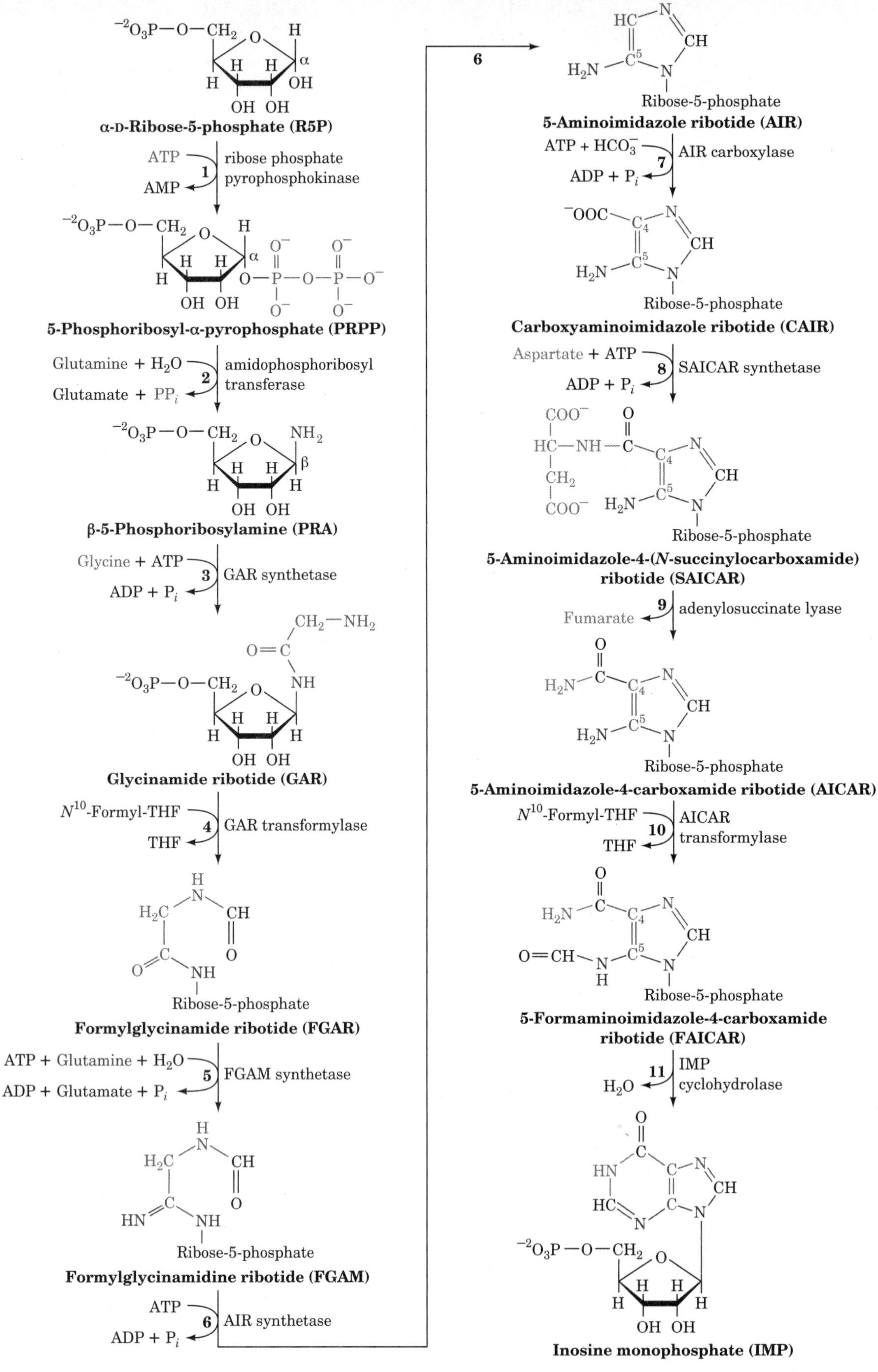

Figure 22-1 *Key to Metabolism.* The metabolic pathway for the *de novo* biosynthesis of IMP.

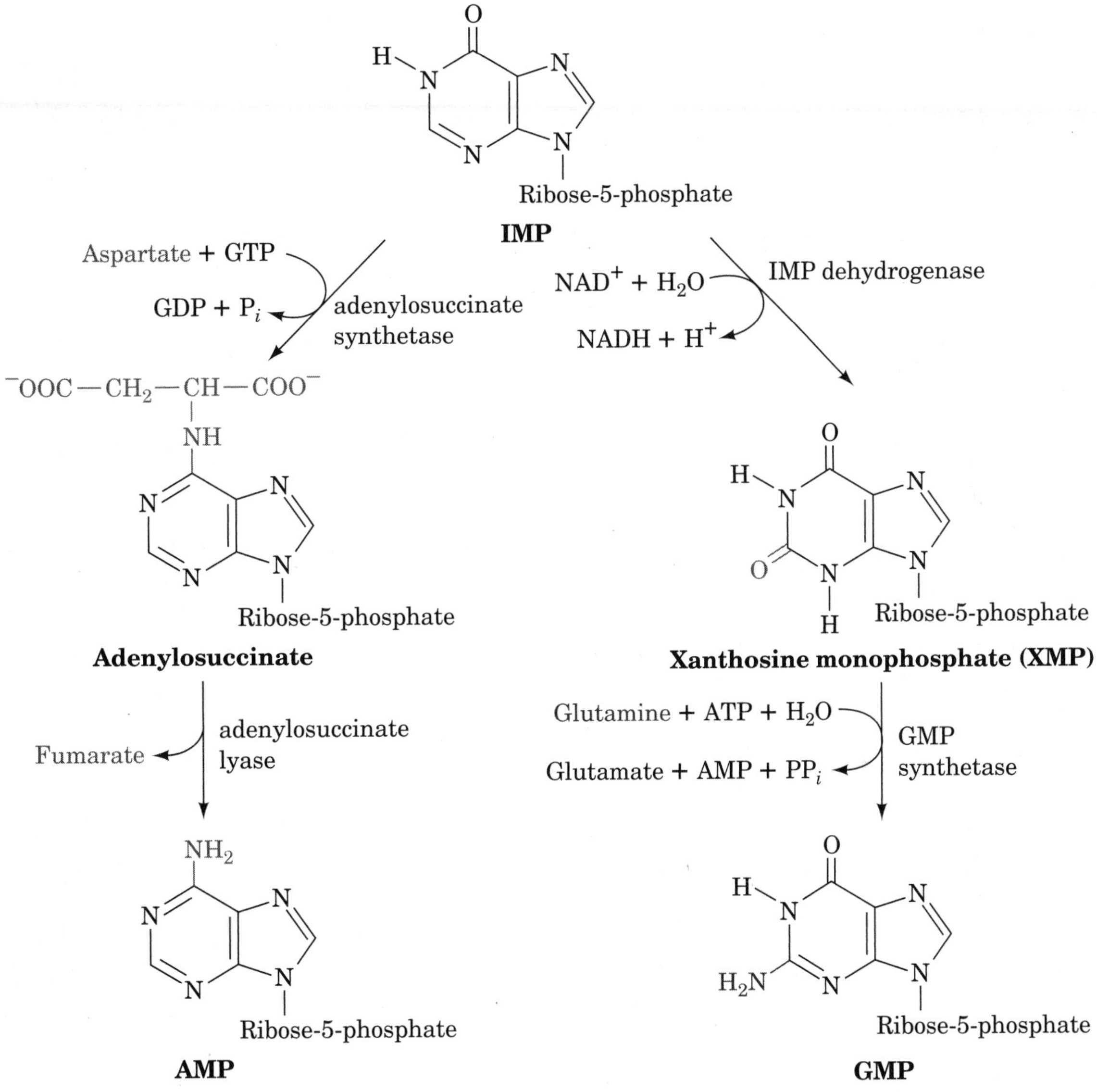

Figure 22-3 Conversion of IMP to AMP or GMP in separate two-reaction pathways.

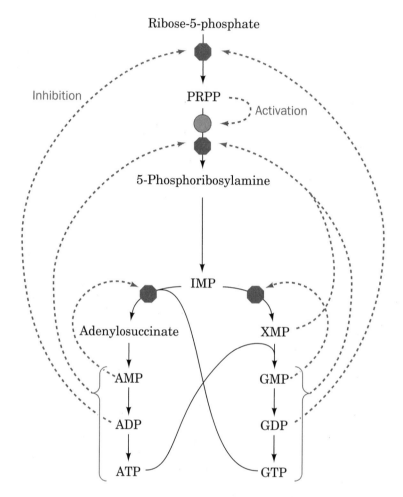

Figure 22-4 Control of the purine biosynthesis pathway.

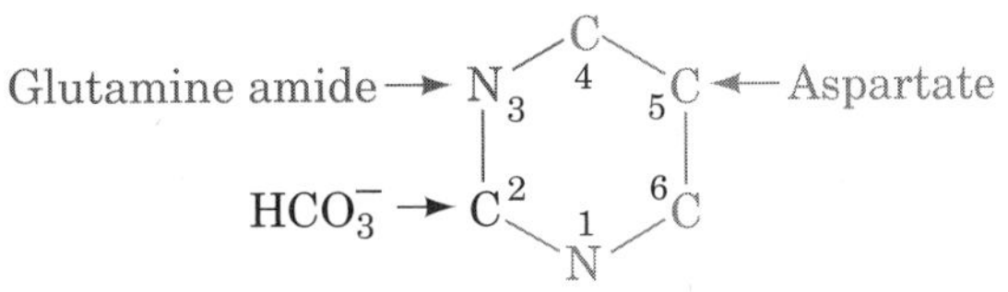

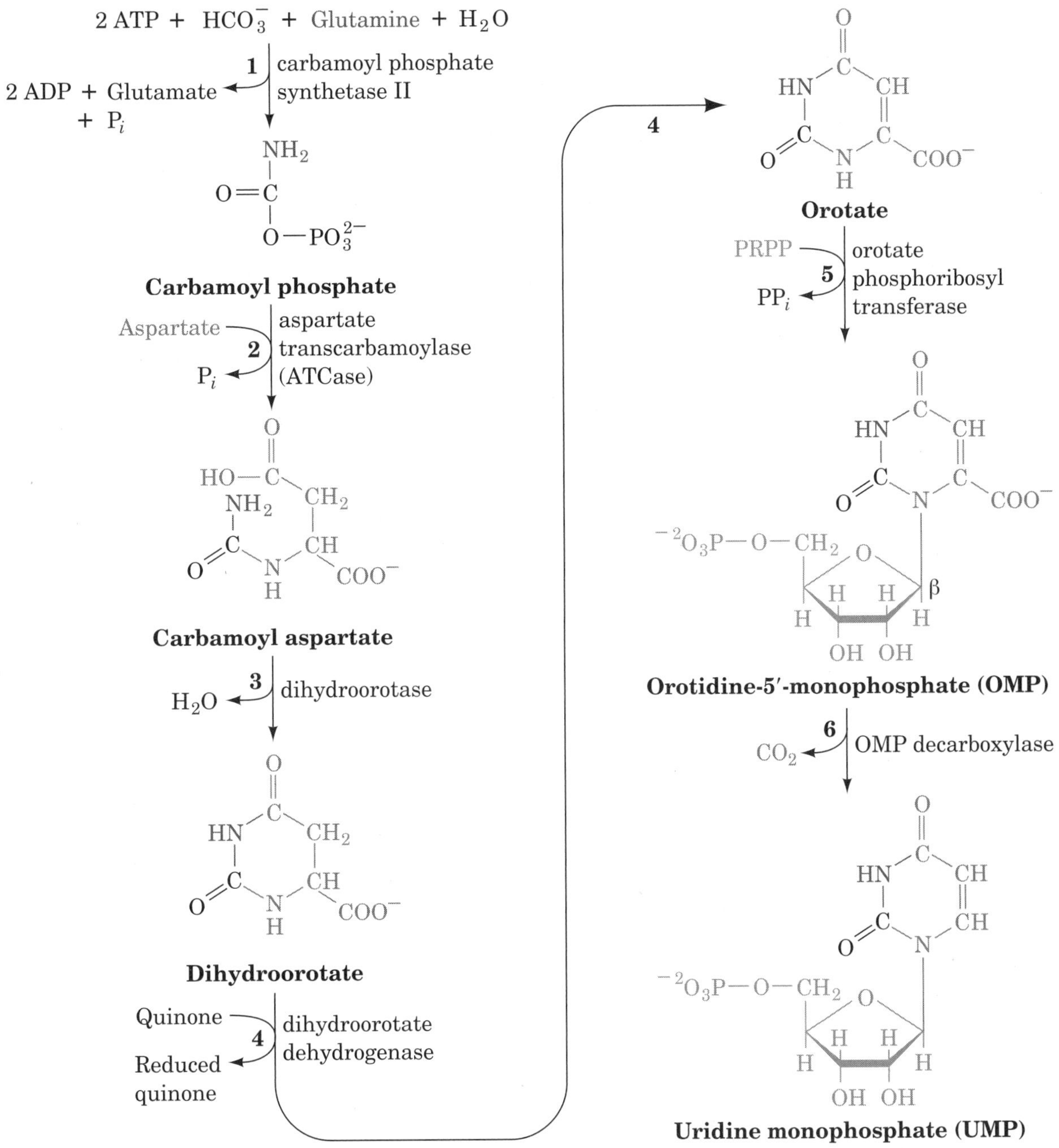

Glutamine amide → N_3

C_4

C_5 ← Aspartate

HCO_3^- → C_2

N_1

C_6

$2 ATP + HCO_3^- + Glutamine + H_2O$

1 carbamoyl phosphate synthetase II

$2 ADP + Glutamate + P_i$

NH_2

$O = C$

$O - PO_3^{2-}$

Carbamoyl phosphate

Aspartate

2 aspartate transcarbamoylase (ATCase)

P_i

$HO - C$

NH_2

CH_2

$O = C$

CH

N H

COO^-

Carbamoyl aspartate

H_2O **3** dihydroorotase

O

C

HN CH_2

$O = C$ CH

N H COO^-

Dihydroorotate

Quinone

4 dihydroorotate dehydrogenase

Reduced quinone

4

O

C

HN CH

$O = C$ C COO^-

N H

Orotate

PRPP **5** orotate phosphoribosyl transferase

PP_i

O

C

HN CH

$O = C$ C COO^-

N

$^{-2}O_3P - O - CH_2$ O

H H β

H H

OH OH

Orotidine-5′-monophosphate (OMP)

CO_2 **6** OMP decarboxylase

O

C

HN CH

$O = C$ CH

N

$^{-2}O_3P - O - CH_2$ O

H H

H H

OH OH

Uridine monophosphate (UMP)

Figure 22-5 *Key to Metabolism.* The *de novo* synthesis of UMP.

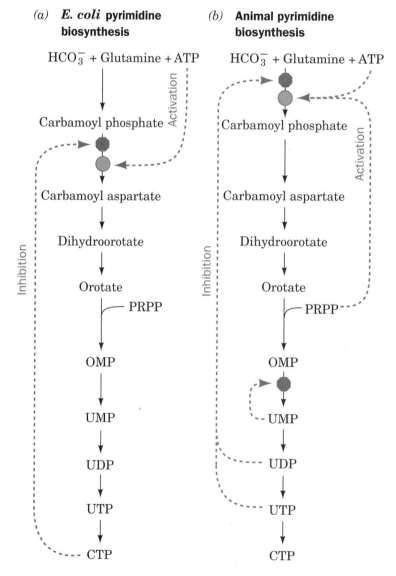

Figure 22-7 The synthesis of CTP from UTP.

(a) *E. coli* **pyrimidine biosynthesis**

HCO$_3^-$ + Glutamine + ATP

↓

Carbamoyl phosphate

↓

Carbamoyl aspartate

↓

Dihydroorotate

↓

Orotate — PRPP

↓

OMP

↓

UMP

↓

UDP

↓

UTP

↓

CTP

(b) **Animal pyrimidine biosynthesis**

HCO$_3^-$ + Glutamine + ATP

↓

Carbamoyl phosphate

↓

Carbamoyl aspartate

↓

Dihydroorotate

↓

Orotate — PRPP

↓

OMP

↓

UMP

↓

UDP

↓

UTP

↓

CTP

Figure 22-8 Regulation of pyrimidine biosynthesis.

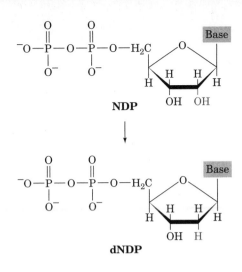

NDP

dNDP

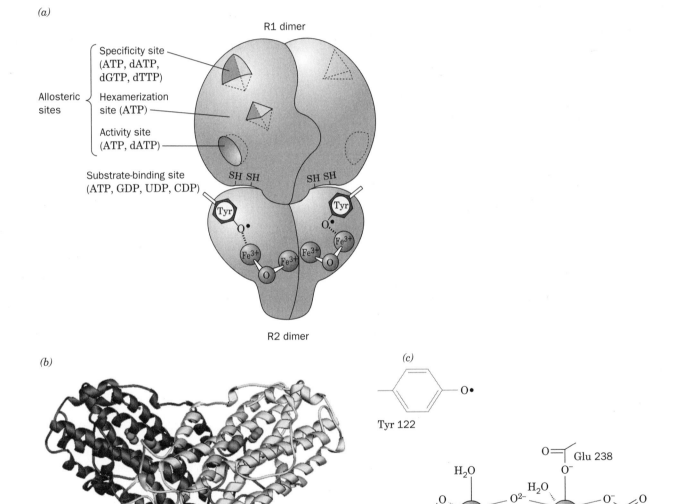

Figure 22-9 Class I ribonucleotide reductase from *E. coli.*

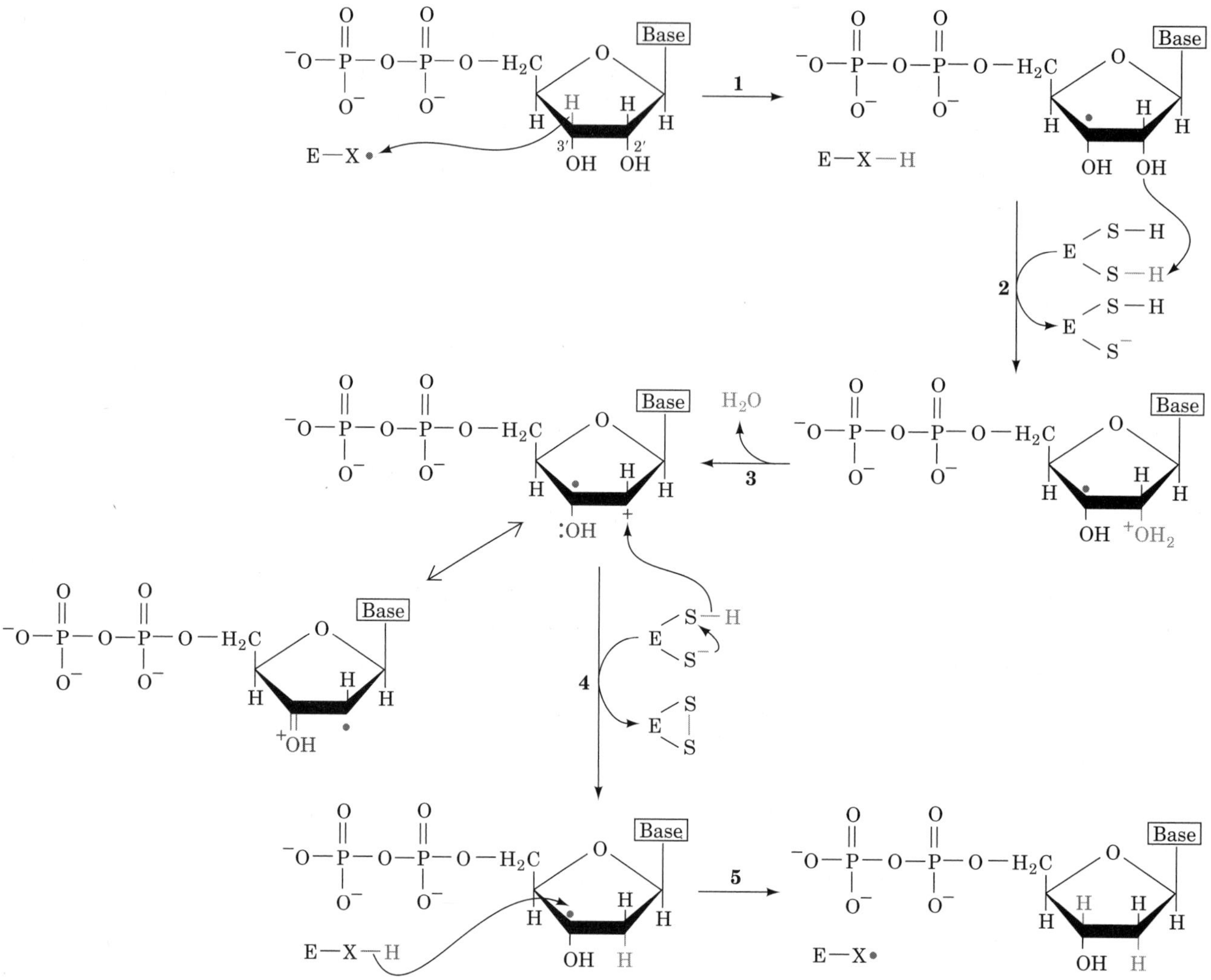

Figure 22-10 Enzymatic mechanism of ribonucleotide reductase.

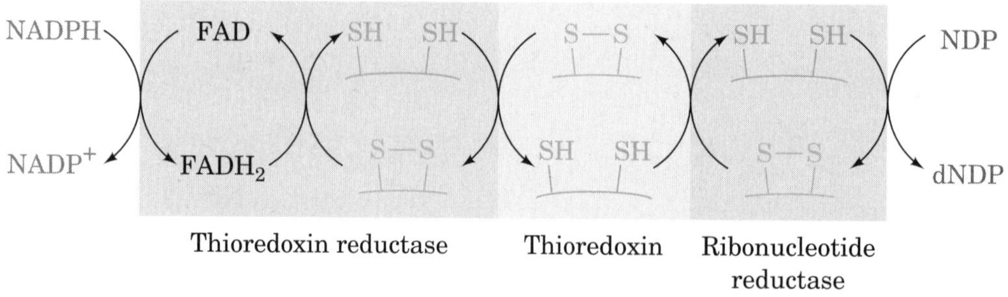

Figure 22-12 An electron-transfer pathway for nucleoside diphosphate (NDP) reduction.

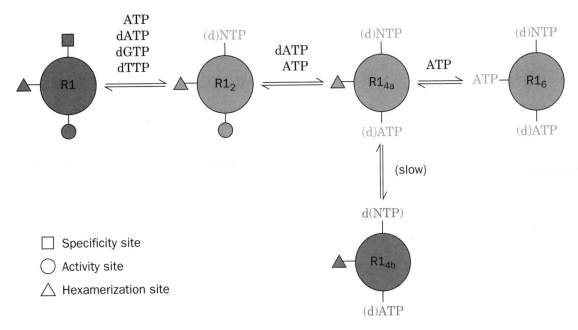

Figure 22-13 Regulation of ribonucleotide reductase.

□ Specificity site
○ Activity site
△ Hexamerization site

dUMP N^5, N^{10}**-Methylenetetrahydrofolate**

dTMP **Dihydrofolate**

$$R = \underset{}{\overset{O}{\underset{}{-\!\!\!\!\!\!\!\!-}}} \; - \; \overset{O}{\underset{}{C}} \left(\overset{H}{\underset{}{N}} - \overset{COO^-}{\underset{}{CH}} - CH_2 - CH_2 - \overset{O}{\underset{}{C}} \right)_n O^- ; \; n = 1\text{--}6$$

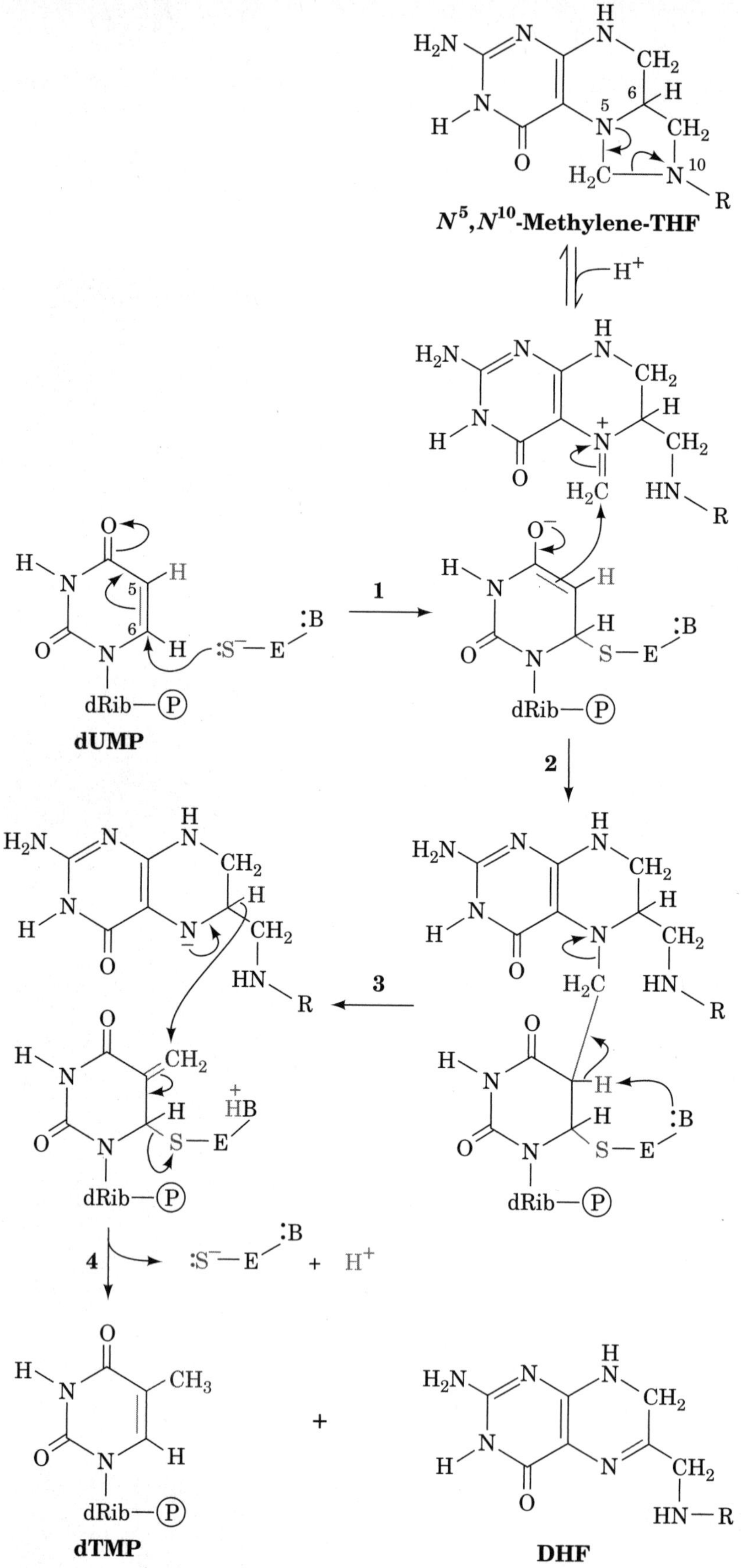

Figure 22-15 Catalytic mechanism of thymidylate synthase.

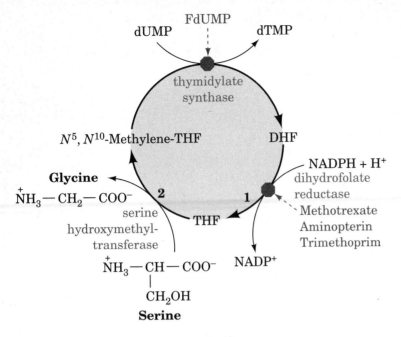

Figure 22-16 Regeneration of N^5,N^{10}-methylenetetrahydrofolate.

5-Fluorodeoxyuridylate (FdUMP)

R = H   **Aminopterin**
R = CH$_3$ **Methotrexate (amethopterin)**

Trimethoprim

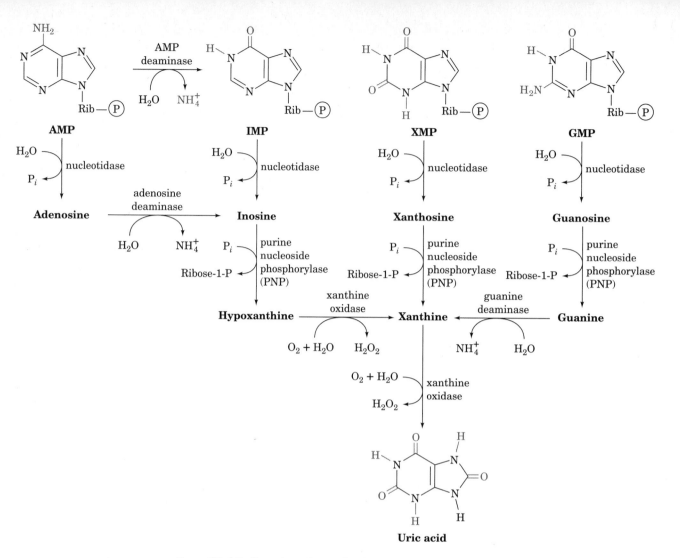

Figure 22-18 The major pathways of purine catabolism in animals.

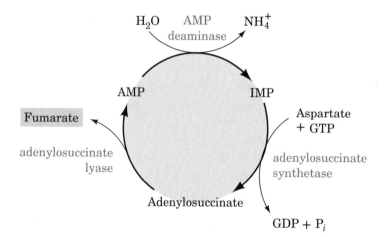

Net: H_2O + Aspartate + GTP $\longrightarrow$ NH_4^+ + GDP + P_i + fumarate

Figure 22-20 The purine nucleotide cycle.

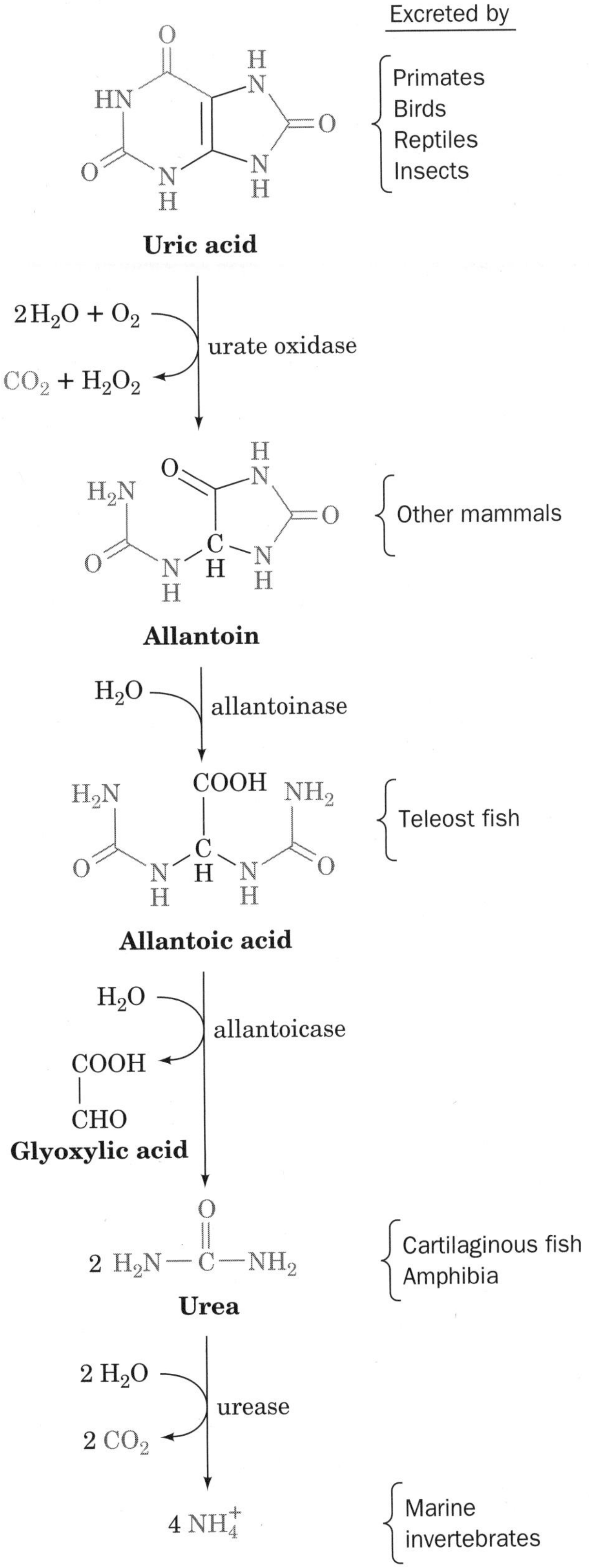

Figure 22-21 Degradation of uric acid to ammonia.

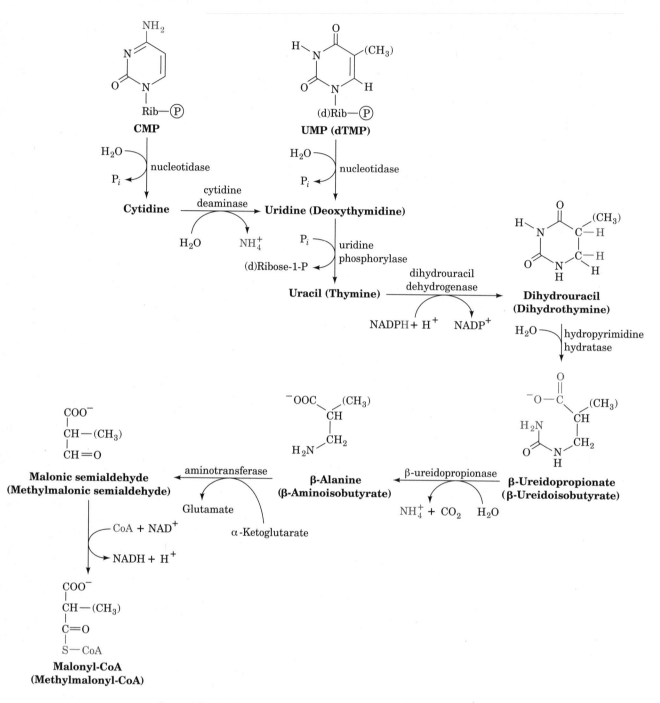

Figure 22-23 The major pathways of pyrimidine catabolism in animals.

Nucleic Acid Structure

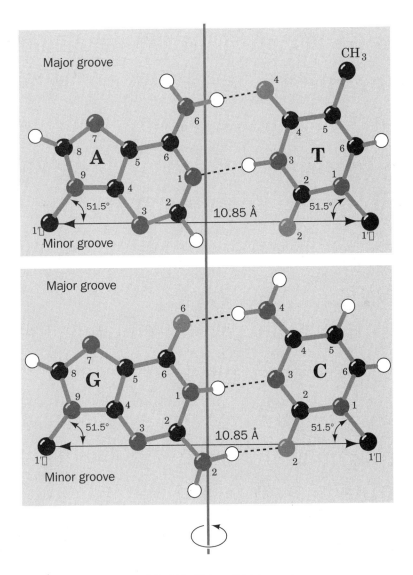

Figure 23-1 The Watson–Crick base pairs.

Table 23-1 *Key to Structure.* Structural Features of Ideal A-, B-, and Z-DNA

	A	B	Z
Helical sense	Right handed	Right handed	Left handed
Diameter	~26 Å	~20 Å	~18 Å
Base pairs per helical turn	11.6	10	12 (6 dimers)
Helical twist per base pair	31°	36°	60° (per dimer)
Helix pitch (rise per turn)	34 Å	34 Å	44 Å
Helix rise per base pair	2.9 Å	3.4 Å	7.4 Å per dimer
Base tilt normal to the helix axis	20°	6°	7°
Major groove	Narrow and deep	Wide and deep	Flat
Minor groove	Wide and shallow	Narrow and deep	Narrow and deep
Sugar pucker	C3'-*endo*	C2'-*endo*	C2'-*endo* for pyrimidines; C3'-*endo* for purines
Glycosidic bond conformation	Anti	Anti	Anti for pyrimidines; syn for purines

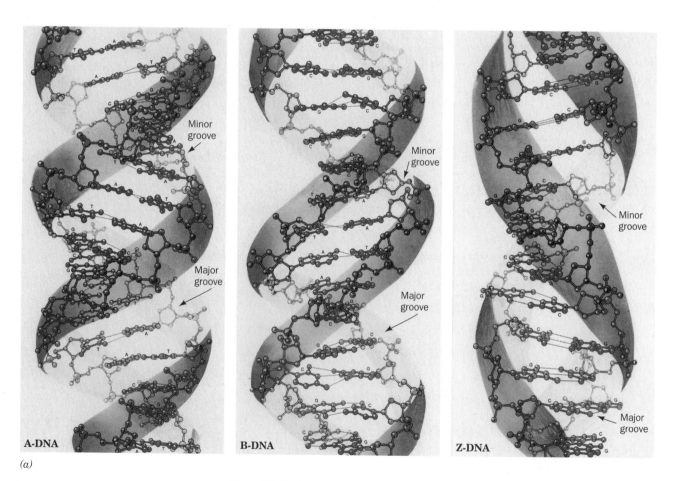

(a)

Figure 23-2 Structures of A-, B-, and Z-DNA.

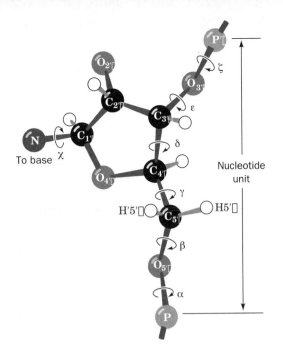

Figure 23-5 The seven torsion angles that determine the conformation of a nucleotide unit.

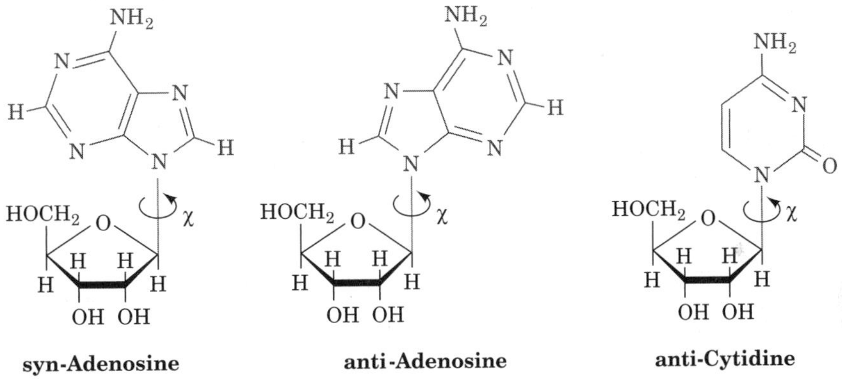

syn-Adenosine   **anti-Adenosine**   **anti-Cytidine**

Figure 23-6 The sterically allowed orientations of purine and pyrimidine bases with respect to their attached ribose units.

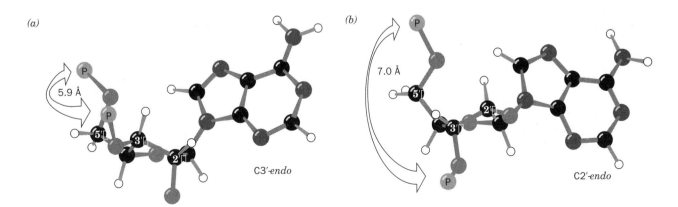

(a)

5.9 Å

C3′-endo

(b)

7.0 Å

C2′-endo

Figure 23-7 Nucleotide sugar conformations.

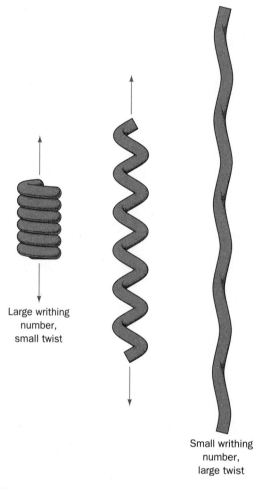

Large writing
number,
small twist

Small writing
number,
large twist

Figure 23-9 The difference between writhing and twist as demonstrated by a coiled telephone cord.

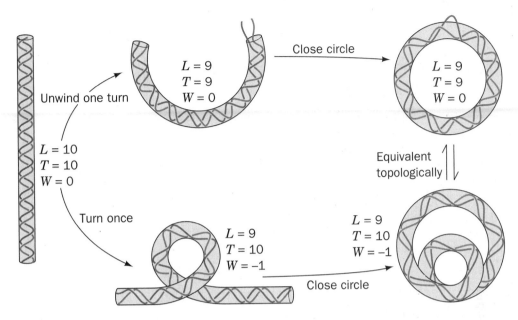

Figure 23-10 Two ways of introducing one supercoil into a DNA that has 10 duplex turns.

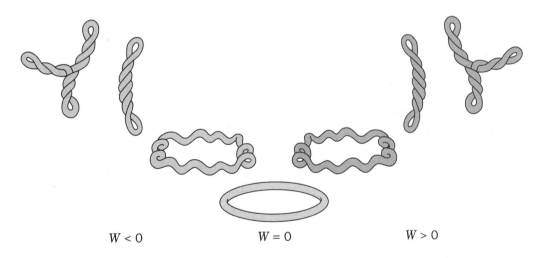

$W < 0$ $W = 0$ $W > 0$

Figure 23-11 Progressive unwinding of a negatively supercoiled DNA molecule.

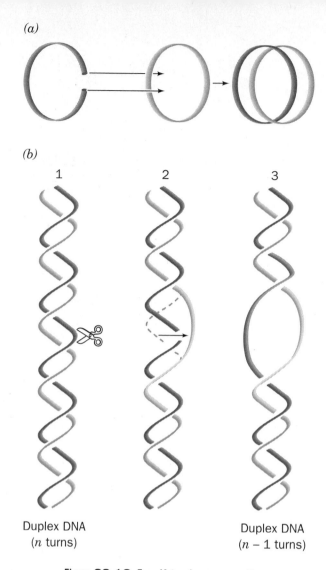

(a)

(b)

1   2   3

Duplex DNA
(n turns)

Duplex DNA
($n - 1$ turns)

Figure 23-12 Type IA topoisomerase action.

Type I topoisomerase

CH_2

Tyr

O^-

$O-P-O-CH_2$ O Base

O

H H

H   H

O H

$^-O-P-O-\cdots-O-P-O-CH_2$ O Base

O   O^-

O   O

H H

H   H

DNA   OH H

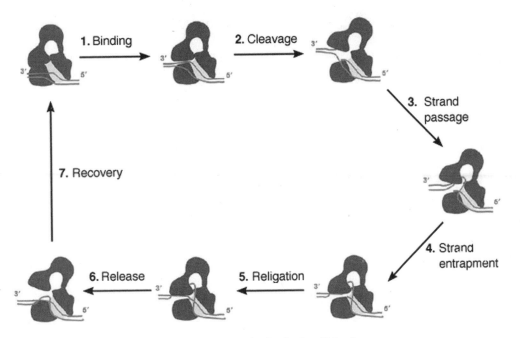

Figure 23-14 Proposed mechanism for type IA topoisomerases.

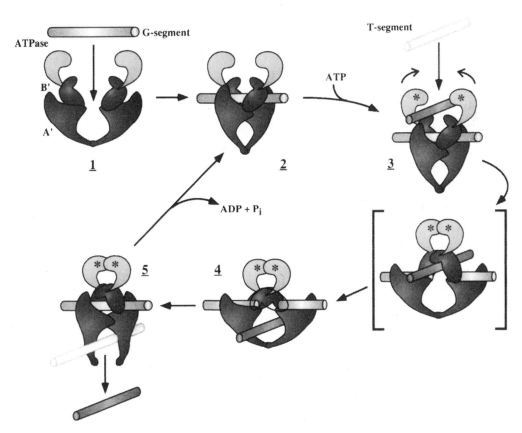

Figure 23-18 Proposed mechanism for Type II topoisomerase.

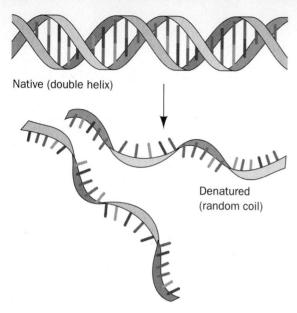

Native (double helix)

Denatured
(random coil)

Figure 23-19 A schematic representation of DNA denaturation.

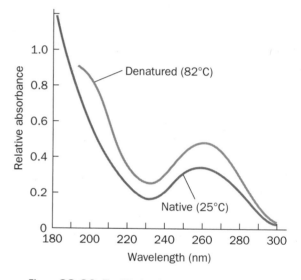

Figure 23-20 The UV absorbance spectra of native and heat-denatured *E. coli* DNA.

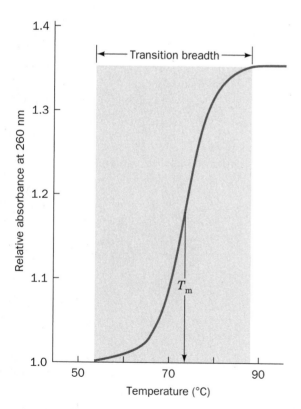

Figure 23-21 An example of a DNA melting curve.

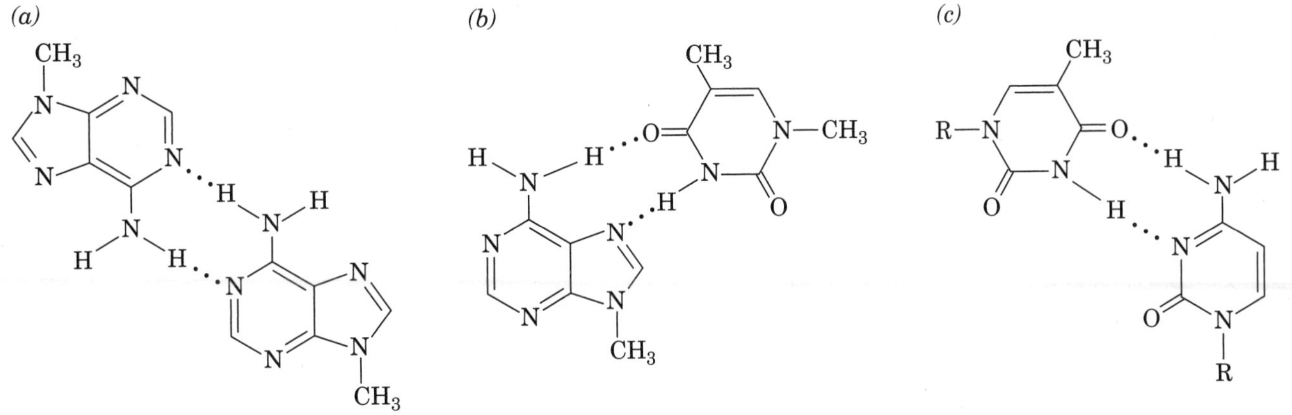

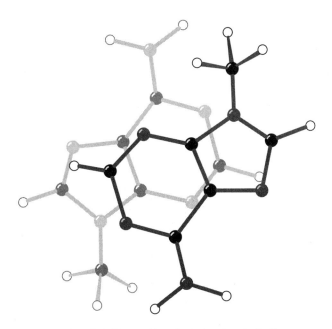

Figure 23-23 Some non-Watson–Crick base pairs.

Figure 23-24 The stacking of adenine rings in the X-ray structure of 9-methyladenine.

Table 23-2 Stacking Energies for the Ten Possible Dimers in B-DNA

Stacked Dimer	Stacking Energy $(kJ \cdot mol^{-1})$
C·G G·C	−61.0
C·G A·T	−44.0
C·G T·A	−41.0
G·C C·G	−40.5
G·C G·C	−34.6
G·C A·T	−28.4
T·A A·T	−27.5
G·C T·A	−27.5
A·T A·T	−22.5
A·T T·A	−16.0

Source: Ornstein, R.L., Rein, R., Breen, D.L., and MacElroy, R.D., *Biopolymers* **17,** 2356 (1978).

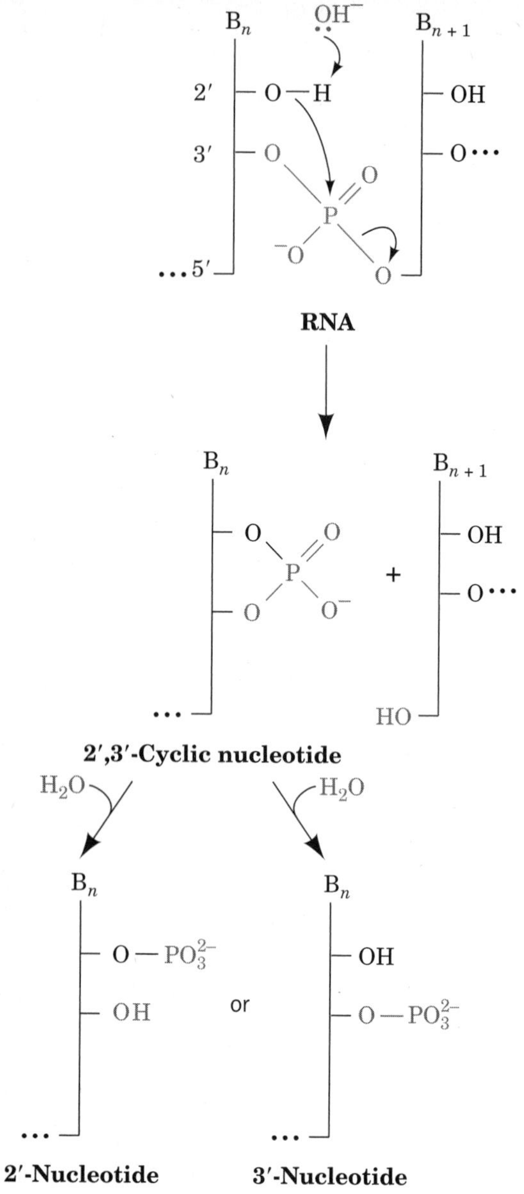

RNA

2′,3′-Cyclic nucleotide

2′-Nucleotide   **3′-Nucleotide**

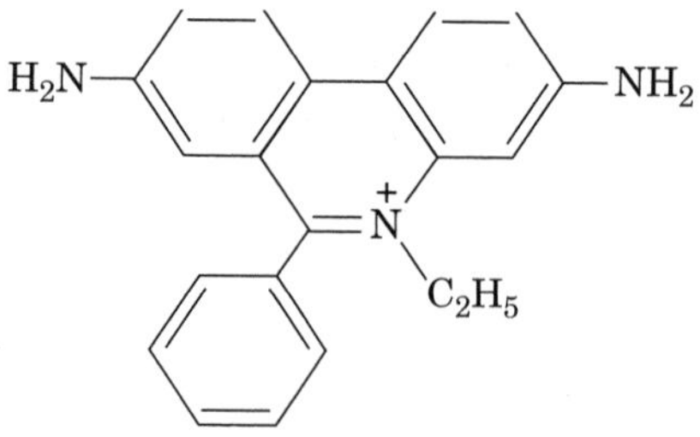

Ethidium

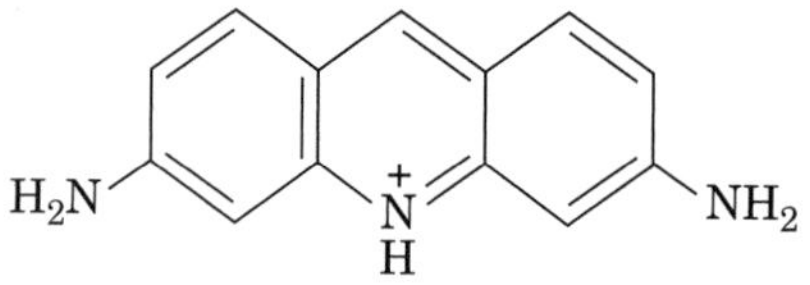

Proflavin 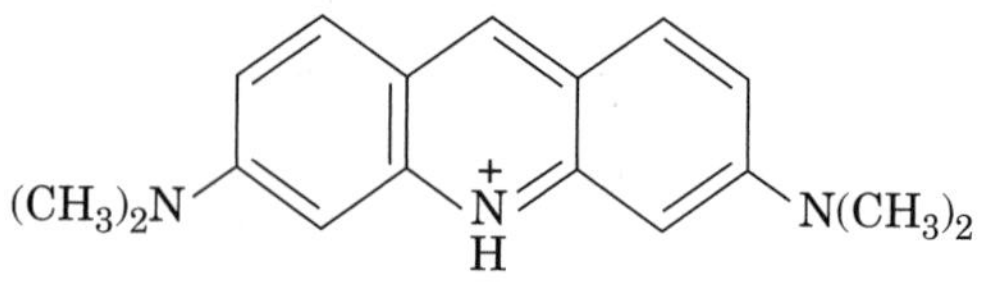   **Acridine orange**

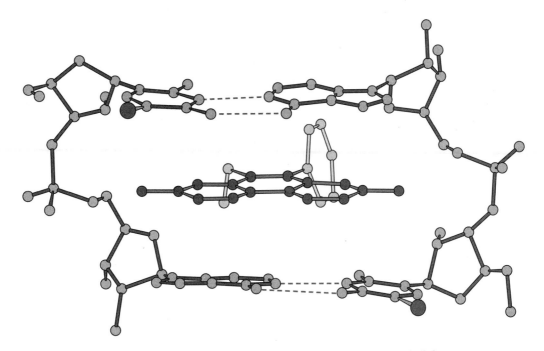

Figure 23-29 X-Ray structure of a complex of ethidium with 5-iodo-UpA.

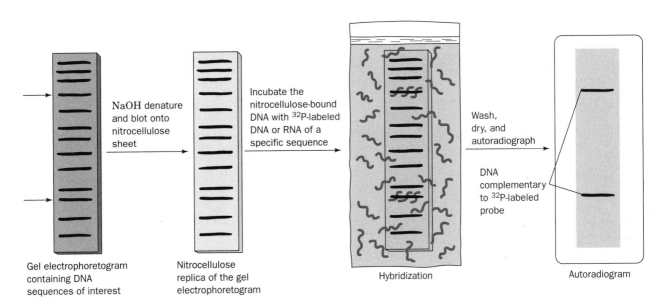

Gel electrophoretogram containing DNA sequences of interest

NaOH denature and blot onto nitrocellulose sheet

Nitrocellulose replica of the gel electrophoretogram

Incubate the nitrocellulose-bound DNA with ^{32}P-labeled DNA or RNA of a specific sequence

Hybridization

Wash, dry, and autoradiograph

DNA complementary to ^{32}P-labeled probe

Autoradiogram

Figure 23-30 The detection of DNAs containing specific base sequences by Southern blotting.

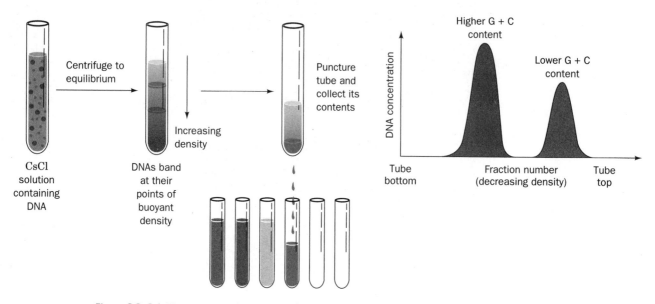

Figure 23-31 The separation of DNAs by equilibrium density gradient ultracentrifugation in CsCl solution.

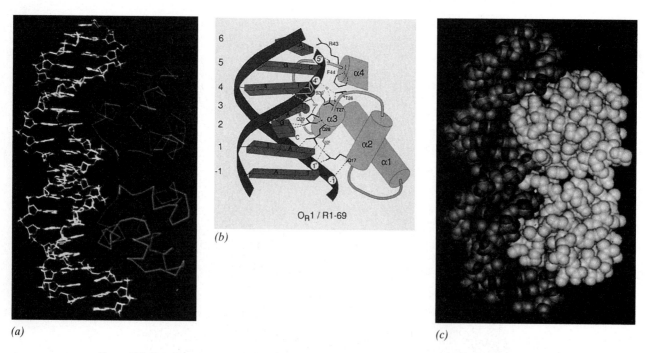

(a)

(b)

O_R1 / R1-69

(c)

Figure 23-34 X-Ray structure of a portion of the 434 phage repressor in complex with its target DNA.

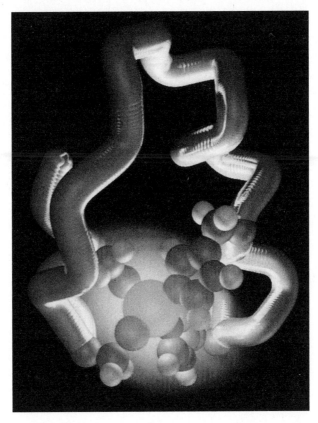

Figure 23-37 NMR structure of a single zinc finger from the *Xenopus* protein Xfin.

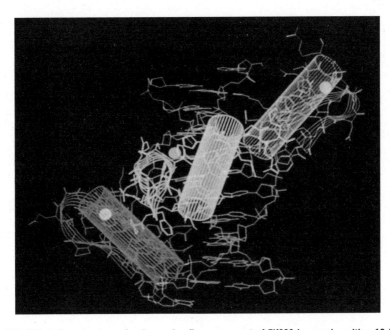

Figure 23-38 X-Ray structure of a three–zinc finger segment of Zif268 in complex with a 10-bp DNA.

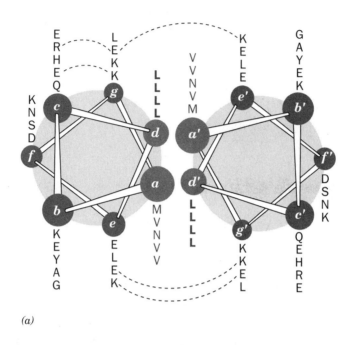

(a)

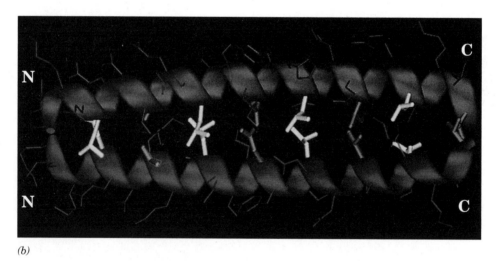

(b)

Figure 23-40 The GCN4 leucine zipper motif.

Table 23-3 Calf Thymus Histones

Histone	Number of Residues	Mass (kD)	% Arg	% Lys
H1	215	23.0	1	29
H2A	129	14.0	9	11
H2B	125	13.8	6	16
H3	135	15.3	13	10
H4	102	11.3	14	11

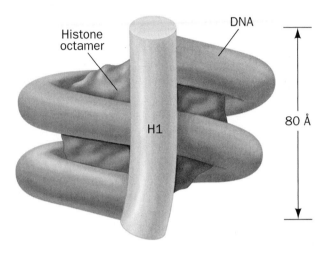

Figure 23-47 Model of histone H1 binding to the DNA of the 166-bp nucleosome.

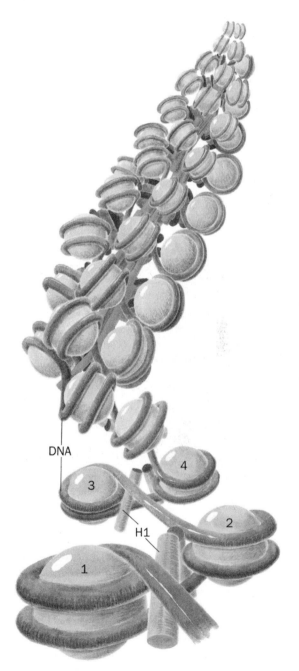

Figure 23-50 Proposed model of the 30-nm-diameter chromatin fiber.

CHAPTER 24

DNA Replication, Repair, and Recombination

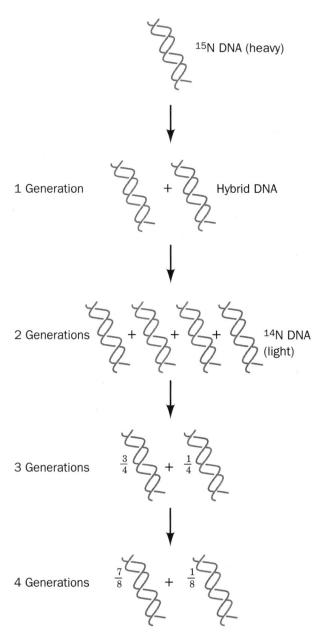

Figure 24-1 The Meselson and Stahl experiment.

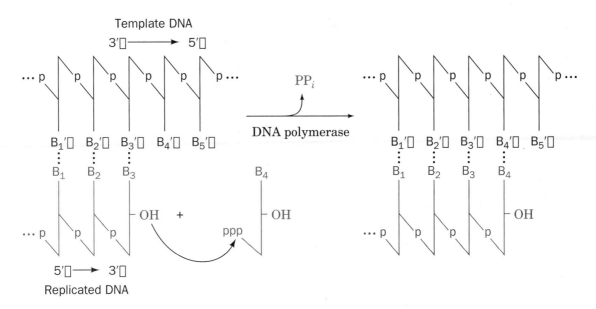

Figure **24-2** Action of DNA polymerases.

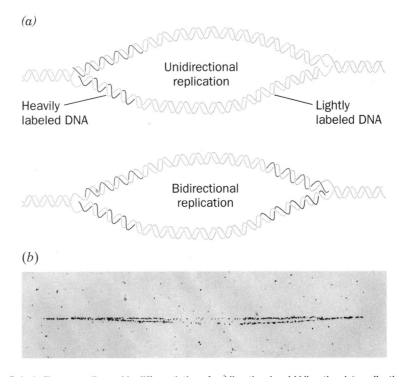

Figure **24-4** The autoradiographic differentiation of unidirectional and bidirectional θ replication of DNA.

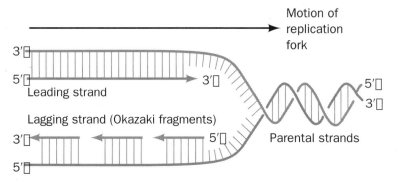

Figure 24-5 Semidiscontinuous DNA replication.

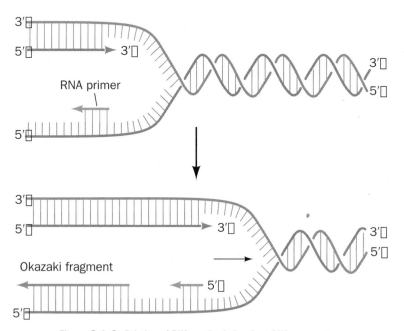

Figure 24-6 Priming of DNA synthesis by short RNA segments.

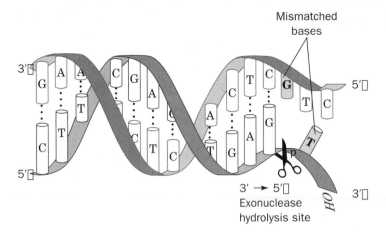

Figure 24-7 The 3′ → 5′ exonuclease function of DNA polymerase I.

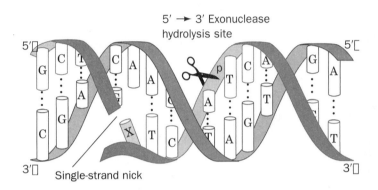

Figure 24-8 The 5′ → 3′ exonuclease function of DNA polymerase I.

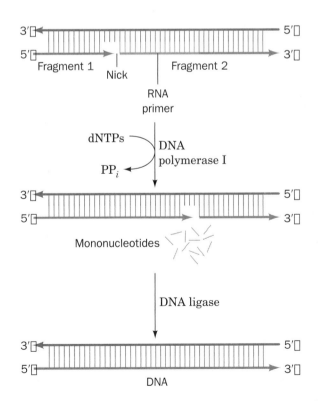

Figure 24-9 The replacement of RNA primers by DNA in lagging strand synthesis.

Table 24-1 Properties of *E. coli* DNA Polymerases

	Pol I	Pol II	Pol III
Mass (kD)	103	90	130
Molecules/cell	400	?	10–20
Turnover number[a]	600	30	9000
Structural gene	*polA*	*polB*	*polC*
Conditionally lethal mutant	+	−	+
Polymerization: $5' \rightarrow 3'$	+	+	+
Exonuclease: $3' \rightarrow 5'$	+	+	+
Exonuclease: $5' \rightarrow 3'$	+	−	−

[a]Nucleotides polymerized $min^{-1} \cdot molecule^{-1}$ at 37°C.

Source: Kornberg, A. and Baker, T.A., *DNA Replication* (2nd ed.), p. 167, Freeman (1992).

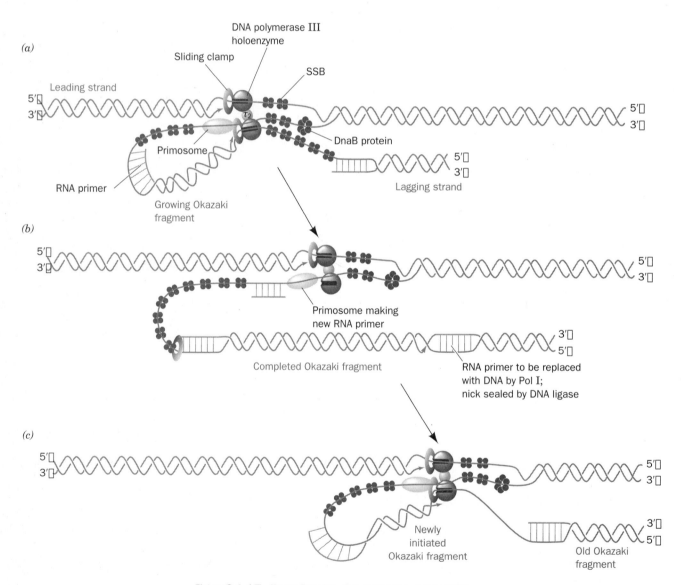

Figure 24-15 *Key to Function.* The replication of *E. coli* DNA.

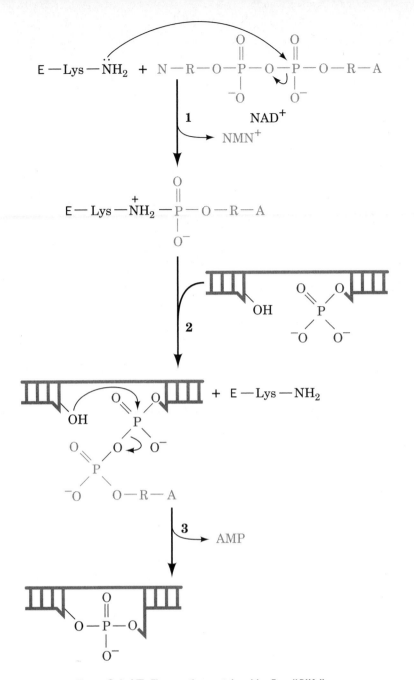

Figure 24-17 The reactions catalyzed by *E. coli* DNA ligase.

Table 24-2 Properties of Some Eukaryotic DNA Polymerases

	α	δ	ε
3′ → 5′ Exonuclease	no	yes	yes
Associates with primase	yes	no	no
Processivity	moderate	high	high
Requires PCNA	no	yes	no

REVERSE TRANSCRIPTASE

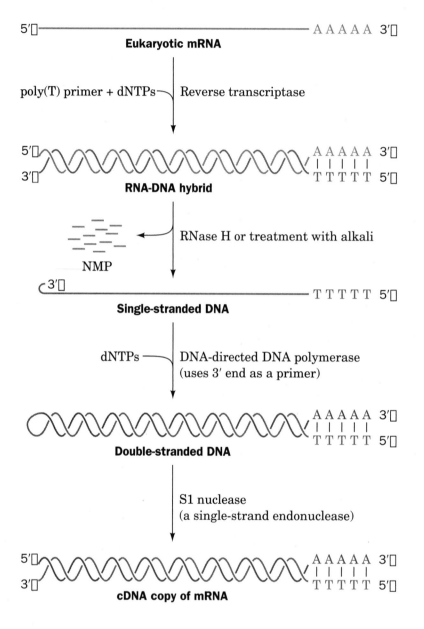

5′ —————————————————— A A A A A 3′
Eukaryotic mRNA

poly(T) primer + dNTPs ⌐ Reverse transcriptase

5′ ～～～～～～～～～～～ A A A A A 3′
                      | | | | |
3′ ～～～～～～～～～～～ T T T T T 5′
RNA-DNA hybrid

⟵ RNase H or treatment with alkali

NMP

3′
C ————————————————— T T T T T 5′
Single-stranded DNA

dNTPs ⌐ DNA-directed DNA polymerase
       (uses 3′ end as a primer)

～～～～～～～～～～～ A A A A A 3′
                    | | | | |
                    T T T T T 5′
Double-stranded DNA

S1 nuclease
(a single-strand endonuclease)

5′ ～～～～～～～～～～～ A A A A A 3′
                      | | | | |
3′ ～～～～～～～～～～～ T T T T T 5′
cDNA copy of mRNA

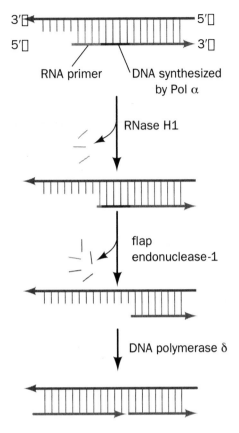

RNA primer   DNA synthesized
                    by Pol α

RNase H1

flap
endonuclease-1

DNA polymerase δ

Figure 24-22   Removal of RNA primers in eukaryotes.

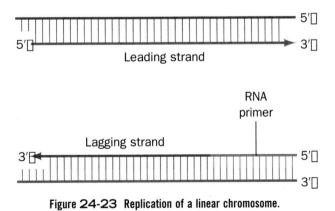

Leading strand

RNA
primer

Lagging strand

Figure 24-23   Replication of a linear chromosome.

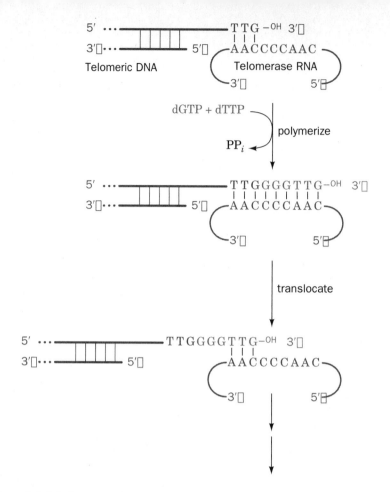

Figure 24-24 Mechanism for the synthesis of telomeric DNA by *Tetrahymena* telomerase.

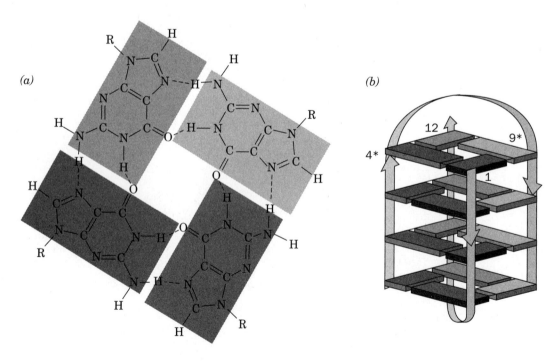

(a)

(b)

Figure 24-25 Structure of the telomeric oligonucleotide d(GGGGTTTTGGGG).

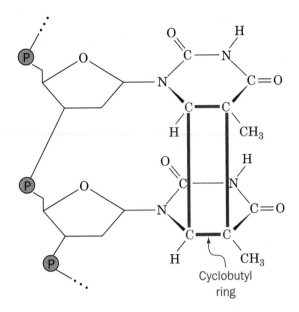

Figure 24-27 The cyclobutylthymine dimer.

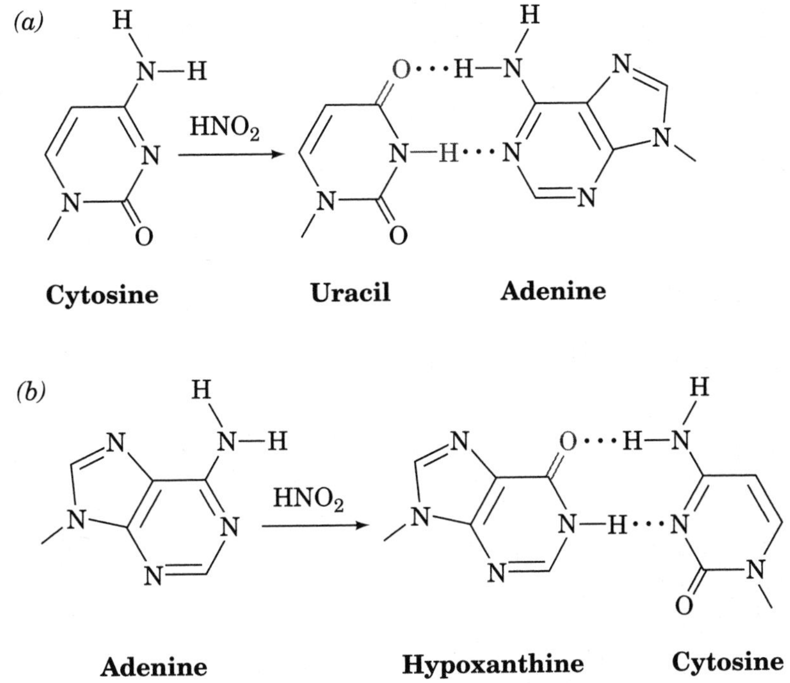

(a)

Cytosine $\xrightarrow{\text{HNO}_2}$ Uracil Adenine

(b)

Adenine $\xrightarrow{\text{HNO}_2}$ Hypoxanthine Cytosine

Figure 24-28 Oxidative deamination by nitrous acid.

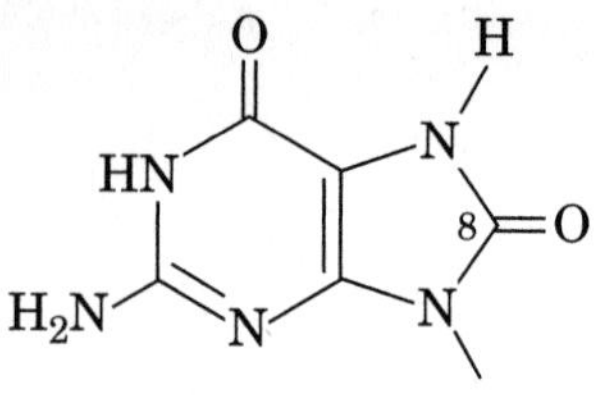

8-Oxoguanine (oxoG)

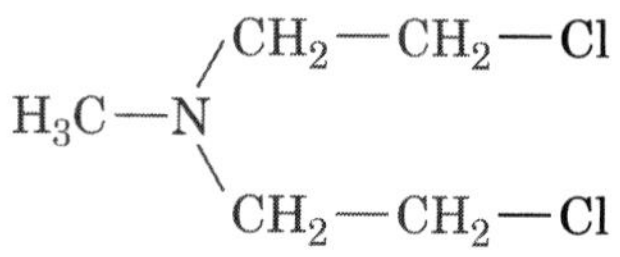

Nitrogen mustard

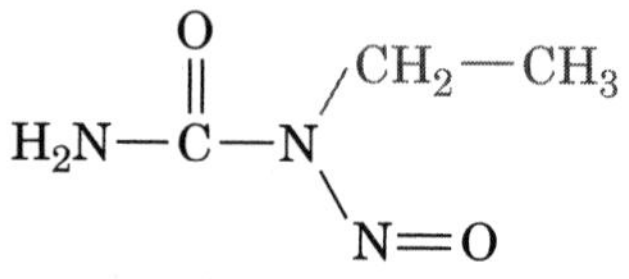

Ethylnitrosourea

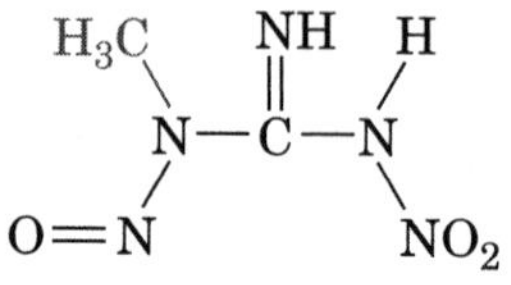

N-Methyl-*N*′-nitro-*N*-
nitrosoguanidine (MNNG)

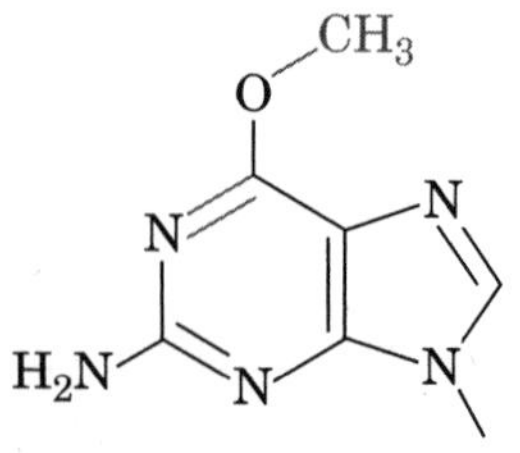

O^6-**Methylguanine residue**

DNA METHYLATION SITES

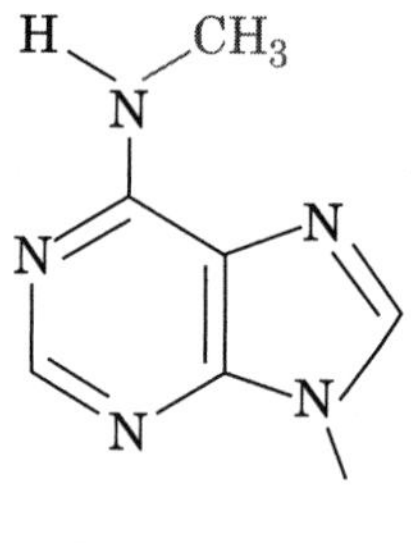

N^6-**Methyladenine (m^{6}A)**
residue

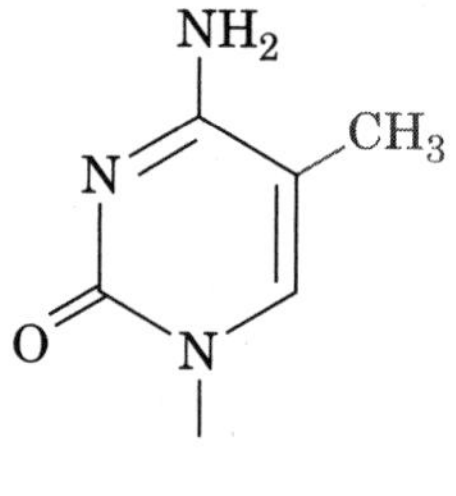

5-Methylcytosine (m^{5}C)
residue

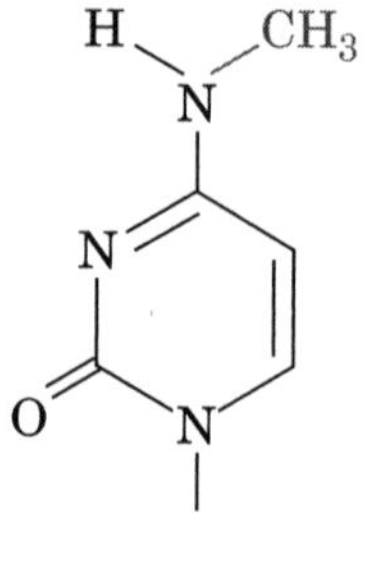

N^4-**Methylcytosine (m^{4}C)**
residue

354

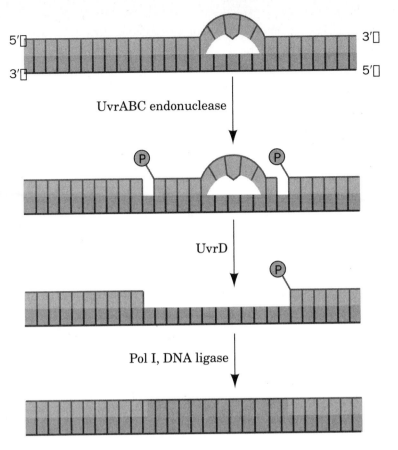

Figure 24-33 The mechanism of nucleotide excision repair (NER) of pyrimidine dimers.

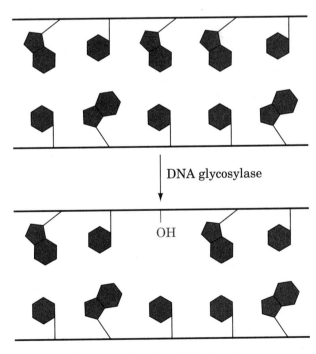

Figure 24-31 Action of DNA glycosylases.

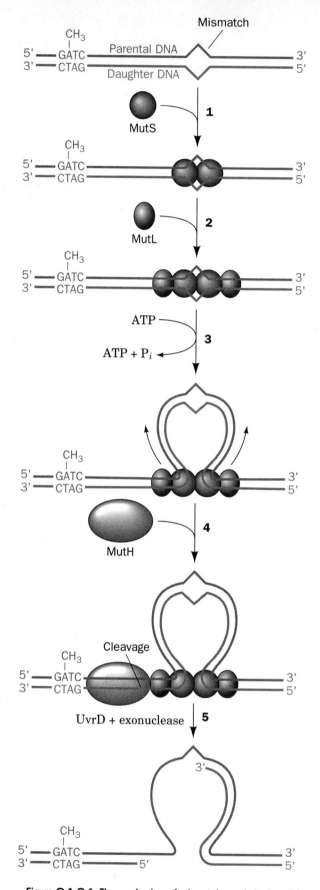

Figure 24-34 The mechanism of mismatch repair in *E. coli.*

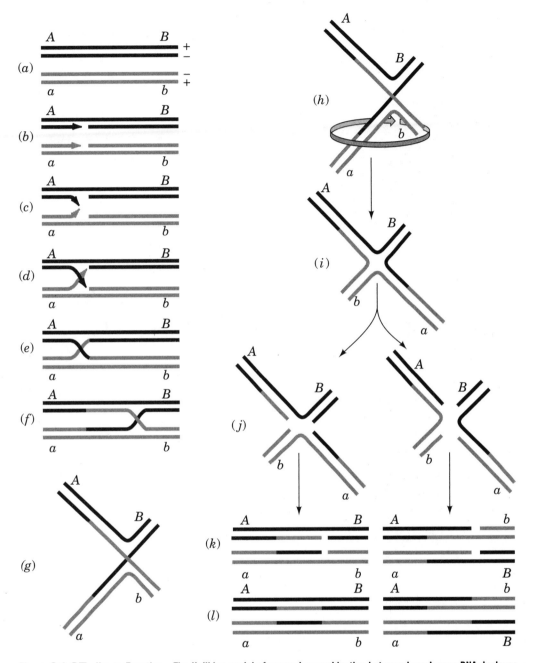

Figure 24-37 **Key to Function.** The Holliday model of general recombination between homologous DNA duplexes.

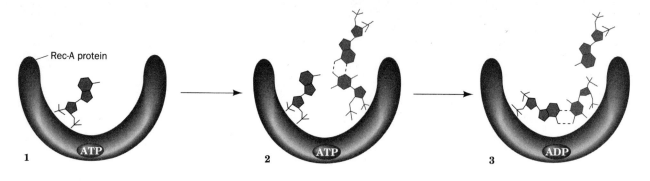

Figure 24-40 A model for RecA-mediated pairing.

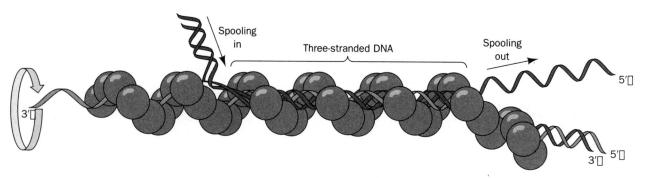

Figure 24-41 Model for RecA-mediated strand exchange.

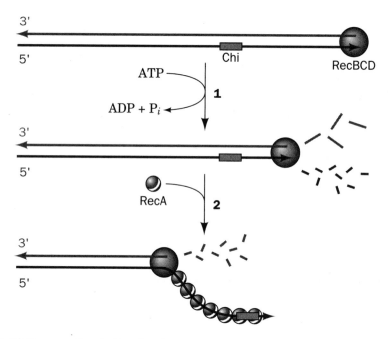

Figure 24-42 The generation of a 3'-ending single-strand DNA segment by RecBCD to initiate recombination.

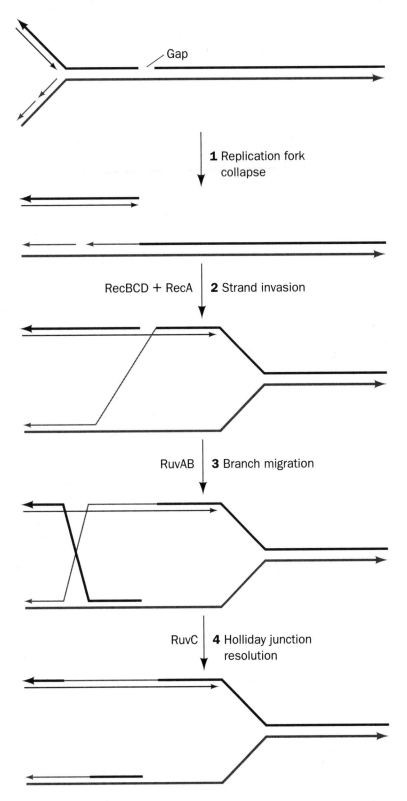

Figure 24-45 The recombination repair of a replication fork that has encountered a single-strand nick.

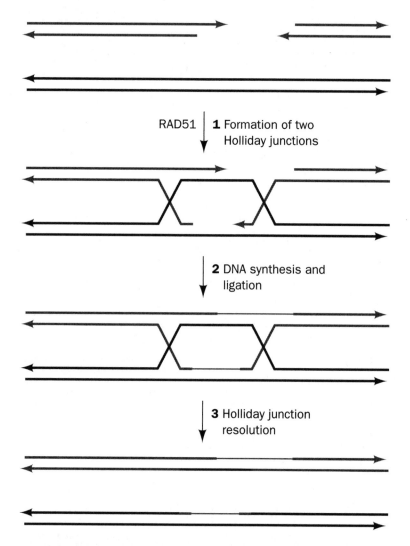

Figure 24-46 The repair of a double-strand break in DNA by homologous end-joining.

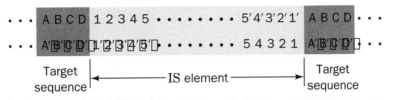

Figure 24-47 Structure of IS elements.

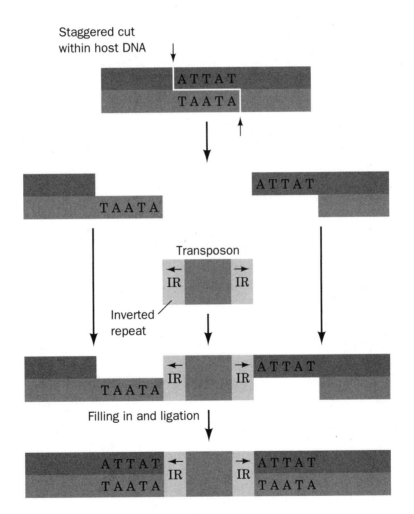

Figure 24-48 A model for transposon insertion.

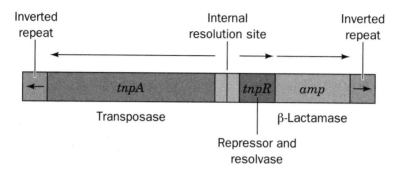

Figure 24-49 A map of transposon Tn3.

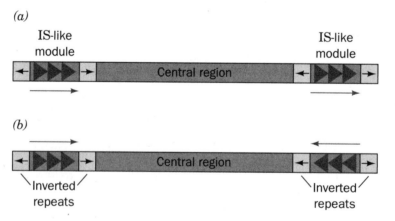

Figure 24-50 A composite transposon.

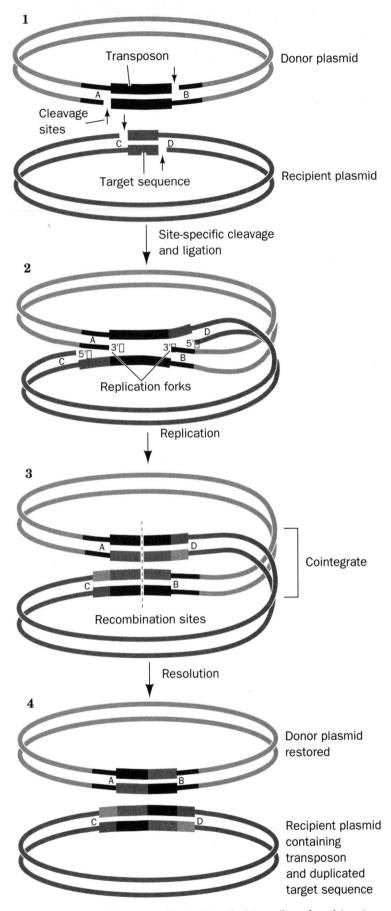

1

Transposon

Donor plasmid

A   B

Cleavage
sites

C   D

Target sequence

Recipient plasmid

Site-specific cleavage
and ligation

2

A   3'☐   3'☐   5'☐   D

C 5'☐   B

Replication forks

Replication

3

A   D

C   B

Cointegrate

Recombination sites

Resolution

4

Donor plasmid
restored

A   B

C   D

Recipient plasmid
containing
transposon
and duplicated
target sequence

Figure 24-51 A model for transposition involving the intermediacy of a cointegrate.

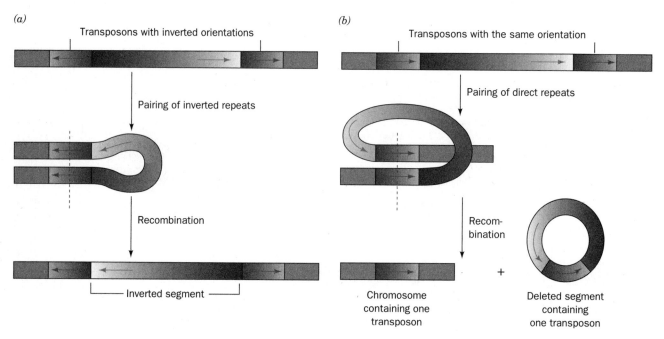

(a)

Transposons with inverted orientations

Pairing of inverted repeats

Recombination

Inverted segment

(b)

Transposons with the same orientation

Pairing of direct repeats

Recombination

Chromosome
containing one
transposon

+

Deleted segment
containing
one transposon

Figure 24-52 Chromosomal rearrangement via recombination.

CHAPTER 25

Transcription and RNA Processing

$$(RNA)_{n\ residues} + NTP \rightleftharpoons (RNA)_{n+1\ residues} + PP_i$$

$$\downarrow\!\!\!\!\curvearrowleft H_2O$$

$$2\,P_i$$

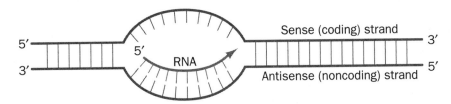

Figure 25-3 Sense and antisense DNA strands.

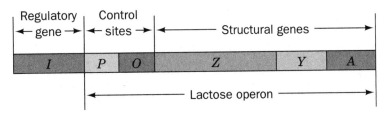

Figure 25-4 The *E. coli lac* operon.

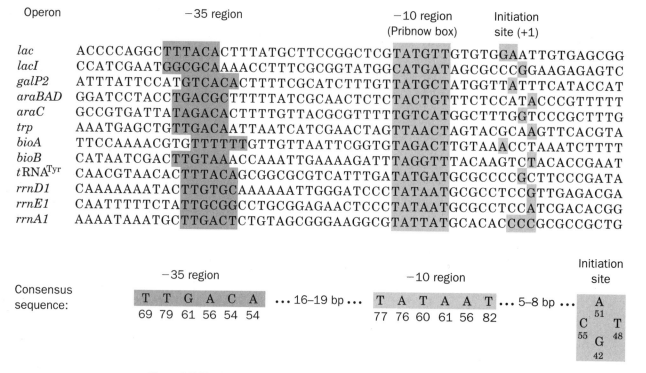

Figure 25-5 The sense (coding) strand sequences of selected *E. coli* promoters.

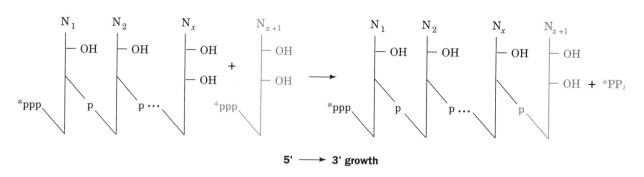

5' ⟶ **3' growth**

Figure 25-6 5'→3' RNA chain growth.

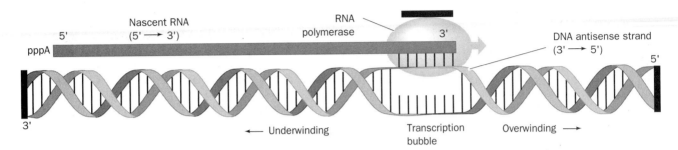

Figure 25-7 DNA supercoiling during transcription.

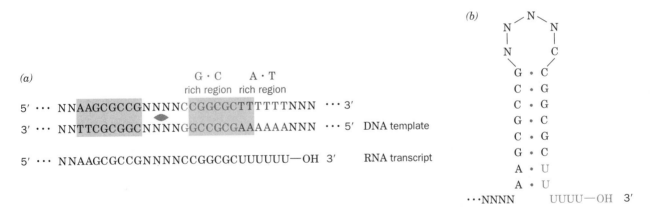

Figure 25-9 A hypothetical strong (efficient) *E. coli* terminator.

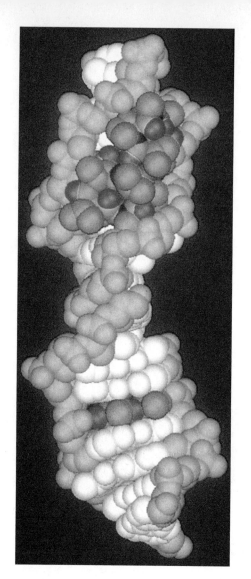

Actinomycin D

Phenoxazone ring system

Methyl-Val

Sarcosine

Pro

D-Val

Thr

Chicken ovalbumin	GAGGCTATATATTCCCCAGGGCTCAGCCAGTGTCTGTACA
Adenovirus late	GGGGCTATAAAAGGGGGTGGGGGCGCGTTCGTCCTCACTC
Rabbit β globin	TTGGGCATAAAAGGCAGAGCAGGGCAGCTGCTGCTAACACT
Mouse β globin major	GAGCATATAAGGTGAGGTAGGATCAGTTGCTCCTCACATTT

$T_{82}A_{97}T_{93}A_{85}\begin{smallmatrix}A_{63}\\T_{37}\end{smallmatrix}A_{83}\begin{smallmatrix}A_{50}\\T_{37}\end{smallmatrix}$

Figure 25-13 The promoter sequences of selected eukaryotic structural genes.

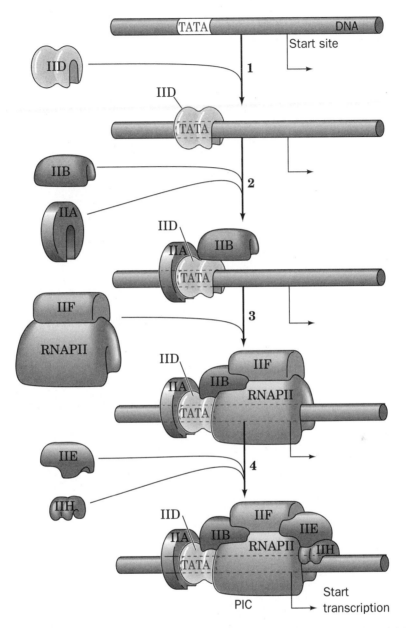

Figure 25-15 *Key to Function.* The assembly of the preinitiation complex (PIC) on a TATA box–containing promoter.

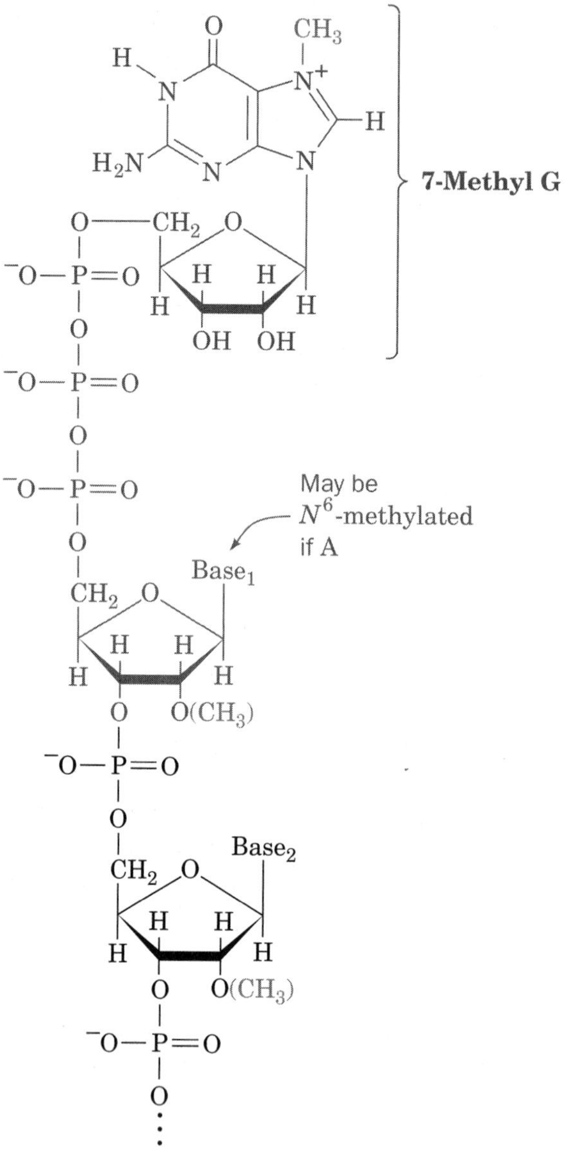

Figure 25-18 Structure of the 5' cap of eukaryotic mRNAs.

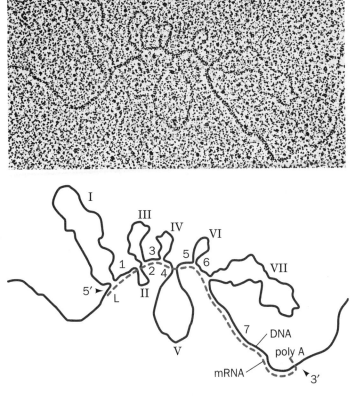

Figure 25-19 The chicken ovalbumin gene and its mRNA.

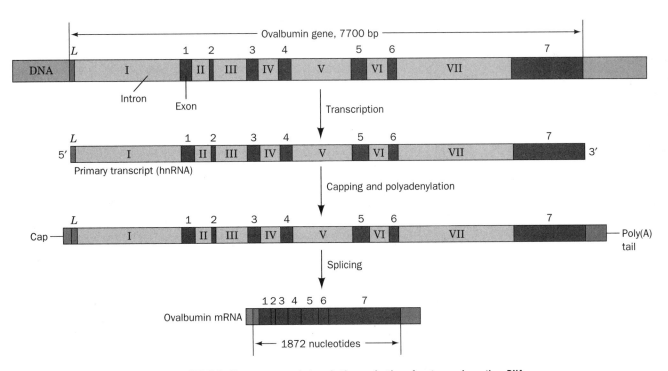

Figure 25-20 The sequence of steps in the production of mature eukaryotic mRNA.

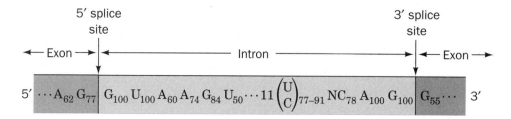

Figure 25-21 The consensus sequences at the exon–intron junctions of eukaryotic pre-mRNAs.

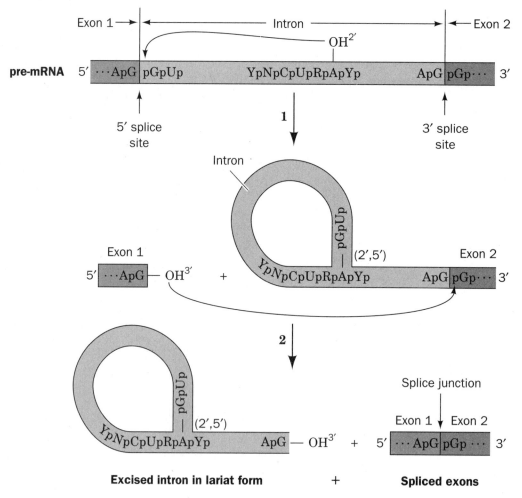

Figure 25-22 *Key to Function.* The splicing reaction.

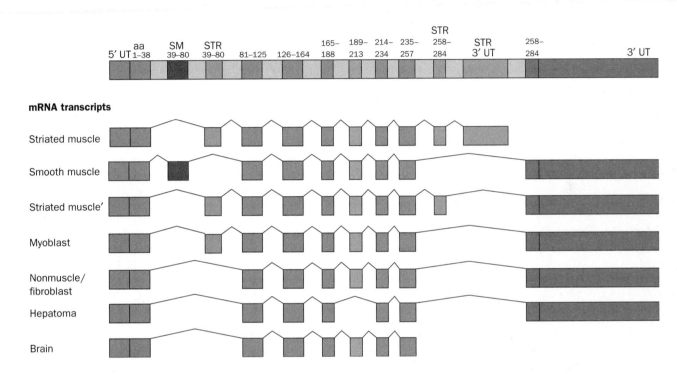

Figure 25-25 Alternative splicing in the rat α-tropomyosin gene.

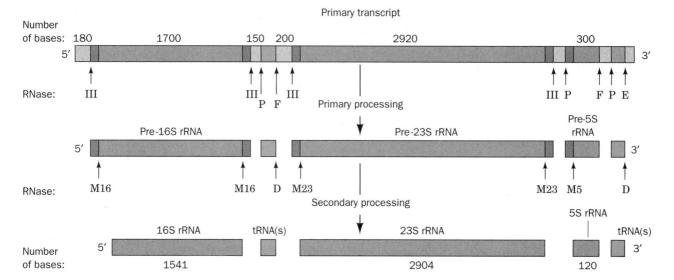

Figure 25-27 The posttranscriptional processing of *E. coli* rRNA.

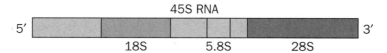

Figure 25-28 The organization of the 45S primary transcript of eukaryotic rRNA.

Uridine   Pseudouridine (ψ)

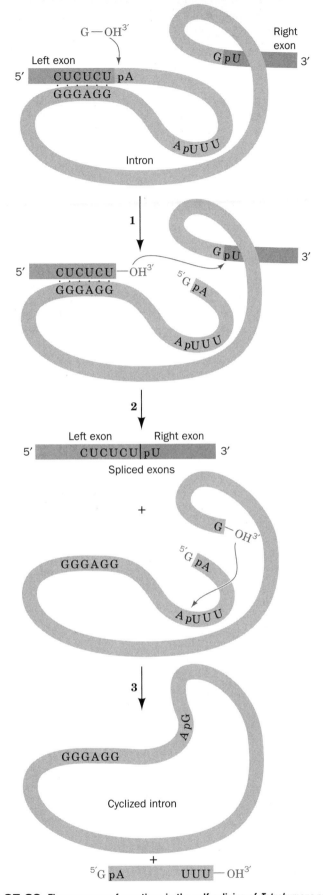

Figure 25-29 The sequence of reactions in the self-splicing of *Tetrahymena* pre-rRNA.

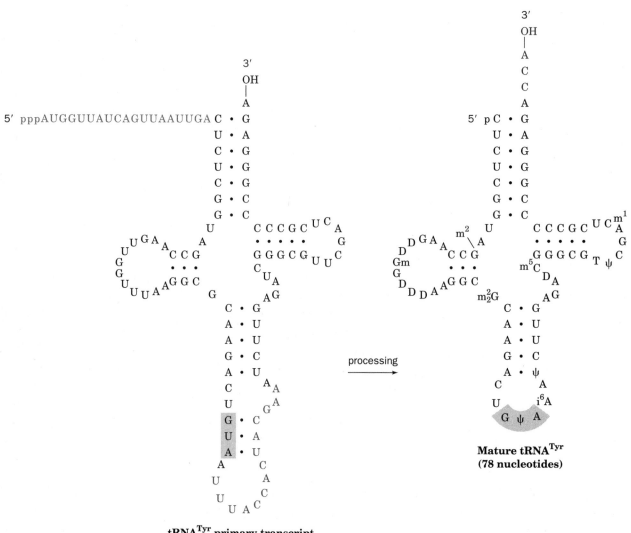

tRNA^{Tyr} primary transcript
(108 nucleotides)

Mature tRNA^{Tyr}
(78 nucleotides)

Figure 25-33 The posttranscriptional processing of yeast tRNA^{Tyr}.

CHAPTER 26

Translation

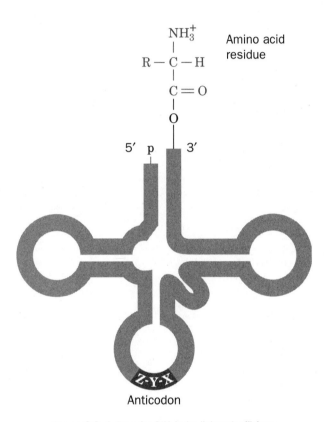

Figure 26-1 Transfer RNA in its "cloverleaf" form.

$$H_3\overset{+}{N}-\overset{\overset{\displaystyle H}{|}}{\underset{\underset{\displaystyle R}{|}}{C}}-\overset{\displaystyle O}{\underset{\displaystyle O^-}{C}}$$

Table 26-1. *Key to Function.* **The "Standard" Genetic Code**[a]

First position (5′ end)	Second position				Third position (3′ end)
	U	**C**	**A**	**G**	
U	UUU Phe	UCU	UAU Tyr	UGU Cys	U
	UUC	UCC	UAC	UGC	C
	UUA Leu	UCA Ser	UAA STOP	UGA STOP	A
	UUG	UCG	UAG	UGG Trp	G
C	CUU	CCU	CAU His	CGU	U
	CUC Leu	CCC	CAC	CGC Arg	C
	CUA	CCA Pro	CAA Gln	CGA	A
	CUG	CCG	CAG	CGG	G
A	AUU Ile	ACU	AAU Asn	AGU Ser	U
	AUC	ACC	AAC	AGC	C
	AUA	ACA Thr	AAA Lys	AGA Arg	A
	AUG Met[b]	ACG	AAG	AGG	G
G	GUU	GCU	GAU Asp	GGU	U
	GUC Val	GCC	GAC	GGC Gly	C
	GUA	GCA Ala	GAA Glu	GGA	A
	GUG	GCG	GAG	GGG	G

[a]Nonpolar amino acid residues are tan, basic residues are blue, acidic residues are red, and polar uncharged residues are purple.

[b]AUG forms part of the initiation signal as well as coding for internal Met residues.

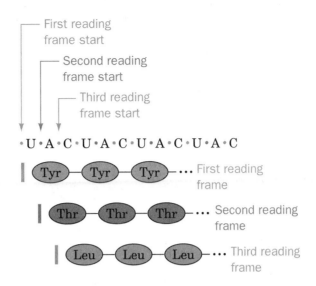

Figure 26-2 The three potential reading frames of an mRNA.

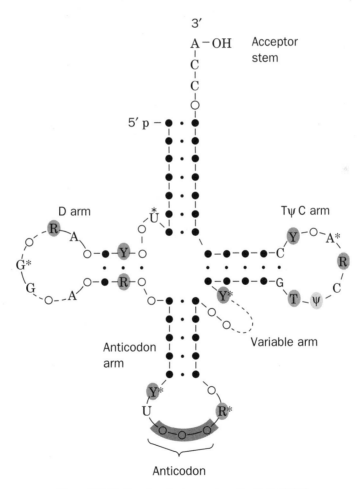

Figure 26-3 The cloverleaf secondary structure of tRNA.

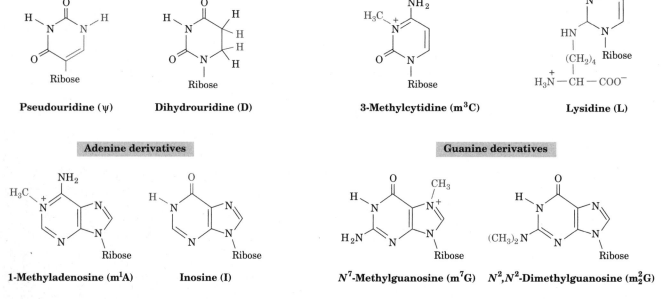

Figure 26-4 A few of the modified nucleosides that occur in tRNAs.

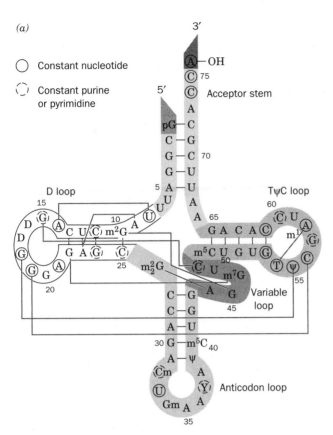

(a)

○ Constant nucleotide

◌ Constant purine or pyrimidine

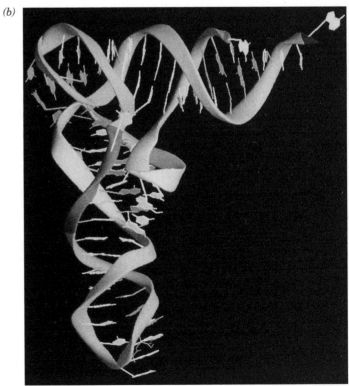

(b)

Figure 26-5 *Key to Structure.* The structure of yeast tRNA^Phe.

tRNA
|
O
|
O=P—O—CH₂
|
O⁻

Adenine

H   H

H₃'   2'H

O   OH
|
C=O
|
H—C—R
|
NH₃⁺

Aminoacyl–tRNA

Figure 26-6 An aminoacyl–tRNA.

$$R-\underset{\underset{NH_3^+}{|}}{\overset{\overset{H}{|}}{C}}-\overset{\overset{O}{\|}}{C}-O^- + ATP \rightleftharpoons R-\underset{\underset{NH_3^+}{|}}{\overset{\overset{H}{|}}{C}}-\overset{\overset{O}{\|}}{C}-O-\underset{\underset{O^-}{|}}{\overset{\overset{O}{\|}}{P}}-O-Ribose-Adenine + PP_i$$

Amino acid

**Aminoacyl–adenylate
(aminoacyl–AMP)**

$$\text{Aminoacyl–AMP} + \text{tRNA} \rightleftharpoons \text{aminoacyl–tRNA} + \text{AMP}$$

$$\text{Amino acid} + \text{tRNA} + \text{ATP} \longrightarrow \text{aminoacyl–tRNA} + \text{AMP} + \text{PP}_i$$

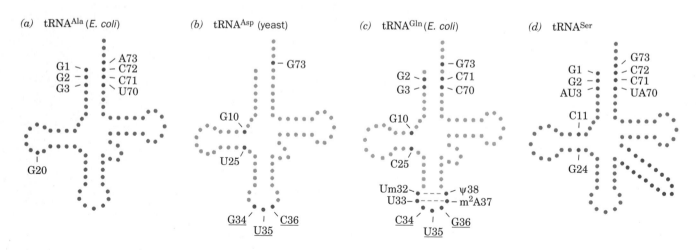

Figure 26-7 Major identity elements in four tRNAs.

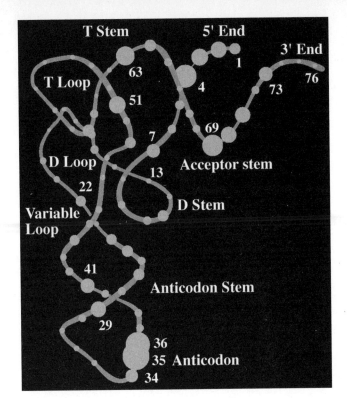

Figure 26-8 Experimentally observed identity elements of tRNAs.

Anticodon: 3′ —A—A—Gm— 5′   3′ —A—A—Gm— 5′
                · · ·   · · ·
                · · ·   · · ·
Codon: 5′ · · · 3′   5′ · · · 3′
            —U—U—C—   —U—U—U—

Anticodon: 3′ —C—G—I— 5′   3′ —C—G—I— 5′
                · · ·   · · ·
                · · ·   · · ·
Codon: 5′ · · · 3′   5′ · · · 3′
            —G—C—U—   —G—C—C—

Anticodon: 3′ —C—G—I— 5′
                · · ·
                · · ·
Codon: 5′ · · · 3′
            —G—C—A—

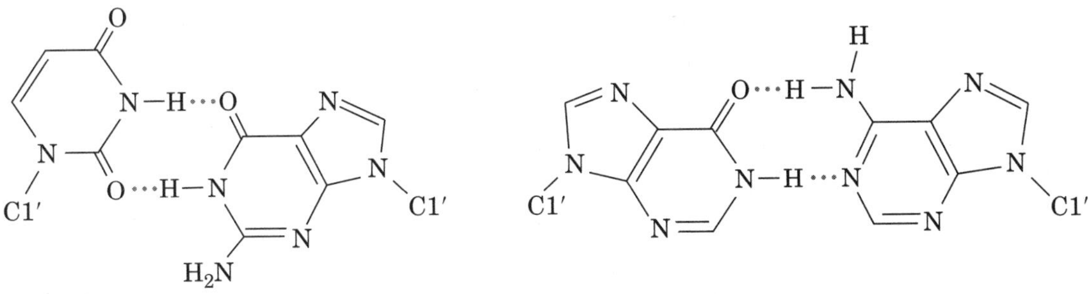

U·G   **I·A**

Figure 26-13 U·G and I·A wobble pairs. Both have been observed in X-ray structures.

Table 26-4 Components of *E. coli* Ribosomes

	Ribosome	Small Subunit	Large Subunit
Sedimentation coefficient	70S	30S	50S
Mass (kD)	2520	930	1590
RNA			
Major		16S, 1542 nucleotides	23S, 2904 nucleotides
Minor			5S, 120 nucleotides
RNA mass (kD)	1664	560	1104
Proportion of mass	66%	60%	70%
Proteins		21 polypeptides	31 polypeptides
Protein mass (kD)	857	370	487
Proportion of mass	34%	40%	30%

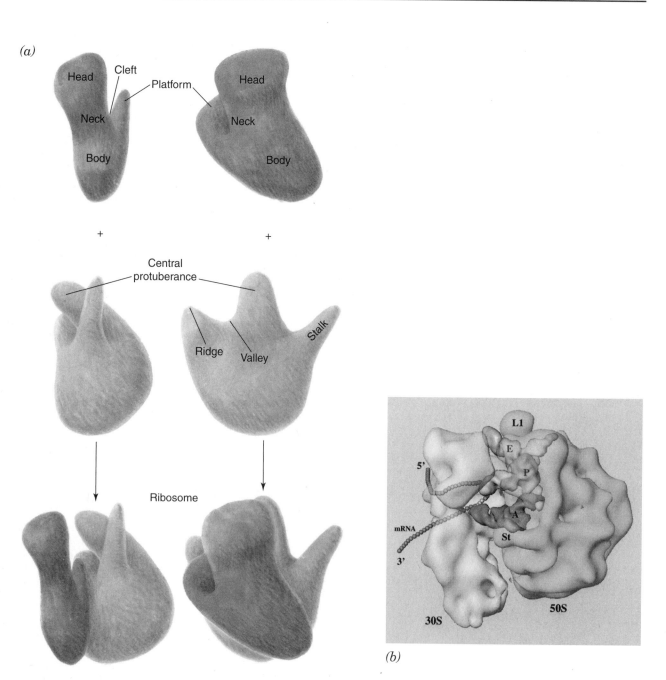

Figure 26-15 Structure of the *E. coli* ribosome.

Table 26-5 Components of Rat Liver Cytoplasmic Ribosomes

	Ribosome	Small Subunit	Large Subunit
Sedimentation coefficient	80S	40S	60S
Mass (kD)	4220	1400	2820
RNA			
Major		18S, 1874 nucleotides	28S, 4718 nucleotides
Minor			5.8S, 160 nucleotides
			5S, 120 nucleotides
RNA mass (kD)	2520	700	1820
Proportion of mass	60%	50%	65%
Proteins		33 polypeptides	49 polypeptides
Protein mass (kD)	1700	700	1000
Proportion of mass	40%	50%	35%

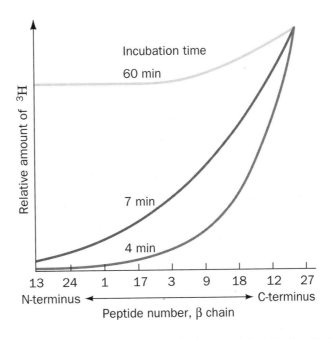

Figure 26-22 Demonstration that polypeptide synthesis proceeds from the N- to the C-terminus.

tRNA
|
O
|
O=P—O—CH$_2$ Adenine
|
O$^-$

H   H

H   H
|
O   OH
|
C=O
|
CH—R$_n$
|
NH
|
C=O
|
CH—R$_{n-1}$
|
NH
|
⋮
|
C=O
|
CH—R$_1$
|
NH$_3^+$

Peptidyl–tRNA

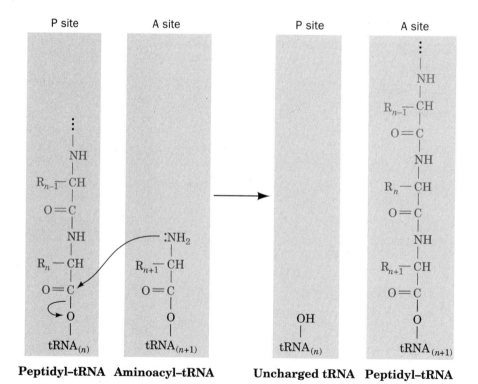

Figure 26-23 The ribosomal peptidyl transferase reaction forming a peptide bond.

araB	– U U U G G A U G G A G U G A A A C G A U G G C G A U U –
galE	– A G C C U A A U G G A G C G A A U U A U G A G A G U U –
lacI	– C A A U U C A G G G U G G U G A U U G U G A A A C C A –
lacZ	– U U C A C A C A G G A A A C A G C U A U G A C C A U G –
Qβ phage replicase	– U A A C U A A G G A U G A A A U G C A U G U C U A A G –
φX174 phage A protein	– A A U C U U G G A G G C U U U U U U A U G G U U C G U –
R17 phage coat protein	– U C A A C C G G G G U U U G A A G C A U G G C U U C U –
Ribosomal S12	– A A A A C C A G G A G C U A U U U A A U G G C A A C A –
Ribosomal L10	– C U A C C A G G A G C A A A G C U A A U G G C U U U A –
trpE	– C A A A A U U A G A G A A U A A C A A U G C A A A C A –
trp leader	– G U A A A A A G G G U A U C G A C A A U G A A A G C A –

3′ end of 16S rRNA 3′ ₍HO₎A U U C C U C C A C U A G – 5′

Figure 26-25 Some translation initiation sequences recognized by *E. coli* ribosomes.

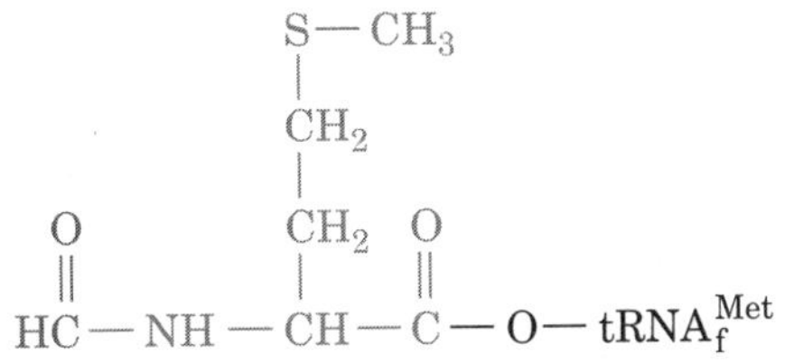

$$
\begin{array}{c}
S\!-\!CH_3 \\
| \\
CH_2 \\
| \\
CH_2 \\
\end{array}
$$

$$
\underset{HC}{\overset{O}{\parallel}}-NH-CH-\underset{C}{\overset{O}{\parallel}}-O-tRNA_f^{Met}
$$

N-Formylmethionine–tRNA$_f^{Met}$
(fMet–tRNA$_f^{Met}$)

Table 26-6 The Soluble Protein Factors of *E. coli* Protein Synthesis

Factor	Number of Residues[a]	Function
Initiation Factors		
IF-1	71	Assists IF-3 binding
IF-2	890	Binds initiator tRNA and GTP
IF-3	180	Releases mRNA and tRNA from recycled 30S subunit and aids new mRNA binding
Elongation Factors		
EF-Tu	393	Binds aminoacyl–tRNA and GTP
EF-Ts	282	Displaces GDP from EF-Tu
EF-G	703	Promotes translocation through GTP binding and hydrolysis
Release Factors		
RF-1	360	Recognizes UAA and UAG Stop codons
RF-2	365	Recognizes UAA and UGA Stop codons
RF-3	528	Stimulates RF-1/RF-2 release via GTP hydrolysis
RRF	185	Together with EF-G, induces ribosomal dissociation to small and large subunits

[a]All *E. coli* translational factors are monomeric proteins.

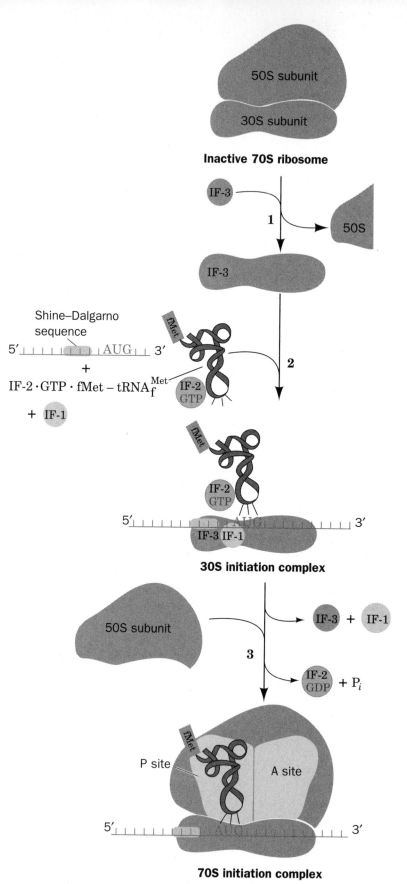

Figure 26-27 Translation initiation pathway in *E. coli.*

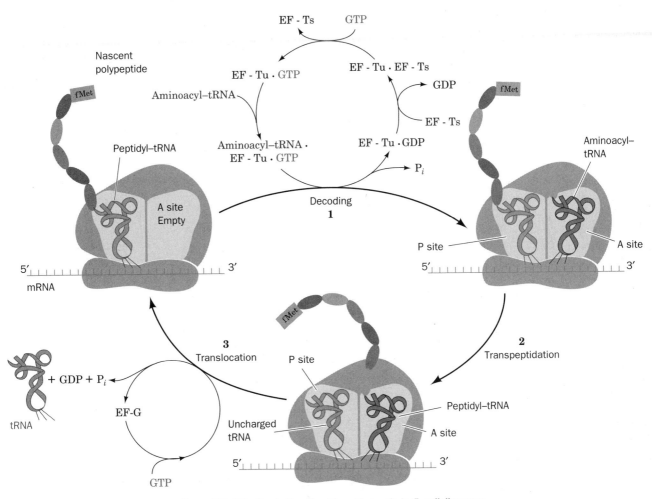

Figure 26-29 *Key to Function.* Elongation cycle in *E. coli* ribosomes.

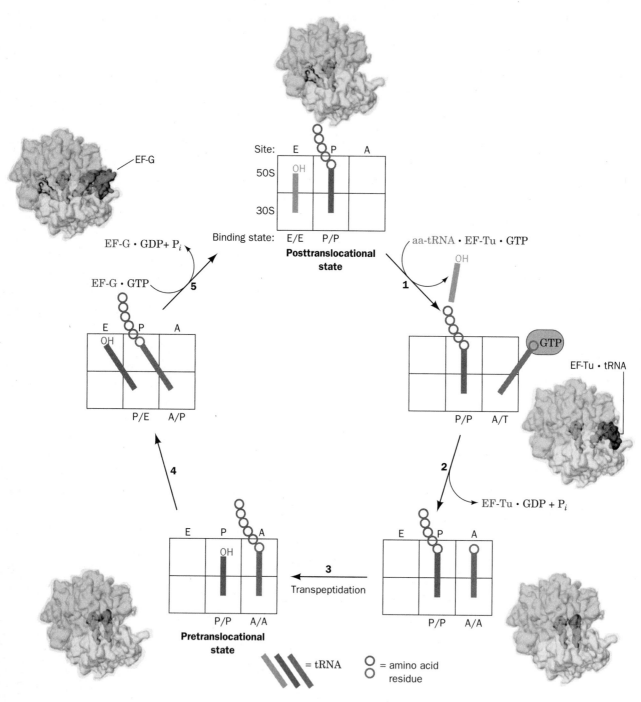

Site:	E	P	A
50S	OH		
30S			
Binding state:	E/E	P/P	

EF-G

EF-G · GDP+ P$_i$

EF-G · GTP

Posttranslational state

aa-tRNA · EF-Tu · GTP

OH

1

GTP

EF-Tu · tRNA

5

E	P	A
OH		
	P/E	A/P

4

E	P	A
	P/P	A/T

2

EF-Tu · GDP + P$_i$

E	P	A
	OH	
	P/P	A/A

Pretranslational state

3

Transpeptidation

E	P	A
	P/P	A/A

= tRNA   = amino acid residue

Figure 26-37 Ribosomal binding states in the elongation cycle.

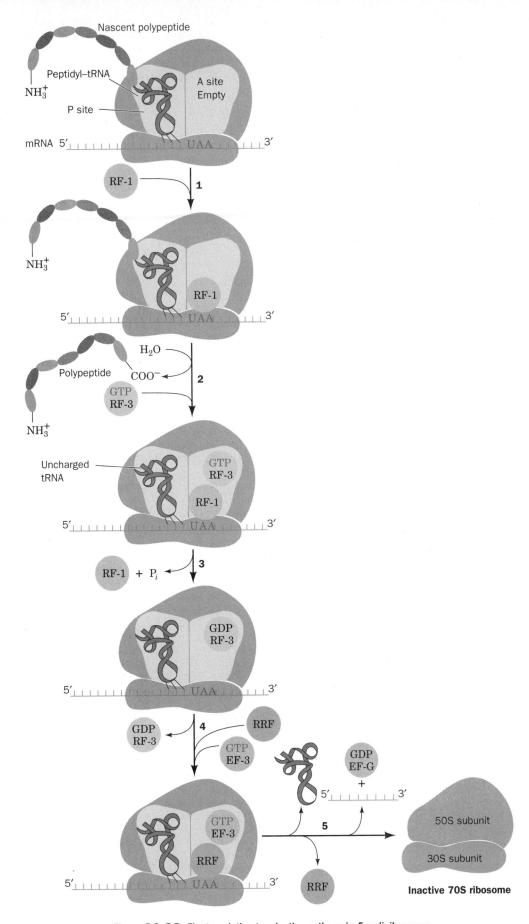

Figure 26-38 The translation termination pathway in *E. coli* ribosomes.

Puromycin

Tyrosyl–tRNA

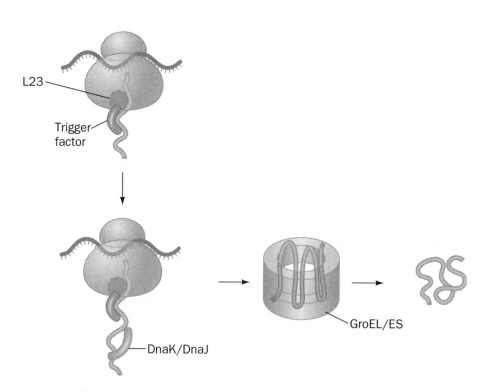

Figure 26-40 Chaperone binding to newly synthesized bacterial proteins.

CHAPTER 27

Regulation of Gene Expression

Table 27-1 Genome Size and Gene Number in Some Organisms

Organism	Genome Size (kb)	Number of Genes
Haemophilus influenzae (bacterium)	1,830	1,740
Escherichia coli (bacterium)	4,639	4,289
Saccharomyces cerevisiae (yeast)	11,700	6,034
Caenorhabditis elegans (nematode)	97,000	19,099
Oryza sativa (rice)	430,000	~35,000
Arabidopsis thaliana (mustard weed)	117,000	~26,000
Drosophila melanogaster (fruit fly)	137,000	13,061
Mus musculus (mouse)	2,500,000	~30,000
Homo sapiens (human)	3,200,000	~30,000

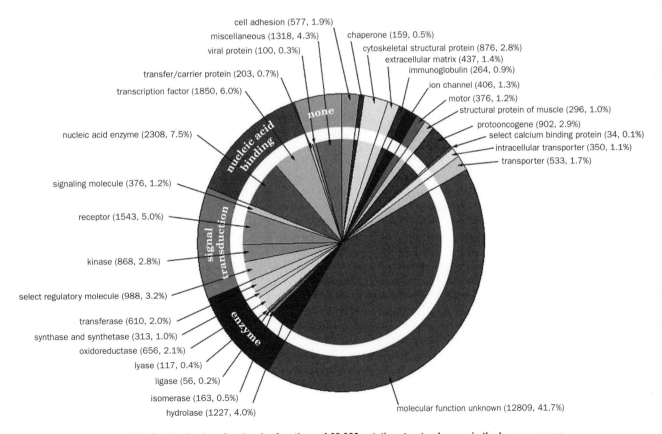

cell adhesion (577, 1.9%)
miscellaneous (1318, 4.3%)
viral protein (100, 0.3%)
transfer/carrier protein (203, 0.7%)
transcription factor (1850, 6.0%)
nucleic acid enzyme (2308, 7.5%)
signaling molecule (376, 1.2%)
receptor (1543, 5.0%)
kinase (868, 2.8%)
select regulatory molecule (988, 3.2%)
transferase (610, 2.0%)
synthase and synthetase (313, 1.0%)
oxidoreductase (656, 2.1%)
lyase (117, 0.4%)
ligase (56, 0.2%)
isomerase (163, 0.5%)
hydrolase (1227, 4.0%)

chaperone (159, 0.5%)
cytoskeletal structural protein (876, 2.8%)
extracellular matrix (437, 1.4%)
immunoglobulin (264, 0.9%)
ion channel (406, 1.3%)
motor (376, 1.2%)
structural protein of muscle (296, 1.0%)
protooncogene (902, 2.9%)
select calcium binding protein (34, 0.1%)
intracellular transporter (350, 1.1%)
transporter (533, 1.7%)

molecular function unknown (12809, 41.7%)

nucleic acid binding

signal transduction

enzyme

Figure 27-3 Distribution of molecular functions of 26,383 putative structural genes in the human genome.

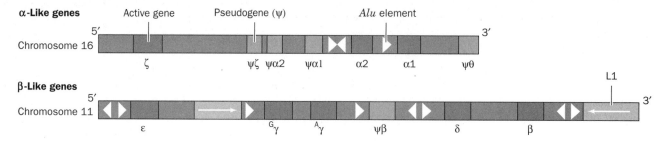

Figure 27-6 The organization of human globin genes.

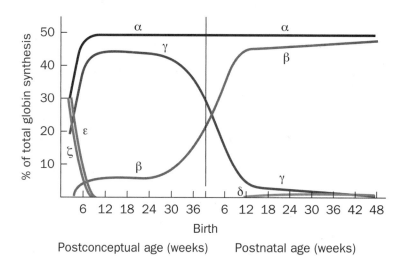

Figure 27-7 The progression of human globin chain synthesis with fetal development.

Table 27-2   Moderately Repetitive Sequences in the Human Genome

Type of Repeat	Length (bp)	Number of Copies (× 1000)	Percentage of Genome
LINEs	6000–8000	868	20.4
SINEs	100–300	1558	13.1
LTR retrotransposons	1500–11,000	443	8.3
DNA transposons	80–3000	294	2.8
Total			44.8

Source: International Human Genome Sequencing Consortium, *Nature* **409,** 800 (2001).

THE LAC OPERON

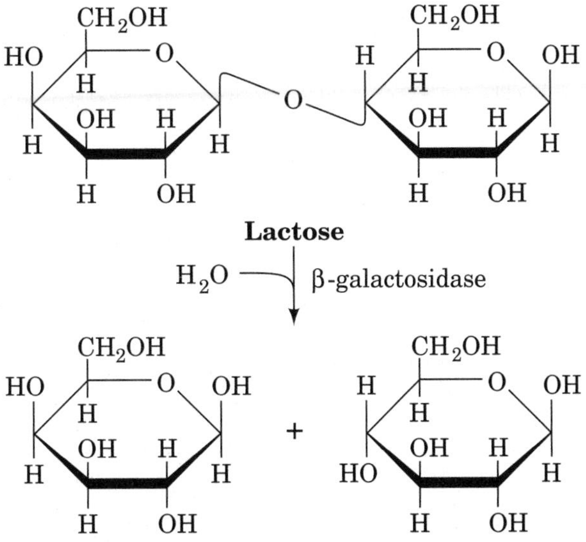

Lactose

H_2O → β-galactosidase

Galactose + **Glucose**

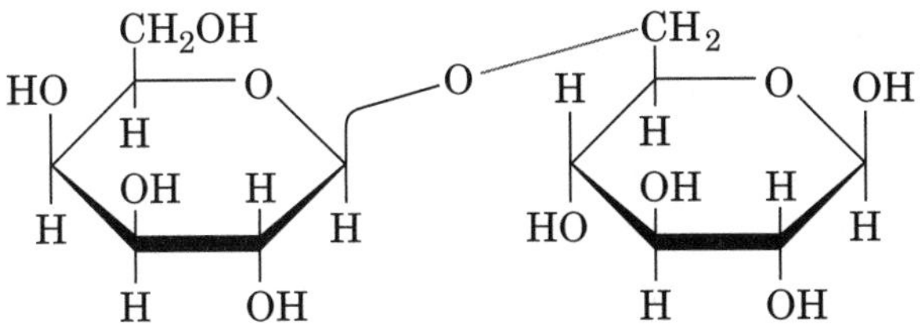

1,6-Allolactose

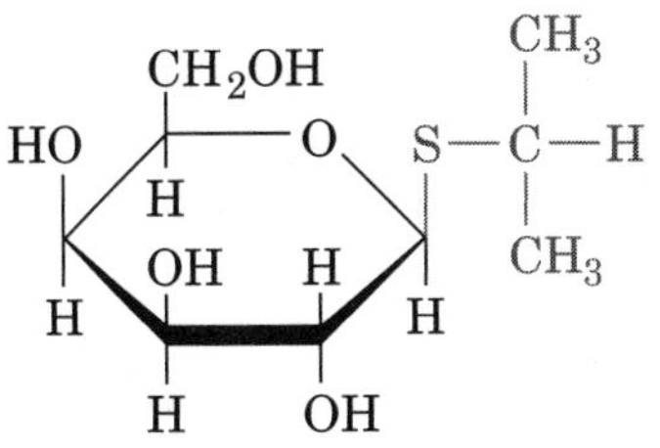

Isopropylthiogalactoside (IPTG)

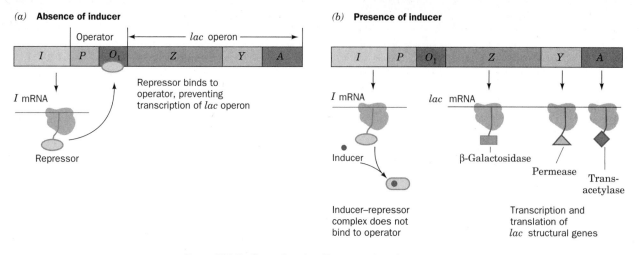

(a) **Absence of inducer**

Operator

lac operon

| I | P | O₁ | Z | Y | A |

I mRNA

Repressor

Repressor binds to operator, preventing transcription of *lac* operon

(b) **Presence of inducer**

| I | P | O₁ | Z | Y | A |

I mRNA

lac mRNA

Inducer

β-Galactosidase

Permease

Trans-acetylase

Inducer–repressor complex does not bind to operator

Transcription and translation of *lac* structural genes

Figure 27-9 *Key to Function.* The expression of the *lac* operon.

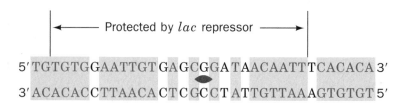

← Protected by *lac* repressor →

5′ TGTGTGGAATTGTGAGCGGATAACAATTTCACACA 3′
3′ ACACACCTTAACACTCGCCTATTGTTAAAGTGTGT 5′

Figure 27-10 The base sequence of the *lac* operator O_1.

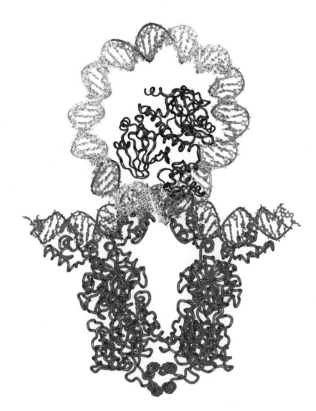

Figure 27-13 Model of the 93-bp loop formed when the *lac* repressor tetramer binds to O_1 and O_3.

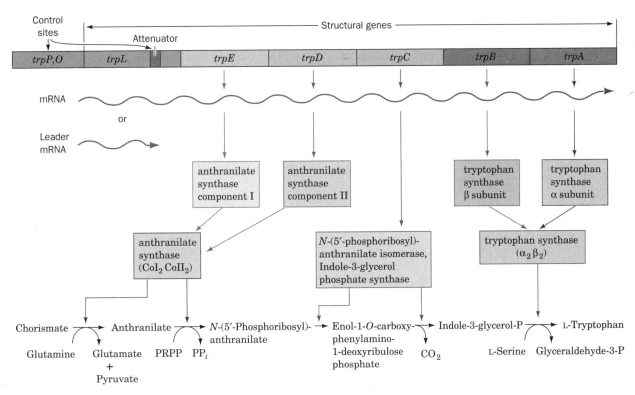

Figure 27-15 Genetic map of the *E. coli trp* operon indicating the enzymes it specifies and the reactions they catalyze.

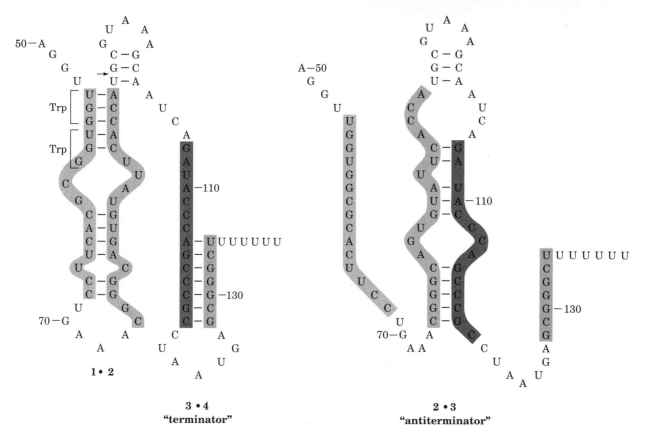

Figure 27-16 Alternative secondary structures of *trpL* mRNA.

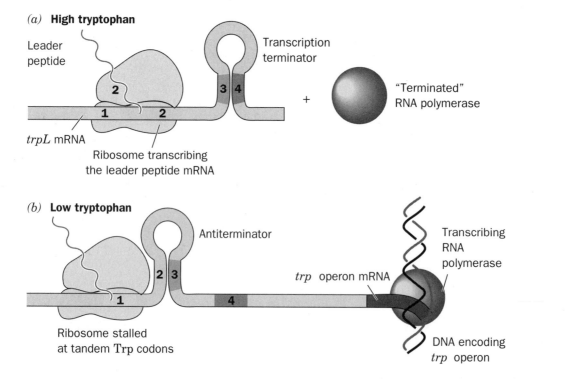

Figure 27-17 Attenuation in the *trp* operon.

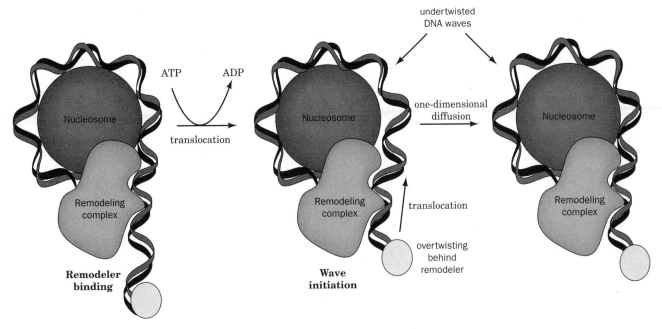

Figure 27-22 Model for nucleosome remodeling by chromatin-remodeling complexes.

```
        |  |
CH₃—C  G
     G  C—CH₃
     |  |
     G  C—CH₃
CH₃—C  G
     |  |
     C  G
     G  C
     |  |

        ↓ replication

     |  |          |  |
CH₃—C  G         C  G
     G  C         G  C—CH₃
     |  |          |  |
     G  C    +    G  C—CH₃
CH₃—C  G         C  G
     |  |          |  |
     C  G         C  G
     G  C         G  C
     |  |          |  |

        ↓ maintenance
          methylation

     |  |                |  |
CH₃—C  G           CH₃—C  G
     G  C—CH₃           G  C—CH₃
     |  |                |  |
     G  C—CH₃     +      G  C—CH₃
CH₃—C  G           CH₃—C  G
     |  |                |  |
     C  G                C  G
     G  C                G  C
     |  |                |  |
```

Figure 27-31 Maintenance methylation.

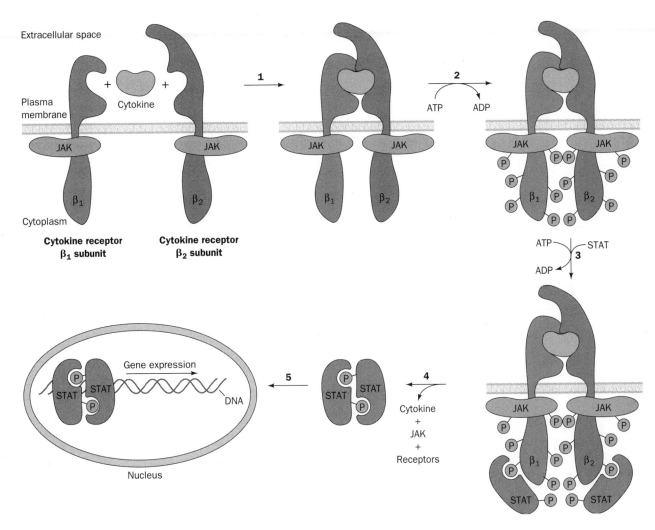

Figure 27-33 The JAK-STAT pathway for the intracellular relaying of cytokine signals.

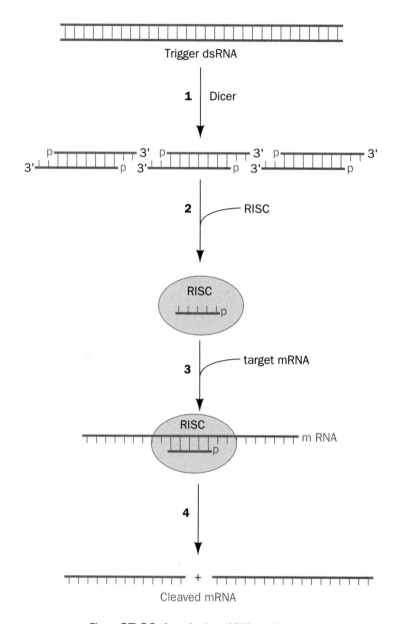

Figure 27-36 A mechanism of RNA interference.

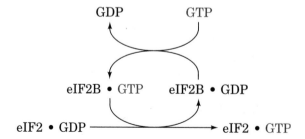

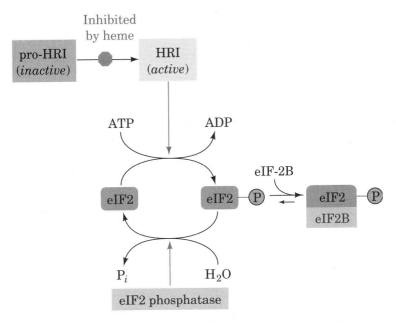

Figure 27-37 A model for heme-controlled protein synthesis in reticulocytes.

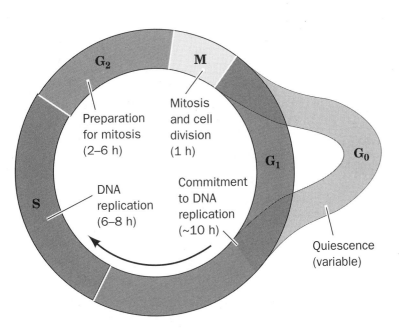

Figure 27-38 The eukaryotic cell cycle.

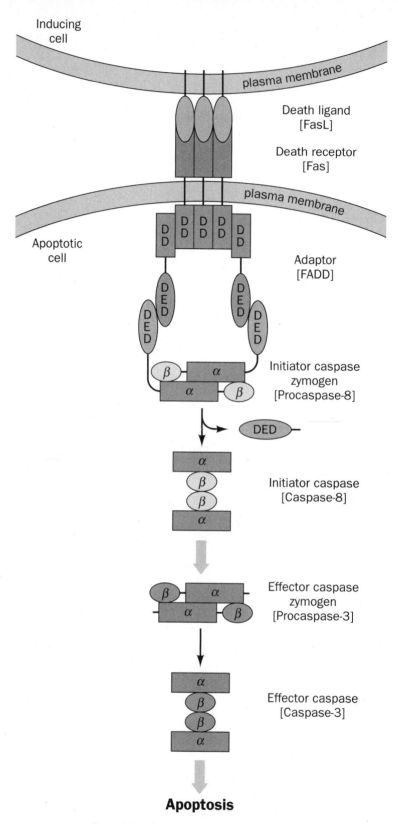

Figure 27-43 The extrinsic pathway of apoptosis.

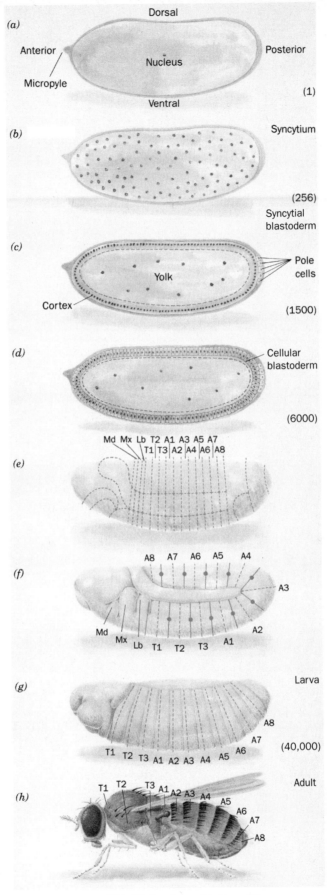

Figure 27-45 Development in Drosophila.

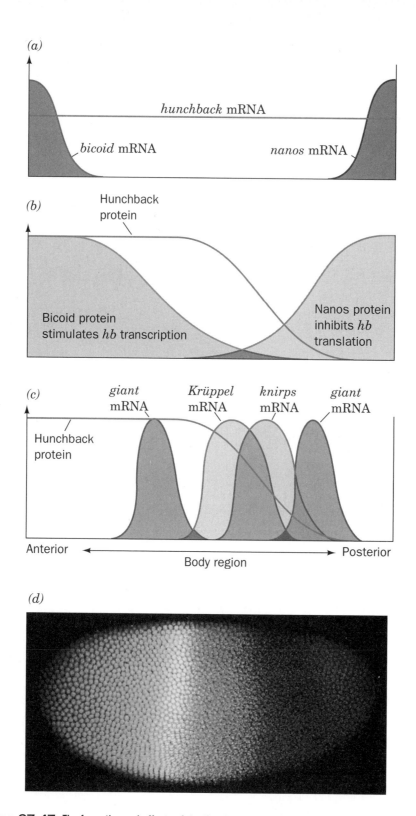

Figure 27-47 The formation and effects of the Hunchback protein gradient in *Drosophila* embryos.

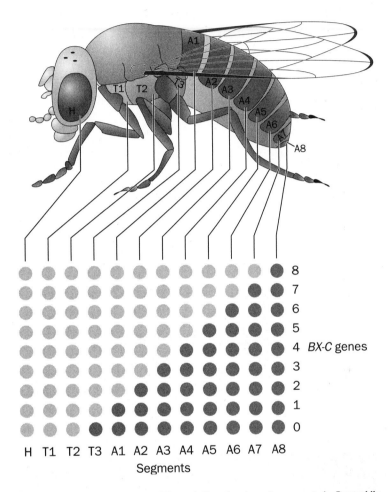

8
7
6
5
4 *BX-C* genes
3
2
1
0

H T1 T2 T3 A1 A2 A3 A4 A5 A6 A7 A8
Segments

Figure 27-50 Model for the differentiation of embryonic segments in *Drosophila*.

CHAPTER 28

Protein Function Part II:
Cytoskeletal and Motor Proteins and Antibodies

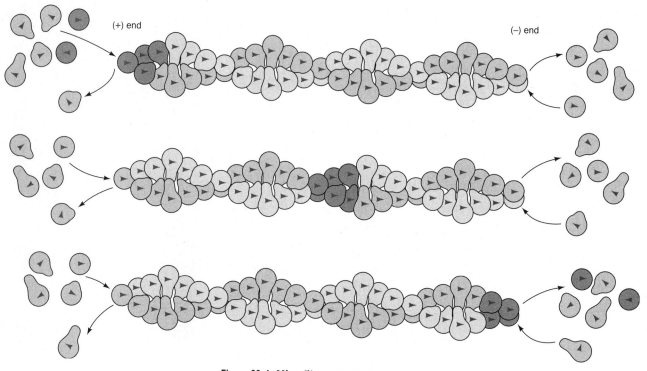

Figure 28-4 Microfilament treadmilling.

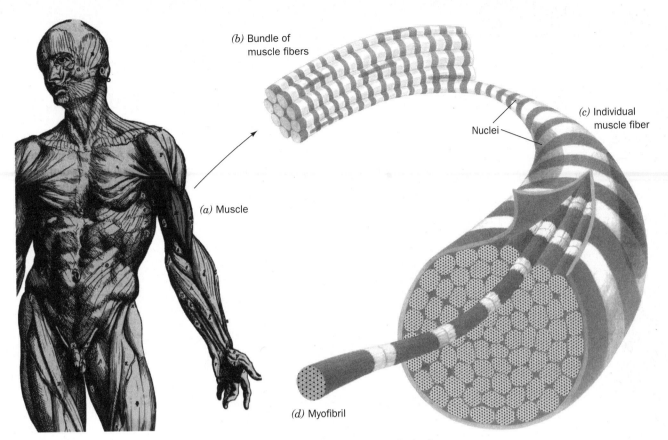

(b) Bundle of
muscle fibers

(c) Individual
muscle fiber

Nuclei

(a) Muscle

(d) Myofibril

Figure 28-10 Skeletal muscle organization.

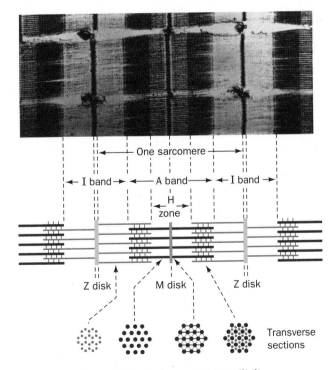

One sarcomere

I band | A band | I band

H
zone

Z disk | M disk | Z disk

Transverse
sections

Figure 28-11 Anatomy of the myofibril.

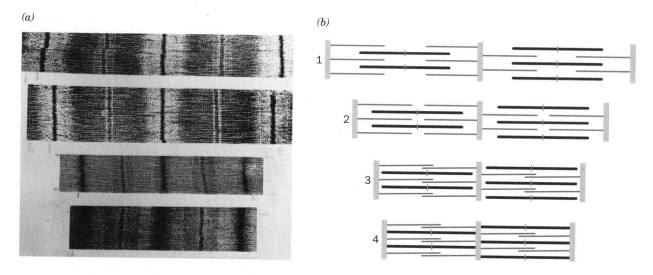

Figure 28-12 Myofibril contraction.

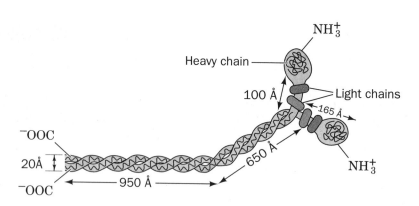

Figure 28-14 The myosin molecule.

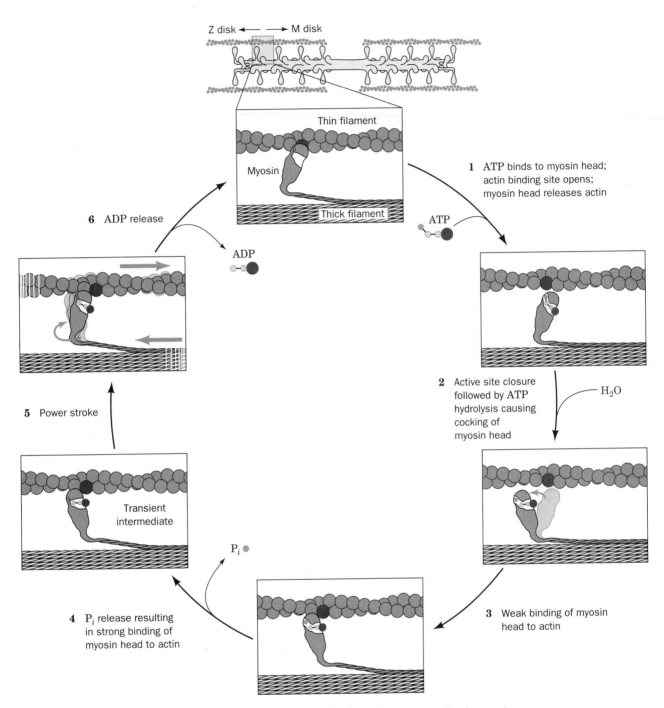

Figure 28-17 *Key to Function.* **Mechanism of force generation in muscle.**

Z disk ← → M disk

Thin filament

Myosin

Thick filament

1 ATP binds to myosin head; actin binding site opens; myosin head releases actin

ATP

6 ADP release

ADP

H_2O

2 Active site closure followed by ATP hydrolysis causing cocking of myosin head

5 Power stroke

Transient intermediate

3 Weak binding of myosin head to actin

P_i

4 P_i release resulting in strong binding of myosin head to actin

411

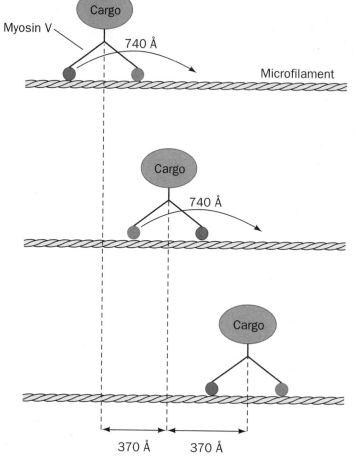

Figure 28-19 Hand-over-hand movement of myosin V.

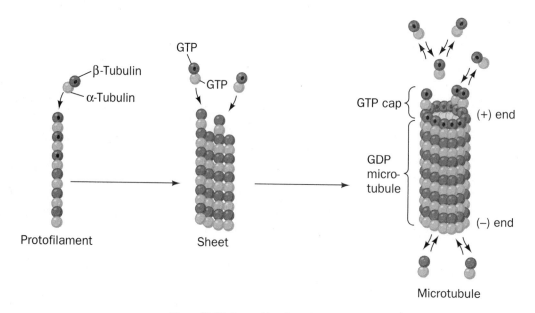

Figure 28-22 Assembly of a microtubule.

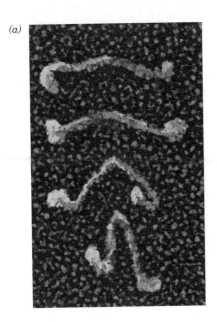

(a)

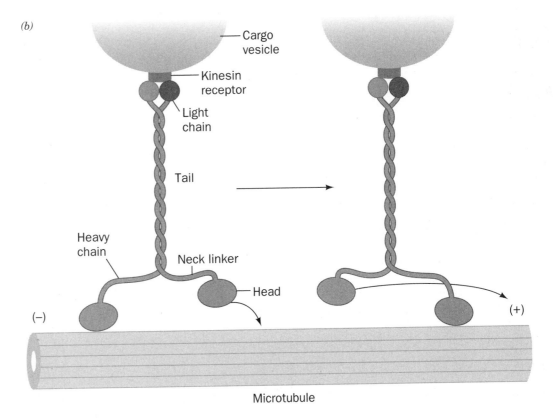

(b)

Figure 28-25 **Conventional kinesin.**

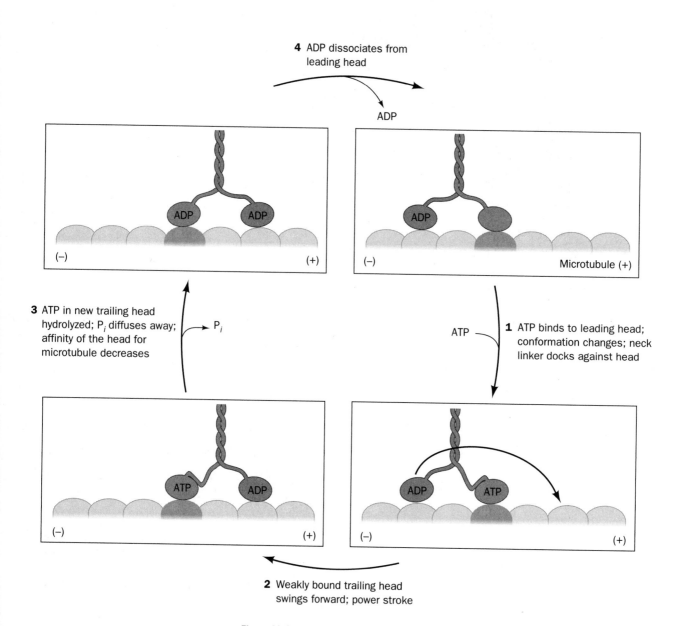

4 ADP dissociates from leading head

ADP

(−)   (+)

(−)   Microtubule (+)

3 ATP in new trailing head hydrolyzed; P_i diffuses away; affinity of the head for microtubule decreases

$\rightarrow P_i$

ATP

1 ATP binds to leading head; conformation changes; neck linker docks against head

(−)   (+)

(−)   (+)

2 Weakly bound trailing head swings forward; power stroke

Figure 28-29 The kinesin reaction cycle.

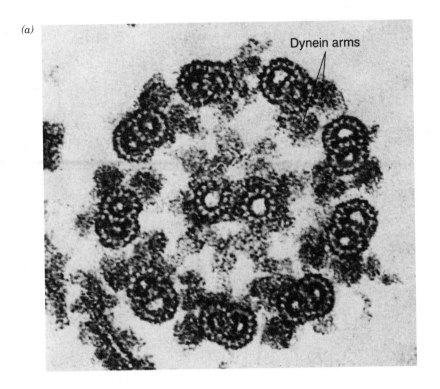

(a)

Dynein arms

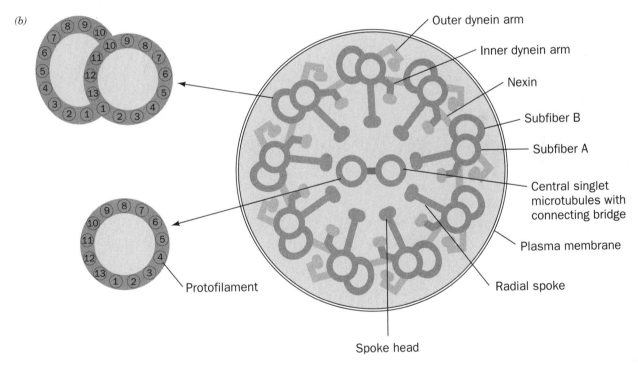

(b)

Outer dynein arm

Inner dynein arm

Nexin

Subfiber B

Subfiber A

Central singlet
microtubules with
connecting bridge

Plasma membrane

Radial spoke

Protofilament

Spoke head

Figure 28-32 Structure of the axoneme.

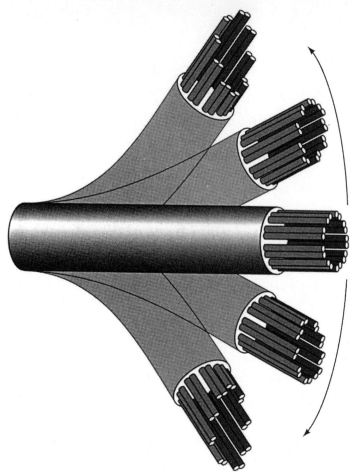

Figure 28-33 The sliding microtubule model of ciliary motion.

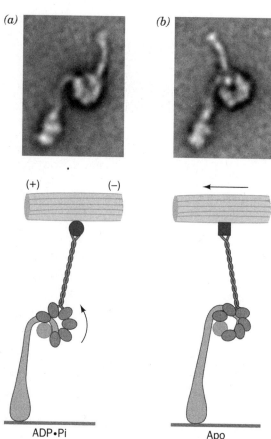

(a) *(b)*

(+)   (−)

ADP•Pi   Apo

Figure 28-35 Conformational changes in dynein.

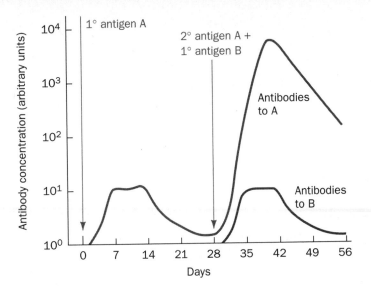

Figure 28-36 Primary and secondary immune responses.

Table 28-1 Classes of Human Immunoglobulins

Class	Heavy Chain	Light Chain	Subunit Structure	Molecular Mass (kD)
IgA	α	κ or λ	$(\alpha_2\kappa_2)_n J^a$ or $(\alpha_2\lambda_2)_n J^a$	360–720
IgD	δ	κ or λ	$\delta_2\kappa_2$ or $\delta_2\lambda_2$	160
IgE	ε	κ or λ	$\varepsilon_2\kappa_2$ or $\varepsilon_2\lambda_2$	190
IgG^b	γ	κ or λ	$\gamma_2\kappa_2$ or $\gamma_2\lambda_2$	150
IgM	μ	κ or λ	$(\mu_2\kappa_2)_5 J$ or $(\mu_2\lambda_2)_5 J$	950

$^a n = 1, 2,$ or 3.

^{b}IgG has four subclasses, IgG1, IgG2, IgG3, and IgG4, which differ in their γ chains.

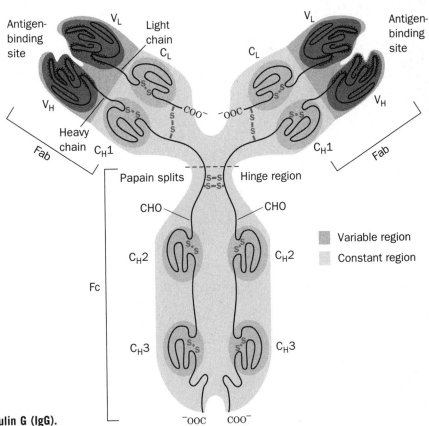

Figure 28-38 *Key to Structure.*
Diagram of human immunoglobulin G (IgG).

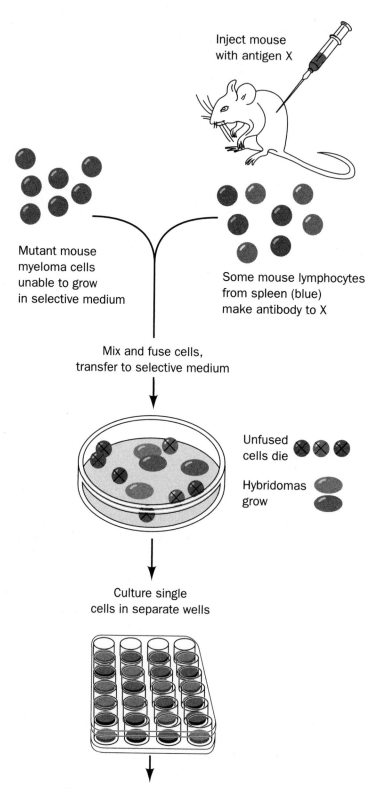

Inject mouse
with antigen X

Mutant mouse
myeloma cells
unable to grow
in selective medium

Some mouse lymphocytes
from spleen (blue)
make antibody to X

Mix and fuse cells,
transfer to selective medium

Unfused
cells die

Hybridomas
grow

Culture single
cells in separate wells

Test each well for antibody to X

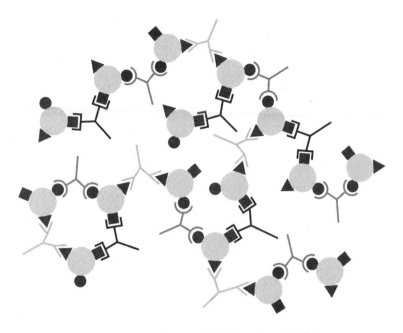

Figure 28-41 Antigen cross-linking by antibodies.

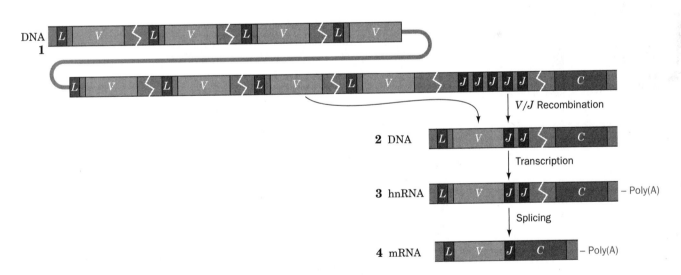

Figure 28-42 The organization and rearrangement of the k chain gene family in mice.

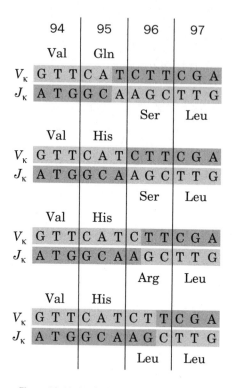

Figure 28-43 Variation at the V_k/J_k joint.

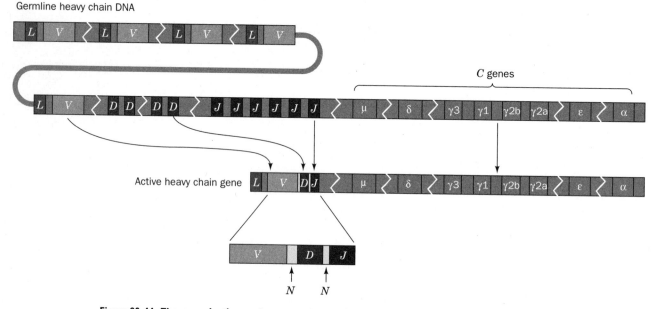

Figure 28-44 The organization and rearrangement of the heavy chain gene family in humans.